Nematode Vectors of Plant Viruses

NATO ADVANCED STUDY INSTITUTES SERIES

A series of edited volumes comprising multifaceted studies of contemporary scientific issues by some of the best scientific minds in the world, assembled in cooperation with NATO Scientific Affairs Division.

Series A: Life Sciences

Volume 1—Fish Vision: New Approaches in Research
edited by M. A. Ali

Volume 2—Nematode Vectors of Plant Viruses
edited by F. Lamberti, C. E. Taylor and J. W. Seinhorst

A Continuation Order Plan may be opened with Plenum for Series A: Life Sciences Subscribers to this scheme receive the same advantages that apply to all our other series: delivery of each new volume immediately upon publication; elimination of unnecessary paper work; and billing only upon actual shipment of the book.

This series is published by an international board of publishers in conjunction with NATO Scientific Affairs Division

A Life Sciences **B Physics**	Plenum Publishing Corporation London and New York
C Mathematical and Physical Sciences	D. Reidel Publishing Company Dordrecht and Boston
D Behavioral and Social Sciences	Sijthoff International Publishing Company Leiden
E Applied Sciences	Noordhoff International Publishing Leiden

Nematode Vectors of Plant Viruses

Edited by

F. Lamberti

Laboratorio di Nematologia Agraria
Bari, Italy

C. E. Taylor

Scottish Horticultural Research Institute
Dundee, Scotland

and

J. W. Seinhorst

Institut voor Planten Ziektenkundig
Wageningen, Netherlands

PLENUM PRESS • LONDON AND NEW YORK
Published in cooperation with NATO Scientific Affairs Division

Lectures presented at the NATO Advanced Study Institute, Riva de I Tessoli, Italy, 19 May to 2 June, 1974

Library of Congress Catalog Card Number 75-2679
ISBN-13: 978-1-4684-0843-0 e-ISBN-13: 978-1-4684-0841-6
DOI: 10.1007/978-1-4684-0841-6

A Division of Plenum Publishing Company, Ltd.
Softcover reprint of the hardcover 1st edition 1975
4a Lower John Street, London W1R 3PD, England
Telephone 01-437 1408

U.S. edition published by Plenum Press, New York,
A Division of Plenum Publishing Corporation
227 West 17th Street, New York, N.Y. 10011

PREFACE

Although nematodes had long been suspected as vectors of soil-borne plant diseases, unequivocal proof of their implication was not forthcoming until 1958 when Professor William Hewitt and his colleagues in California demonstrated experimentally that *Xiphinema index* was the vector of grapevine fanleaf virus. This opened up a new and exciting field in plant pathology and discoveries quickly followed of other nematode species associated with soil-borne diseases of many different crops and in several countries. After the initial enthusiasm of discovering new vectors and new viruses there followed a period of consolidation in which research workers sought answers to tantalising questions about the location of the virus within the nematode, the factors governing the close specificity between virus and vector; and more mundane but equally important and compelling questions about life cycles, geographical distribution, host relations, morphology and taxonomy. No other group of nematodes has attracted such a concentrated effort involving many different scientific specialisations and yielding so much progress in a relatively short time.

The NATO Advanced Study Institute held at Riva dei Tessali, Italy, during 19 May to 2 June, 1974, provided the forum for a critical discussion of all aspects of biology of virus vector nematodes. Indeed, the occasion brought together most of the nematologists and several of the virologists who are working in this particular field and these published Proceedings therefore represent a comprehensive and authoritative account of the subject, with the summarised discussions providing further critical appraisals and suggestions for future investigations.

The Institute was an undoubted success due mainly to the enthusiasm and interest of all those who participated, and we hope that this is reflected in this book to the enjoyment of all who read it. We thank the many helpers from the University of Bari who assisted with the administration and organisation of the Institute at Riva dei Tessali, and typists, photographers and graphic artists at the Scottish Horticultural Research Institute who were involved in the preparation of typescripts for publication.

C. E. Taylor
F. Lamberti
J. W. Seinhorst

CONTENTS

VIRUS VECTOR NEMATODES - TAXONOMY AND GENERAL INTRODUCTION

DAVID J. HOOPER

Nematology Department, Rothamsted Experimental Station

Harpenden, Herts., England.

INTRODUCTION AND HISTORICAL REVIEW

Nematodes that transmit viruses to plants are known to occur only in two suborders of the order Dorylaimida class Adenophorea whereas most other plant parasitic nematodes are in the order Tylenchida class Secernentea. Plant parasitism seems to have developed independently in these two classes. Maggenti (18) suggests that the Adenophorea are the older group and that within the Dorylaimida there are two relatively recent divergent lines; one containing the genera Longidorus, Paralongidorus and Xiphinema, some species of which transmit nepoviruses (see Taylor and Robertson in these Proceedings), is in the family Longidoridae, suborder Dorylaimina. The other line containing the genera Trichodorus and Paratrichodorus, some species of which transmit tobraviruses, is in the family Trichodoridae, suborder Diphtherophorina. Although species of Paralongidorus have not so far been implicated as virus vectors their morphology is so similar to Longidorus that it is probably prudent to treat them as potential virus vectors.

Dorylaimus elongatus de Man, 1876 was made the type of the subgenus Longidorus by Micoletzky (1922) which was recognised at generic level by Filipjey, 1934. The genus Xiphinema was erected by Cobb (1913) with the type species X. americanum. However, it has been mainly in the last fifteen years that nematodes in these two genera and the closely related genus Paralongidorus Siddiqi, Hooper and Khan, 1963, have received much attention. The discovery of Hewitt, Raski and Goheen (10) that X. index was a vector of grape vine fanleaf virus instigated further studies of longidorid nematodes and several species have now been implicated as virus vectors. The biological studies have been paralleled by taxonomic interest and many new species have been described (see papers by Lamberti and by Luc in these Proceedings).

The genus Trichodorus was erected by Cobb (1913) on a single species T. obtusus. However, Micoletzky (1922) placed Dorylaimus primitivus de Man, 1880 in the genus Trichodorus and made T. obtusus a junior synonym of T. primitivus. This synonymy has been questioned by Seinhorst (20) but expediently ignored by most other workers so that T. primitivus has generally been accepted as being the name of the type species of the genus Trichodorus (but see Loof in these Proceedings). The comprehensive review by Allen (1) of the genus Trichodorus with descriptions of ten new species was very important as it established the criteria for differentiating species to the benefit of many subsequent workers. Although some species, especially Trichodorus christiei (now Paratrichodorus (Nanidorus) christiei), were already known to be serious root ectoparasites, the implication of some species as vectors of tobraviruses (see Taylor and Robertson in these Proceedings) stimulated further interest in this group and, as with the Longidoridae, many new species have been described (see Loof in these Proceedings).

The higher classification of both these groups has been steadily raised, almost in sympathy with their biological importance, so that now longidorids are ranked as the family Longidoridae and the trichodorids as a superfamily - Trichodoroidea.

THE LONGIDORIDAE

Longidorid nematodes have a typical Dorylaimoid oesophagus with a long narrow anterior part and a prominent muscular cylindrical bulb. The oesophageal bulb has a prominent dorsal oesophageal gland duct opening near the anterior end and an associated gland nucleus; there is a pair of subventral gland ducts and gland nuclei about half way along the bulb and sometimes a suggestion of a second pair of subventral gland duct openings, but no gland nuclei, towards the base of the bulb. The absence of the second pair of subventral gland nuclei helps to separate longidorids from other Dorylaims most of which have two pairs of subventral glands (see Coomans and Loof (4); Loof and Coomans (17); Coomans in these Proceedings).

Longidorids are readily recognised from most other Dorylaims by their long (2-12 mm) rather narrow body and their elongated axial mouth spear (odontostyle) plus an extension (odontophore) about half the odontostyle length. Juveniles are recognised by having a functional and replacement odontostyle, the latter being located in the wall of the anterior oesophagus, first stage juveniles are distinguished from others in that the anterior part of the replacement odontostyle is embedded within the wall of the functional odontophore. Longidorus spp. (Needle nematodes) and Xiphinema spp. (Dagger nematodes) are root ectoparasites as are also Paralongidorus spp. and they can cause serious root damage whether or not they may also be transmitting virus to the plant.

Classification of the LONGIDORIDAE

Class ADENOPHOREA (von Linstow, 1905) Chitwood, 1958

Order DORYLAIMIDA Pearse 1942

Suborder DORYLAIMINA Pearse 1936

Oesophagus consisting of a narrow anterior part and a wider posterior one. Stoma with a protrusible tooth or odontostyle. Dorsal oesophageal gland nucleus a short distance behind its gland duct orifice, well anterior to first pair of subventral gland nuclei which are about at the same level as their gland duct orifices. Second pair of subventral gland nuclei (if present) slightly anterior to their gland duct orifices.

Superfamily DORYLAIMOIDEA (de Man, 1876) Thorne, 1934.

Stoma with axial odontostyle. Usually no free cardiac gland cells. Gubernaculum mostly absent. First pair of subventral gland nuclei usually different in size from the dorsal nucleus and located far behind it.

Family LONGIDORIDAE (Thorne, 1935) Meyl, 1961 - Dorylaimoidea. Large nematodes 2-12 mm long, spear (odontostyle) greatly attenuated, 50-200μm long, plus a long extension (odontophore) which is plain or heavily flanged. Anterior oesophagus a narrow tube; posterior part a wider, muscular cylinder with a dorsal and two subventral gland nuclei. Tails of sexes somewhat similar where both are known. Males have large arcuate spicules with lateral guiding pieces but no gubernaculum; testes paired, opposed; a series of prominent preanal supplementary papillae present including an adanal pair. Lateral cords relatively broad with lateral pores; dorsal and ventral body-pores also sometimes present especially at anterior end. Lip region smoothly rounded, continuous with, or set off from the body, with six amalgamated lips bearing 16 distinct papillae arranged in two circles, 6 inner, 10 outer. Amphid pouches large, extending back from base of lip region.

Subfamily LONGIDORINAE Thorne, 1935.

Longidoridae. Guide ring single, prominent, around anterior part of odontostyle. Odontostyle-odontophore junction simple, without prongs. Odontophore plain, without basal flanges. Vulva about median; genital branches paired, opposed with reflexed ovaries.

Tails bluntly rounded to roundly conoid. Dorsal oesophageal gland nucleus well posterior to its orifice.
Type genus : Longidorus (Micoletzky, 1922) Filipjev, 1934.
Other genus : Paralongidorus Siddiqi, Hooper and Khan, 1963.

Subfamily XIPHINEMINAE Dalmasso, 1969 (5).

Longidoridae. Amphid openings wide, slit-like, almost extending across the base of the lip region in lateral view; amphid pouches usually inverted-stirrup (goblet) shape. Base of odontostyle forked (2-3 prongs) at its junction with the odontophore. Odontophore with prominent basal flanges. Guiding sheath around base of odontostyle showing as two guide rings, the posterior being the more prominent. Vulva position variable; genital tract sometimes having paired, opposed, reflexed branches (didelphic). Some species have a more anterior vulva with a posterior genital branch only (monodelphic) and some also a vestigial anterior branch lacking an ovary (pseudomonodelphic). Tails vary between species - bluntly rounded, rounded with a ventral peg, elongate conoid to filiform. Dorsal oesophageal gland just posterior to its orifice.
Type and only genus : Xiphinema Cobb, 1913.

THE TRICHODORIDAE

Trichodorid spp. (Stubby root nematodes) have a characteristically shaped, elongate, ventrally curved mouth spear (onchiostyle) which readily separates them from other dorylaimid nematodes and particularly from those in their own suborder (genera Diphtherophora and Triplonchium) which tend to have a shorter more complicated stomal armature. The method of feeding is exceptional in that after the solid tipped onchiostyle has punctured a cell a feeding tube is established through which food is pumped to the oesophagus via a pharyngeal lumen that lies ventrally adjacent to the spear (see Robertson and Taylor in these Proceedings). Adults vary in length from 0.5 to just under 1 mm and the females, which usually have a subterminal anus, often appear plump and cigar shaped especially when moribund and even more so when killed and fixed.

Classification of the TRICHODORIDAE

Suborder DIPHTHEROPHORINA Coomans and Loof, 1970 (4) emend Siddiqi, 1974 (22)

Dorylaimida. Body plump usually less than 2 mm long; cuticle appears thick and loose. Occasional lateral body pores and caudal pores may be present; caudal glands and spinnerets absent. Amphids pouch shaped, post labial with ellipsoidal apertures, separated from sensilla sac by a constriction. Stoma with a complex spear (onchiostyle) composed of dorsal tooth and supporting structures. New onchiostyle tip formed in buccal capsule. Oesophagus consisting of a narrow anterior part swelling posteriorly to form a pyriform to elongate basal bulb that contains one dorsal and at least two subventral uninucleate glands. Excretory pore present. No distinct pre-rectum. Females usually with paired opposed genital tracts the ovaries being reflexed. Males with single testis, no ejaculatory muscles and no adanal pair of supplementary papillae. Spicules paired, tripyloid, without lateral guiding pieces but with a gubernaculum. Each spicule surrounded by a capsule of suspensor muscles that replaces the form and function of the typical Dorylaimoid protractor muscles.

Superfamily TRICHODOROIDEA (Thorne 1935) Siddiqi, 1974.

Diphtherophorina. Spear (onchiostyle) is an elongate dorsal tooth with an attenuated ventrally arcuate extension without basal knobs. Female genital tracts paired except in one species. Males with or without a bursa (caudal alae). Spicules almost straight to ventrally curved with or without fine transverse striations. Tail length not more than one anal body width long.

Family TRICHODORIDAE (Thorne 1935) Clarke 1961
Diagnosis as for superfamily.
Type genus *Trichodorus* Cobb, 1913.
Other genus *Paratrichodorus* Siddiqi 1974.

LONGIDORELLA AND XIPHINEMELLA

Longidorella Thorne 1939 and *Xiphinemella* Loos, 1950. As their names imply these genera have at times been closely linked with the longidorids. However, although *Longidorella* has an odontostyle and odontophore similar in appearance to *Longidorus* its guiding ring is relatively weaker, the body length is usually less than 1 mm (15) and the amphids are stirrup shaped with wide transverse apertures. Loof and Coomans (17) note that it has two pairs of subventral gland nuclei (only one pair in *Longidorus*) and is therefore better placed in the Dorylaimidae. *Xiphinemella* has an odontostyle and odontophore somewhat similar to *Xiphinema* but the body length is less than 2 mm, the lip region has a distinct terminal disc and the guide ring tends to be around the anterior half of the odontostyle. The taxonomic position of *Xiphinemella* and the closely related genus *Botalium* Heyns, 1963 is discussed by Ferris (8) who leaves them in the subfamily Tylencholaiminae, family Tylencholaimidae.

NEW APPROACHES AND TAXONOMIC AIDS

Although the longidorid and trichodorid nematodes fall into

two readily recognisable groups, opinions as to the ranking of these groups relative to each other and to other closely related groups have varied considerably in the last decade.

The observations by Coomans and Loof (4) and Loof and Coomans (17) regarding the position, size and shape of oesophageal gland nuclei and their ducts have made a significant contribution to the understanding of relationships between various groups in the Dorylaimida. In particular their observations supported the division, by Dalmasso (5), of the Longidoridae into the two subfamilies, Longidorinae and Xiphineminae, while eliminating *Longidorella* from the Longidoridae because it has two pairs of subventral oesophageal glands.

Ultrastructure studies by various workers (see Taylor and Robertson in these Proceedings) also confirmed a fundamental difference in the morphology of the stoma and pharynx between Longidorinae and Xiphineminae. Ultrastructure studies have also helped to elucidate the unique structure of the trichodorid spear (11, 19, 2) and have shown that the musculature in trichodorids is of a more primitive type than in the longidorids, supporting the separation of these two groups into separate suborders (12). Ultrastructure studies also helped Siddiqi (22) to interpret the spicule musculature in trichodorids.

Scanning electron microscopy also greatly helps in displaying structures such as amphid openings and head papillae which are very important characters in the Longidoridae (see Lamberti, Luc in these Proceedings).

Cytogenetic studies, especially by Dalmasso (6), have shown basic differences in chromosome numbers between *Longidorus* and *Xiphinema* spp. (see Dalmasso in these Proceedings).

Polyacrylamide gel electrophoresis has been useful in appraising "species" in some nematode groups. Provisional tests with *Longidorus* spp. (13) showed promise although results tended to parallel traditional specific classification in that 'large' and 'small' forms of *Longidorus leptocephalus* gave the same gel water-soluble protein patterns which were somewhat similar to *L. elongatus*, whereas *L. macrosoma* had a very different pattern.

Specific identification can often be aided by knowledge of ecological and geographical distribution (see several short contributions in these Proceedings). Dalmasso (7) and Taylor (24) in particular note that climate, soil type and longitude greatly affect the distribution of some longidorid species. Some species are closely associated with a particular host, e.g. *X. index* with grapevines and *L. laevicapitatus* with sugar cane occurring in various parts of the world where their host crop is grown. Most of the Paralongidorus spp. have been described from Africa, India or Australia. Likewise some Trichodorus spp. seem confined to temperate latitudes whereas others like T. christiei/T. minor usually occur in warmer soils.

Numerical taxonomy has been used successfully by Lima (16) to appraise species groups in the genus Xiphinema and helped to justify the recognition of seven species in the X. americanum complex. Likewise Bird (2) appraised the species in the genus Trichodorus and indicated groupings that have been followed to some extent in the review by Siddiqi (22). However, much detailed information is required to get worthwhile results and computer systems are not always readily available to those requiring help in the identification of individual species.

Mainly because of the pioneer work of Allen (1) species of Trichodorus are usually well defined, especially those with males, and their characters tend to remain remarkably consistent. Unfortunately, the identification of longidorid species is much more subjective; some species have very inadequate descriptions sometimes erected on very few specimens. Intraspecific morphometric and morphological variation is considerable; tail shape, which is often used as a distinguishing character, is often very variable. Lip region shape, also much used to separate species, is a very subjective character which is difficult to evaluate especially when comparing drawings of different species made in different styles by different specialists. Although supplementary help may be given by information obtained by one of the aforementioned methods, one has to accept that intraspecific variants do occur and that convincing specific identification is not always possible with individual specimens. There is also the possibility of undescribed species being encountered and if their description is deemed necessary as much information as possible should be given based at least on several specimens preferably from more than one population.

SOIL SAMPLING

Trichodorid species tend to aggregate about plant roots. Therefore it is advisable to take soil samples from the rhizosphere. Bor and Kuiper (3) noted that some species do not like 'rough handling' and that more specimens are recovered if the samples are taken in chunks (with a spade or trowel) or with a large diameter corer rather than with a small one (up to 85% loss of specimens). Also, samples should be treated gently rather than being dropped or sent through the post (up to 89% may be killed).

The large size of longidorids (up to 1 cm long) renders them susceptible to damage by soil sampling. Therefore large diameter cores (2.5 cm or more) or 'chunk' samples are advocated. Some longidorids seem to prefer relatively undisturbed soil and may occur below 30 cm, and allowance should be made for this when sampling.

EXTRACTION FROM SOIL

The generally large size of longidorid nematodes means that they can easily be extracted from most soils by the Cobb sieving and decanting method. About 200–400 ml of soil is mixed with water in a bowl, stirred vigorously at the same time breaking down any lumps of soil. The resultant slurry, except for heavy particles, is then poured through a graded series of sieves. A sieve with 2–4 mm apertures is useful to remove coarse debris and then a 250 μ aperture mesh sieve is used which should retain a good proportion of adult longidorids; more specimens can be obtained by passing the filtrate through finer mesh sieves. The method can be made quantitative by using standard amounts of soil mixed with standard volumes of water as advocated by Flegg (9). To avoid damaging specimens, especially if they are required for virus transmission tests, the sieve should be kept just submerged in water while the suspension is being poured onto it. Extracts from many soils when washed from the sieves can be examined, at a suitable dilution, in a counting dish viewed under a binocular dissecting microscope. However, it is difficult to see nematodes in extracts that have much organic debris and it is then best to pour the extract on to a suitable filter e.g. a coarse milk filter or nylon cloth with about 90 μ apertures. The filter is then left just submerged in clean water for several hours (overnight) in a suitable funnel or dish so that the nematodes can migrate from the debris. Relatively clean extracts can be obtained from larger samples (up to 1 litre) by using an elutriator such as the Oostenbrink Mark III in which the water flow is run at about 2 l/min. The extract is caught on large diameter sieves with mesh in the 250 to 150 μ range. However, depending on soil type, the extract may need cleaning by filtering.

Trichodorid nematodes can also be extracted by the Cobb sieving method but a sieve with about 100 μ apertures is required to retain adults of most species and finer mesh sieves down to about 50 μ apertures are necessary to obtain juveniles and adults of some species. The extracts from such fine sieves are generally rather dirty and further cleaning by filtering usually is required. The Seinhorst (20) two Erlenmeyer (conical) flask method is a simple way of obtaining much cleaner extracts and is recommended for soil samples of up to about 300 ml.

Although longidorids and trichodorids seem to filter through extracts obtained by sieving or elutriation they do not readily emerge from raw soil immersed in water. Therefore the Baermann funnel type extraction technique cannot be recommended unless the sample has already been partially extracted. There are many modifications and alternatives to the above mentioned methods and further details are given in Hooper and Flegg (14).

Dorylaimid nematodes may be inactivated or even killed by low concentrations of metal ions especially copper. Metal sieves and gauzes should not therefore be used in small quantities of static water. Sieves to be used for filtering extracts can be made from inert plastic rings with plastic or muslin cloth gauze attached.

MOUNTING AND MEASURING

Measurements can be made on specimens killed by heat (at about 65°C) and mounted temporarily, with coverslip supports, in fixative. Refractive structures such as spear, guide ring, spicules, body pores and to some extent amphids are usually most easily seen in specimens freshly fixed and mounted in TAF (formalin 7 ml; triethanolamine 2 ml; water 91 ml). More acid fixatives such as dilute formalin or formal acetic tend to instantly make the body contents rather dark and thus obscure structures. Although TAF is a good short term fixative/preservative, specimens should be transferred to one of the other more acid fixatives if they are to be stored for a long time. Details of genital tract structure are more easily seen in specimens that have been fixed and subsequently cleared in hot lactophenol and, if desired, processed to glycerol by the Baker rapid method. Better results are often obtained if specimens are killed and fixed with hot formaldehyde: propionic acid 4:1 and processed to glycerol by the Seinhorst rapid method (see Hooper in Southey (23)). Either method seems to give good results with longidorid nematodes but unfortunately trichodorids do not always make good permanent mounts.

Measurements usually include the standard de Man indices (L, a, b, c, V) with certain supplementary measurements and ratios (see Hooper _in_ Southey (23)). However, some measurements are more useful than others and for longidorids it is recommended, in the first instance, that the measurements L (mm), a, c' (tail length divided by anal body width), V and lengths of odontostyle and odontophore be made as routine. Short straight measurements can be made with a calibrated eyepiece scale. Other measurements can be obtained from drawings made with the aid of a camera lucida or drawing tube attached to the microscope; scale lines for comparison should always be added, at the time of making the drawing, from a stage micrometer.

With trichodorid species the relative position of pores and papillae, especially in males, and the spicule shape is very important, and it is often worthwhile making sketch outline drawings of the appropriate regions, with a drawing aid, so that specimens can be readily compared.

REFERENCES

1. Allen, M.W. (1957). A review of the nematode genus Trichodorus with descriptions of ten new species. Nematologica 2, 32-62.

2. Bird, G.W. (1971). Taxonomy: the science of classification. In: Zuckerman, B.M., W.F. Mai & R.A. Rohde eds., Plant Parasitic Nematodes. Vol.I. New York & Lond.: Academic Press. pp. 117-138.

3. Bor, N.A. & Kuiper, K. (1966). Gevoeligheid van Trichodorus teres en T. pachydermus voor uitwendige invloeden. Meded. LandbHoogesch. OpzoekStns Gent 31, 609-616.

4. Coomans, A. & Loof, P.A.A. (1970). Morphology and taxonomy of Bathyodontina (Dorylaimida). Nematologica 16, 180-196.

5. Dalmasso, A. (1969). Etude anatomique et taxonomique des genres Xiphinema, Longidorus et Paralongidorus (Nematoda: Dorylaimidae). Mem. Mus. natn. Hist. nat., Paris, Ser. A. Zool. 61, 33-82.

6. Dalmasso, A. (1970a). La gametogenese des genres Xiphinema et Longidorus (Nematoda: Dorylaimida). C.R. Acad. Sci. Paris 270, 824-827.

7. Dalmasso, A. (1970b). Influence directe de quelques facteurs ecologiques sur l'activite biologique et la distribution des especes Francaises de la famille des Longidoridae (Nematoda: Dorylaimida). Ann. Zool. Ecol. anim. 2, 163-200.

8. Ferris, V.R. (1971). Taxonomy of the Dorylaimida. In: Zuckerman, B.M., W.F. Mai & R.A. Rohde eds. Plant Parasitic Nematodes Vol. I. New York & Lond.: Academic Press pp. 164-189.

9. Flegg, J.J.M. (1967). Extraction of Xiphinema and Longidorus species from soil by a modification of Cobb's decanting and sieving technique. Ann. appl. Biol. 60, 429-437.

10. Hewitt, W.B., Raski, D.J. & Goheen, A.C. (1958). Nematode vector of soil-borne fan leaf virus of grape vines. Phytopathology 48, 586–595.

11. Hirumi, H., Chen, T.A., Lee, K.J. & Maramorosch, K. (1968). Ultrastructure of the feeding apparatus of the nematode Trichodorus christiei. J. Ultrastruct. Res. 24, 434–453.

12. Hirumi, H., Raski, D.J. & Jones, N.O. (1971). Primitive muscle cells of nematodes: morphological aspects of platymyarian and shallow coelomyarian muscles in two plant parasitic nematodes, Trichodorus christiei and Longidorus elongatus. J. Ultrastruct. Res. 34, 517–543.

13. Hooper, D.J., Pike, K. & Trudgill, D.L. (1973). [Soluble proteins of several vermiform nematodes]. In: Rep. Rothamsted exp. Stn for 1972, Pt. 1. p.161.

14. Hooper, D.J. & Flegg, J.J.M. (1970). Extraction of free living stages from soil. In: Southey, J.F. ed. Laboratory methods for work with plant and soil nematodes. Minist. Agric. Fish & Food. Tech. Bull. 2 London: H.M.S.O. pp. 5–22.

15. Jairajpuri, M.S. & Hooper, D.J. (1969). The genus Longidorella Thorne (Nematoda). Nematologica 15, 275–284.

16. Lima, M.B. (1968). A numerical approach to the Xiphinema americanum complex. C.r. 8e Symp. Int. Nematol. Antibes, Sept. 1965. p.30.

17. Loof, P.A.A. & Coomans, A. (1972). The oesophageal gland nuclei of Longidoridae (Dorylaimida). Nematologica 18, 213–233.

18. Maggenti, A.R. (1971). Nemic relationships and origins of plant parasitic nematodes. In: Zuckerman, B.M., W.F. Mai & R.A. Rohde, eds., Plant Parasitic Nematodes Vol. I. New York & Lond.: Academic Press. pp. 65–81.

19. Raski, D.J., Jones, N.O. & Roggen, D.R. (1969). On the morphology and ultrastructure of the oesophageal region of Trichodorus allius Jensen. Proc. helm. Soc. Wash. 36, 106–118.

20. Seinhorst, J.W. (1955). Een eenvoudige methode voor het afscheiden van aaltjes uit grond. Tijdschr. PlZiekt. 61, 188–190.

21. Seinhorst, J.W. (1963). A redescription of the male of Trichodorus primitivus (de Man), and the description of a new species T. similis. Nematologica 9, 125-130.

22. Siddiqi, M.R. (1974). Systematics of the genus Trichodorus Cobb 1913 (Nematoda: Dorylaimida), with descriptions of three new species. Nematologica 19, 259-278. (Year 1973).

23. Southey, J.F. ed. (1970). Laboratory methods for work with plant and soil nematodes. Minist. Agric. Fish. & Food. Tech. Bull. 2. London: H.M.S.O. 148 pp.

24. Taylor, C.E. (1972). Nematode transmission of plant viruses Pest. Art. News Summ. 18, 269-282.

DISCUSSION

Extraction methods

It was generally agreed that extraction methods varied between laboratories, modifications in standard methods reflecting particular local needs in relation to soil type, the kinds of nematodes to be extracted, or the way in which the nematodes were to be treated afterwards e.g. for fixation, for inoculation of plants, or for virus transmission. Indeed, Loof concluded that there was no single extraction method suitable for all situations, and several methods should be tried to ensure that the best can be selected, or adapted, for each nematode species.

Several points were made about the suitability of the modified Baermann funnel method which is frequently used for final separation of nematodes from soil debris. Seinhorst had found that Longidorus macrosoma does not pass readily through Ederol filter paper (N66) but L. caespiticola does and concluded this is not due to pore size and therefore must be due to some action of the paper on the nematodes. Cotten had found that L. macrosoma stuck in filters with 95 µ pores but he overcame the problem by increasing the pore size; Yassin said that previously he had found that a single layer of 'Kleenex' tissue gave better extraction of L. elongatus than when used double, and questioned whether pore size was a limiting factor, but Hooper pointed out that with the double filter several of the strands in one layer would lie across pores in the other direction and thus effectively reduce pore size. Cooper suggested that the surface charge of a nematode might play some role in its ability to move through filters. Pitcher suspected that many Dorylaimoids may be inhibited or killed by copper in solution, and reported that this had been found to be the case when phosphor-bronze screens were used for supporting filters. He said that Flegg (J.J.M. Flegg, Ann. appl.Biol., 60, 429-437 (1967)) had shown that separation was quicker, and with a better yield of nematodes, if a nylon screen of suitable mesh size was used rather than tissue or paper filters. McElroy reported that he had excellent results for Xiphinema bakeri, X. americanum and L. elongatus using the modified Flegg extraction procedure. However, he had found that nematodes allowed to remain in contact with soil debris collected on a 60 mesh screen were viable for several days whereas those in water alone were viable for only 24 hours.

Lippens said that he had found that chilling (below 10°C) soil prior to extraction affected the yields of Xiphinema, but the affect was only temporary. Thomason pointed out that nematode species differ in their sensitivity to low temperatures, for example, Meloidogyne javanica is affected at 10°C, but M. hapla not until 2°C, and Aphelenchus avenae can be stored without affect at 0°C and remains viable even when thawed out from ice. The depth of water in the tube was also important when storing Trichodorus, several cm depth being better than a shallow covering (Seinhorst).

The suitability of the centrifugal sugar flotation method for Xiphinema or Longidorus was discussed. Hooper thought that the technique was rarely used except for small samples, and said there was a problem of osmosis particularly with Trichodorus. Hackney said he had observed that the efficiency of the method was related to soil type; it was adequate for light, sandy soils but problems were encountered with clay soils. Jatala said that Separan 2610 (Dow Chem.Co.) was used in N. Carolina following the rapid technique of Byrd et al (D.W. Byrd, Jr., C.J. Nusbaum and K. R. Barker, Pl.Dis.Reptr., 50, 954-957 (1966)). Hooper said that clay was always found to be a problem in extraction. He also pointed out that damage to nematodes on the screens could partially be avoided by submerging the screens in water during the sieving procedure. He also said it was worthwhile looking at the raw extract collected on the screens to get an estimate of the numbers and kinds of nematodes present before proceeding further with the extraction process.

Taxonomy

In discussing the taxonomy of Trichodorus the question was raised about the family Ailaimidae, which Clark (W.C. Clark, Nematologica 7, 119-121 (1962)) had included with the family Diphtherophoridae in the suborder Alaimina; Kirjanova and Krall (E.S. Kirjanova and E.L. Krall, Parasitic nematodes of plants and their control. Vol. 1. Leningrad: Izdatelstvo "Nauka", 447 pp) grouped in Alaimina the families Alaimidae, Diphtherophoridae and Trichodoridae without assigning them to superfamilies. In the discussion Coomans pointed out that the Alaimids have been moved to the Oxystominidae by Gerlach (S.A. Gerlach. Mitt.biol.BundAnst. Ld-u.Forstw. 118, 25-39 (1966)) and that they seem to fit quite well in that group; hence Coomans and Loof (A. Coomans and P.A.A. Loof, Nematologica 16, 180-196 (1970)) considered only Diphtherophoridae and Trichodoridae under Diphtherophorina. Coomans stated that Nematoda have been considered a Class of the Phylum Aschelminthes or Nemathelminthes mainly on the basis of negative evidence. Taking into consideration all the groups belonging to the so-called Aschelminthes, only the Gastrotricha and the Nematoda show a relationship based on the general characters of embryology and comparative morphology although the Nematomorpha also show resemblances; but such relationships do not suggest that these forms belong in the same group. Taking account of available evidence, without entering into the realms of speculation, Coomans considered nematodes to be in a separately defined group i.e. Phylum.

NOTE: "stoma" in the longidorid section of this paper is equivalent to pharynx ss (= modified anterior stomodaeum) as proposed by Coomans and used by Hooper in following papers of these Proceedings.

MORPHOLOGY OF LONGIDORIDAE

A. COOMANS

Instituut voor Dierkunde

Rijksuniversiteit, Gent, Belgium.

INTRODUCTION

Longidorids are moderate to long (L varies from about 1.5 mm to about 13 mm), relatively slender, sluggish nematodes. Their general body shape, upon relaxation by gentle heat, is usually ventrally arcuate to C-shaped or spiral, rarely straight; the curvation is more pronounced towards the tail, especially in males. The anterior end (lip region) may be continuous with the body or offset to a variable extent. The tail is usually short, hemispheroid or conoid, with or without a peg; it may also be elongate conoid and even long and filiform. The variation in tail shape is largest in the genus Xiphinema. All representatives possess a very long spear used for puncturing plant cells on the contents of which they feed.

In describing the morphology of Longidoridae we have to rely on general principles of nematode morphology and on comparative data from related forms in order to understand the specializations that occurred during evolution.

A nematode in general can be considered as being composed of an outer tube - the body wall, and an inner tube - the digestive tract. The inner tube only makes contact with the outer one at at both its ends, while throughout its length it is separated from the outer tube by a fluid-filled space, usually called a pseudocoelome. In this body cavity lies also another tubular system, the genital tract, with an independent opening to the exterior in females, but joining with the digestive tract in males. The nervous system has its components mainly in the outer, but also in the inner tube, while its most obvious parts are lying in the body cavity.

BODY WALL

The body wall is composed of a cuticle, a hypodermis or epidermis, and the somatic musculature.

Under the light microscope 2 to 4 main layers are visible in the cuticle of longidorids. Internally there are radial striations that are oblique-anteriad in the head region and oblique-posteriad in the tail region, but transverse throughout the greatest part of the body; these striations form a criss-cross pattern of minute dots in surface view. The external layer shows faint to very faint superficial transverse striations which are often difficult to see under the light microscope, hence the cuticular surface is frequently described as "smooth". Usually the internal layers are thicker than the outer ones near both ends of the body. Throughout the body cuticular canals open at the surface as so-called body pores.

The cuticle is secreted by the hypodermis that consists of a very thin layer except mediodorsal, medioventral and lateral where it forms respectively the dorsal, ventral and lateral chords. The chords have a narrow basis in the anterior body region, the main part of it bulging out into the body cavity; in the rest of the body the lateral chords have a wider basis.

The hypodermal chords divide the somatic musculature in four fields with oblique longitudinal muscles. Longidorids, as most dorylaims, are coelomyarian.

DIGESTIVE TRACT

The digestive tract consists of mouth opening, vestibulum, stomodaeum, intestine, rectum and anus.

Cheilostome

Typically the mouth opening is surrounded by six lips and thus shows a hexaradiate symmetry. In longidorids the mouth opening is very small and the lips are amalgamated.

The vestibulum or cheilosome, originating from a secondary embryological invagination, is lined with cuticle which is continuous with that of the lips and hence this part also typically shows a hexaradiate symmetry. In longidorids and many other dorylaims this symmetry is less obvious.

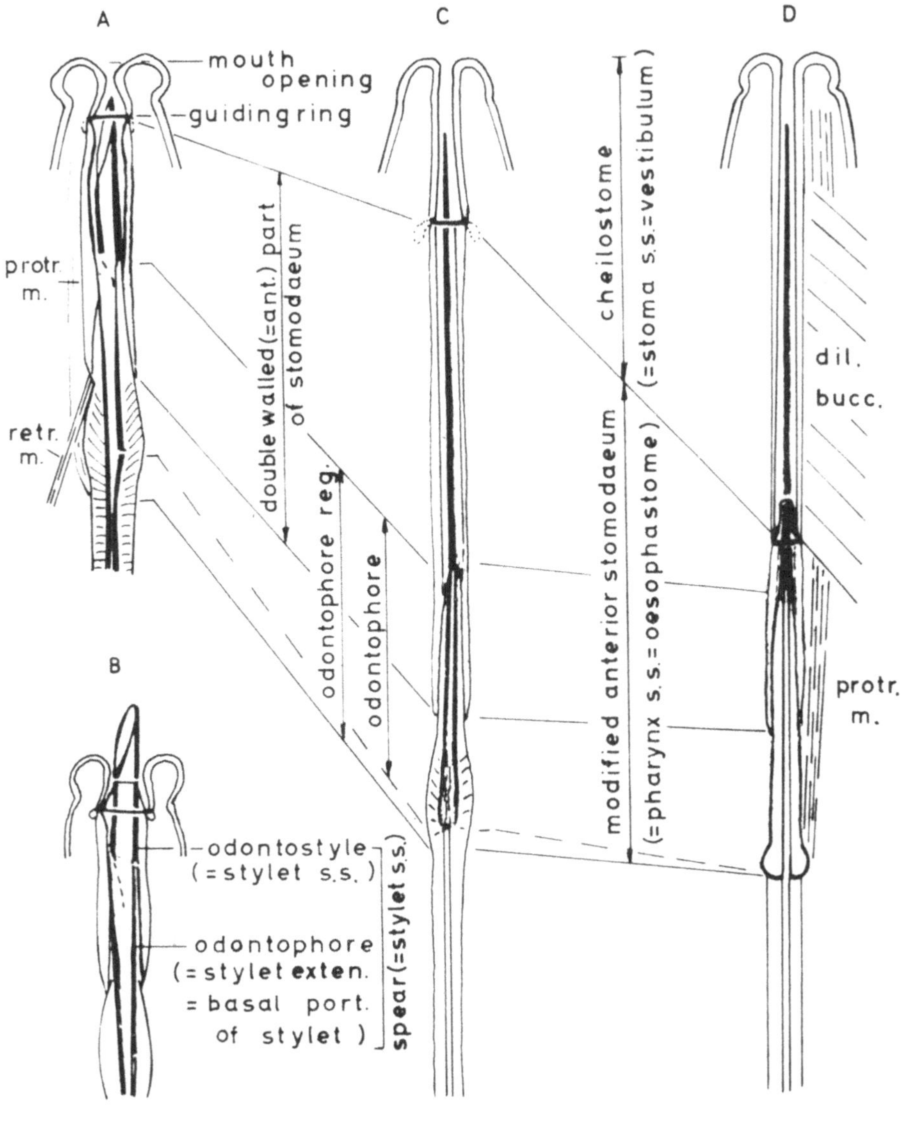

Fig.1. Terminology of the different parts of the anterior feeding apparatus in dorylaims. A: Primitive representative with spear in resting position. B: Same with spear protruded. C: Longidorus - Paralongidorus. D: Xiphinema. ant. = anterior, dil.bucc. = dilatores buccae, exten. = extension, port. = portion, protr.m. protractor muscle, reg. = region, retr.m. = retractor muscle, s.s. = sensu stricto.

Originally the cheilostome was about as long as the lip region height, but most present day dorylaims show an elongation of the cheilostome. This is also true for longidorids with the extreme in the genus Xiphinema.

Stomodaeum

The stomodaeum originated from a primary embryological invagination and typically shows a triradiate symmetry. It is lined with a different type of cuticle secreted by the invaginated hypoderm.

Where cheilostome and stomodaeum meet, a guiding ring is formed. Originally the ring was irregular and not sclerotized, but in longidorids, as well as in many other dorylaims, this ring became a sclerotized structure, easily recognisable under the light microscope.

The anterior stomodaeum is specialised in relation to food-uptake; it is also called oesophastome (12) or pharynx s.s. It consists of an anterior and a posterior part. The anterior part appears as double walled and partly surrounds a tube-like structure, the odontostyle. The latter originally was a rather short and wide cylinder, implanted on one sector of the stomodeal wall. In longidorids a considerable change in both the odontostyle and its associated structures has occurred: it has become very long (from about 50 to over 200 µm) and narrow and is implanted on the three fused sectors of the stomodeal wall.

The specialised anterior stomodaeum of longidorids can be better understood by comparing it with that of less specialised forms. The odontostyle can be derived from a U-shaped or grooved tooth implanted on a ventrosublateral stomodeal sector (either left or right). Such a tooth is still present in bathyodontids and nygolaims. A comparison of bathydontids and nygolaims shows an elongation and narrowing of the tooth in the latter (7); this is also true for the whole stomodaeum (from cylindrical to more flask-shaped). Together with this evolutionary change, the first part of the anterior stomodeal lining became thinner and less rigid, so that it could be folded upon protraction of the tooth (4).

Nygolaims are carnivorous animals that puncture their prey (Enchytraeids), withdraw the tooth and suck the body contents. In other forms (most present day dorylaims) the margins of the grooved tooth meet each other dorsally, thus forming a hollow cylinder, the odontostyle, which is closely surrounded by the stomodeal lining. The odontostyle is used to puncture the food material (either animal or plant according to the species) and remains protruded while feeding, so that the food passes through it.

When the odontostyle protrudes, the anterior stomodeal lining folds along it thus forming a guiding sheath. This could only be achieved when the inner cuticular lining became detached from the rest of the wall (the hypodermis that secreted the cuticle), except at its anterior (guiding ring) and posterior end and when the cavity in between became a fluid filled cavity acting as a separate hydrostatic system [cf. "hydrostatic tissue" of Taylor & Robertson (15) or hydrostatic substance]. This organisation already occurs in nygolaims and is further developed in dorylaims. The second part of the anterior stomodaeum acts as a supporting structure, the odontophore. In nygolaims it is formed by a ventrosublateral sector; in dorylaims this originally ventrosublateral sector becomes larger and ventral in position.

In longidorids the anterior stomodaeum became extremely elongated and very narrow. The odontostyle is a needle-like tube with a minute lumen through which plant sap can be sucked. Also the odontophore is now built up by the three stomodeal sectors thus giving a better support for the long odontostyle. Here, more than in other dorylaims, the terms spear or stylet are applicable to the functional unit of odontostyle + odontophore.

There is quite a difference between Longidorus-Paralongidorus on the one hand and Xiphinema on the other. In the latter genus the cheilostome is much longer, resulting in a more posteriorly situated guiding ring and a guiding sheath which is already partly folded beyond the guiding ring when the spear is in resting position. In Longidorus and Paralongidorus the cheilostome is shorter, the guiding ring more anterior and the guiding sheath straight and completely behind the guiding ring when the spear is in a resting position. The odontophore of Xiphinema is more sclerotized than in Longidorus and Paralongidorus and is posteriorly "flanged", but basically the structure is similar.

The very long cheilostome in Xiphinema apparently creates a special problem: upon protraction of the stylet a forwardly directed force is exerted upon the cheilostomial wall tending to compress it longitudinally (15). A special set of obliquely orientated, ligamentory muscles, the dilatores buccae, counteract this effect of stylet protraction in that they attach the cheilostomial wall to the body wall.

Protraction and retraction of the feeding apparatus is mediated by corresponding sets of muscles. Typically eight stylet protractor muscles are present in dorylaims, attached posteriorly to the stomodeal wall near the base of the odontophore and anteriorly to the lip region, thus running almost parallel with the feeding apparatus. This system also exists in longidorids. In most dorylaims there are eight stylet retractor muscles arranged in four pairs (one per

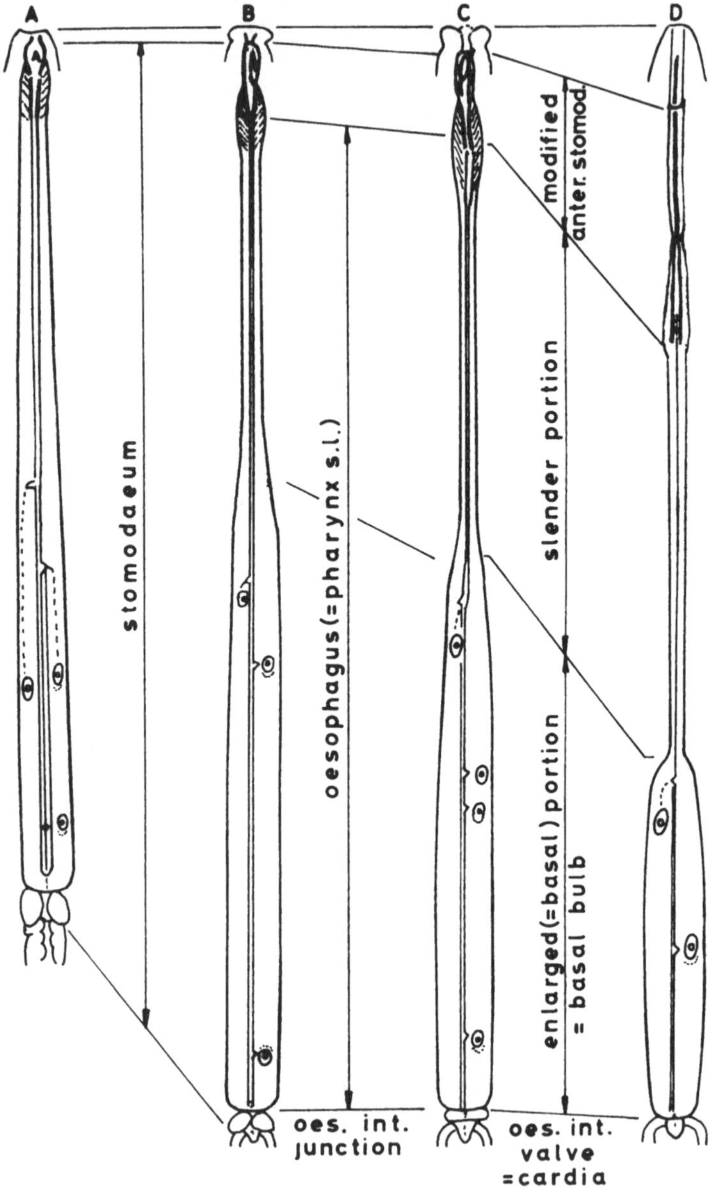

Fig. 2. Terminology and evolution of the dorylaimid feeding apparatus. A: Mononchulidae (Bathyodontina). B: Nygolaimidae. C: Dorylaimidae. D: Longidoridae. anter. = anterior, oes.int. = oesophago-intestinal, s.l. = sensu lato.

quadrant), running from the stomodeal wall (near the junction of anterior and posterior part of the oesophastome) posteriad to the body wall.

In longidorids there is some variation in the retractor system and true stylet retractors may even be absent (16).

Oesophagus

The next part of the stomodaeum is usually called the oesophagus, sometimes the pharynx, and it could also be called the posterior stomodaeum. In dorylaims it is typically flask-shaped with a relatively narrow anterior and wider posterior portion. Its cuticular lining shows a longitudinal series of triangular plates arranged in cross section as three pairs, one pair per sector. Radial muscles connect these platelets with the outer wall of the oesophagus. Upon contraction these muscles cause the lumen to open.

In longidorids and some other dorylaims the platelets are present only in the posterior, enlarged part of the oesophagus, which thus is the only part that can exert a pumping action. The narrow anterior part of the oesophagus shows a circular lumen which apparently cannot change in diameter. Because of this, the radial muscles of the narrow part no longer serve their original function, and either disappear almost completely or become involved in a different functional pattern in relation with a set of retractor muscles that connect the oesophagus with the body wall (16). The latter muscles may be responsible for the formation of a loop at the posterior end of the narrow part of the oesophagus in resting position.

The anterior portion of the oesophagus, together with the spear, forms a long narrow tube through which the food is pumped by the action of the enlarged part or basal bulb. Apart from being well muscularized this bulb also contains the oesophageal glands. Dorylaims usually possess five such glands, one dorsal and two pairs of ventrosublateral ones. In longidorids, however, the second pair has disappeared. This loss seems to be a recent one as testified by the fact that the outlets of this pair are often still visible and occasionally specimens are found possessing small nuclei. The dorsal gland opens into the lumen near the anterior end, just before the series of platelets starts; the ventrosublateral ones open about halfway along the bulb. This means that the secretion produced by the dorsal gland could flow towards the feeding apparatus while the main part of the lumen in the basal bulb remains closed.

There is a distinct difference between Xiphinema on the one hand and Longidorus-Paralongidorus on the other, with regard to location, size and shape of the dorsal gland nucleus (13).

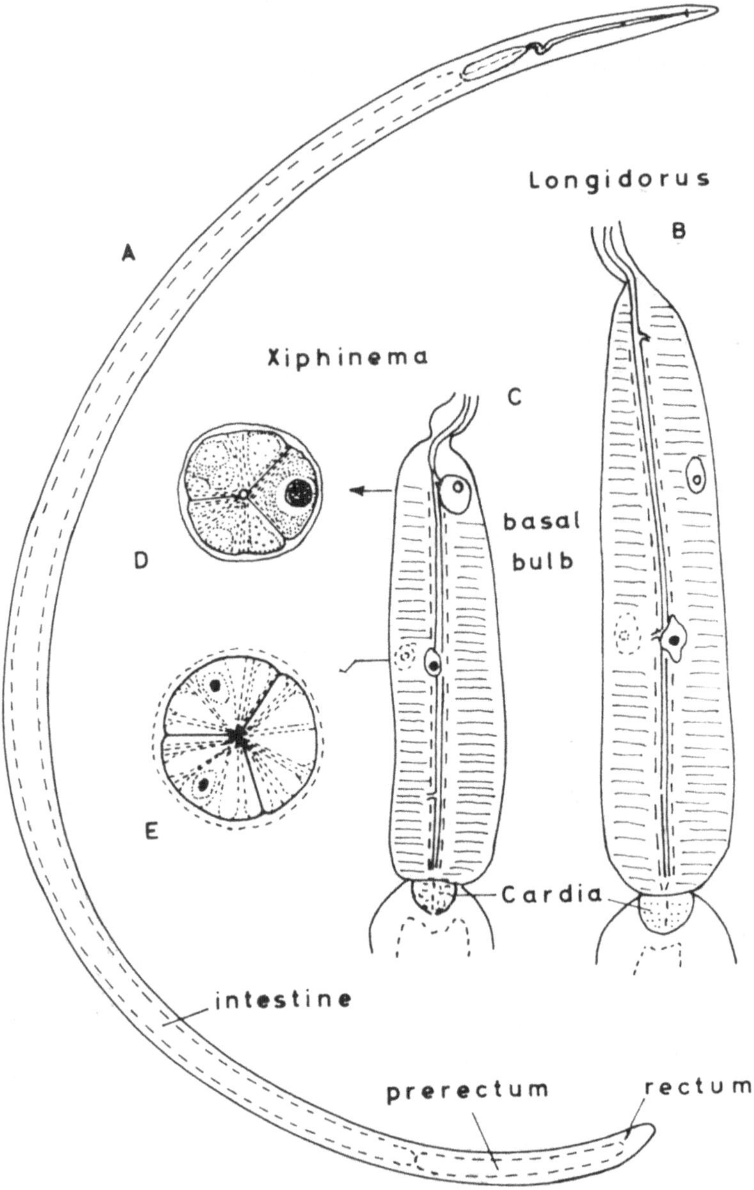

Fig. 3. Alimentary canal of Longidoridae. A: General structure. B: Basal bulb of Longidorus macrosoma. C: Basal bulb of Xiphinema diversicaudatum. D: Cross section at the level of the dorsal gland nucleus. E: Cross section at the level of the ventrosublateral gland nuclei.

In Xiphinema this nucleus is larger than the ventrosublateral ones, almost circular and close to the outlet; in the other genera it is smaller than the ventrosublateral ones, elongate and situated further backward. The basal bulb occupies a variable position in relation to the body wall, which indicates that it (and probably also other parts of the stomodaeum) rotates along its longitudinal axis during forward and backward movements of the spear.

The oesophagus is connected with the intestine by means of a one way oesophago-intestinal valve, often called cardia, through which the food is injected into the intestine. This valve represents the terminal part of the stomodaeum and typically it is also triradiate.

Intestine

The intestine is a long endodermal tube, the wall of which consists of one layer of epithelial cells provided with microvilli that project into the lumen. The cytoplasm often contains many granules with reserve food material.

The terminal part of the intestine is called the prerectum. It is several body widths long, usually less opaque and separated from the intestine by a valve-like structure formed by columnar cells. The prerectum leads to the rectum which is again ectodermal and lined with cuticle. The junction between prerectum and rectum is guarded by a sphincter muscle. Defaecation is mediated by another muscle, the H-shaped "depressor ani", extending from the dorsal wall of the rectum and the posterior lip of the anus to the latero- or sub-dorsal body wall. Since it acts in dilation rather than depression, the muscle should be better called "dilator ani".

BODY CAVITY

The body cavity is quite large and is usually called a pseudocoelome, but it is difficult to find embryological evidence for this, since the blastocoel disappears completely during development and the body cavity originates in fact as a kind of schizocoel. The fluid-filled cavity provides the nematodes with a hydrostatic skeleton which is unique because of its high turgor pressure (10). The implications of this system are considerable and account for the uniformity of the nematode body plan; variations are apparently only possible when they do not interfere with the turgor pressure system.

A number of structures protrude in or run through the body cavity: the hypodermal chords in the stomodeal region, several elements of the nervous system, the reproductive system and possibly the "excretory" system.

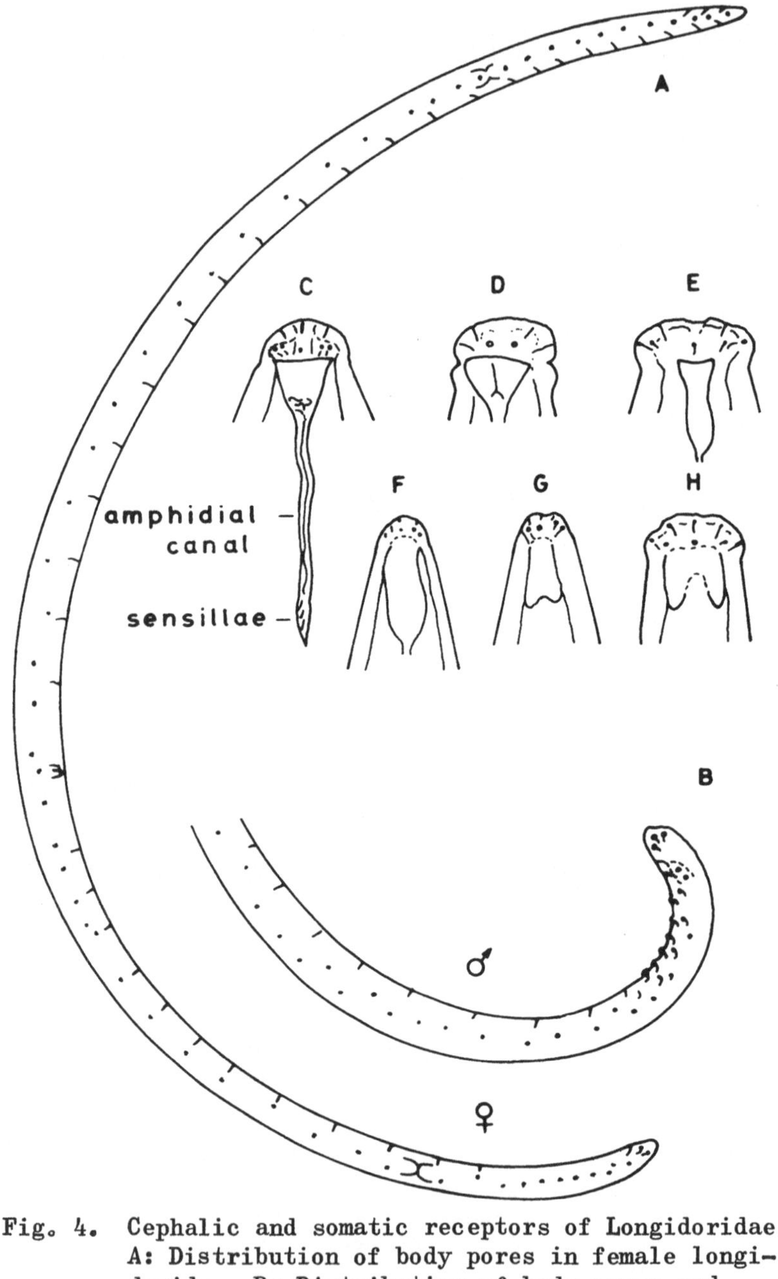

Fig. 4. Cephalic and somatic receptors of Longidoridae. A: Distribution of body pores in female longidorids. B: Distribution of body pores and supplements in the posterior body region of male longidorids. C–H: Shape of amphid in *Xiphinema* (C), *Paralongidorus* (D, E), and *Longidorus* (F–H).

Nervous system

The nervous system consists of a nerve ring (= circumoesophageal commissure) with associated ganglia as the main centre from which longitudinal nerves run forward and backward. The ganglia at the anterior side of the nerve ring are those connected with the cephalic sense organs, those at the posterior side are mainly connected with the dorsal, lateral and ventral nerves that run through the respective hypodermal chords. Ganglia also occur in the anal region. The motor nerves that control the body movements lie in the dorsal and ventral chords, the dorsal nerve controlling the somatic muscles of the dorsal half and the ventral nerve those of the ventral half. The connection between muscles and nerves are formed by processes of the muscle cells.

The only parts of the nervous system readily recognisable under the light microscope are the nerve ring, the large ventral and lateral ganglia, the small dorsal ganglion and the receptor organs. The latter can be divided in cephalic and somatic receptor organs.

Cephalic sensory system. This comprises six internal labial, six external labial and four cephalic mechanoreceptors, all situated on the lips, and two lateral chemoreceptors, the amphids, just postlabial in position. In most dorylaims the mechanoreceptors are arranged in two circles (6+10) and are papilliform. In all of them the amphid is pocket-like. Amongst longidorids, Xiphinema shows a typical goblet- or stirrup-shaped amphid with a large, slit-like opening. In Paralongidorus the amphid opening is also a transverse slit but the pouch is variable in shape. In Longidorus the amphid opening is a minute pore and the pouch is very large with or without a bilobed base. The sensillae are situated rather far behind the amphidial opening.

Somatic sensory system. This comprises longitudinal series of receptors that are in connection with the environment by the so-called body pores. These occur along the lateral, dorsal and ventral body sides. Dorsal body pores are median in position and occur near the anterior end only; ventral body pores are also in median position, but are present throughout the greatest part of the body. In females, ventral pores may be lacking in the vicinity of the vulva and in the posterior prerectal region; in males they continue till the region of the supplements (= male genital papillae), which may be considered as modified and more elaborate ventral receptors. The lateral body pores are distributed in a single, somewhat irregular, line in the stomodeal region and in two or three lines in the rest of the body (this is correlated with the width of the lateral chord-basis). In females the number of lateral body pores diminishes posteriad, but in males there is again an increase in number in the region of the supplements. In this area a number of lateral body pores shift to a ventrosublateral position.

Excretory systems

Whether or not longidorids possess an excretory system is still doubtful. One of the ventral pores in the region of the nerve ring has often a somewhat different appearance and has been described as a rudimentary excretory pore. Also, a glandular structure with two cells has been tentatively described as an excretory system (1), but this structure is situated at the same place as the ventral ganglia and it could also be a part of these.

Reproductive system

The reproductive system consists of the gonads and the gonoducts, which are continuous with each other.

Females. In females there are typically two branches extending in opposite directions from an almost equatorial vulva, a condition called amphidelphic. In some cases a reduction of one of the branches occurs; this condition is called monodelphic (complete loss of one of the branches) or pseudomonodelphic (partial reduction of one of the branches). The monodelphic condition is usually associated with a shift in the position of the vulva towards the side of the reduced branch.

Most longidorids are amphidelphic, but some species of *Xiphinema* are pseudomono- or monodelphic (see Luc in these Proceedings). The reduction occurs in the anterior branch. Although all these species have an anteriorly placed vulva, a number of didelphic species also show a pre-equatorial vulva; this is apparently related with the rather small area occupied by the reproductive system in the large body.

Each complete genital branch consists of ovary, oviduct and uterus. The ovary is reflexed at its junction with the oviduct. In the ovary the developing gametes are enclosed within a sac-like structure, which can extend considerably when large oocytes are produced. At its ventral side this ovarial sac is connected with the oviduct which consists first of a narrow portion becoming gradually or more abruptly enlarged near the junction with the uterus. This enlarged portion may act as a spermatheca (= receptaculum seminis). The uterus is separated from the oviduct by a narrow portion provided with a sphincter muscle. In *Longidorus* and *Paralongidorus* the uterus is a simple, moderately long tube, whereas in *Xiphinema* it may be very short (about one body width in *X. americanum* to very long and with a complicated structure (5, 14). In the most complex form it consists of: (1) an enlarged distal portion with gland cells secreting a substance which plays a role in egg shell formation and eventually a part which may act as a

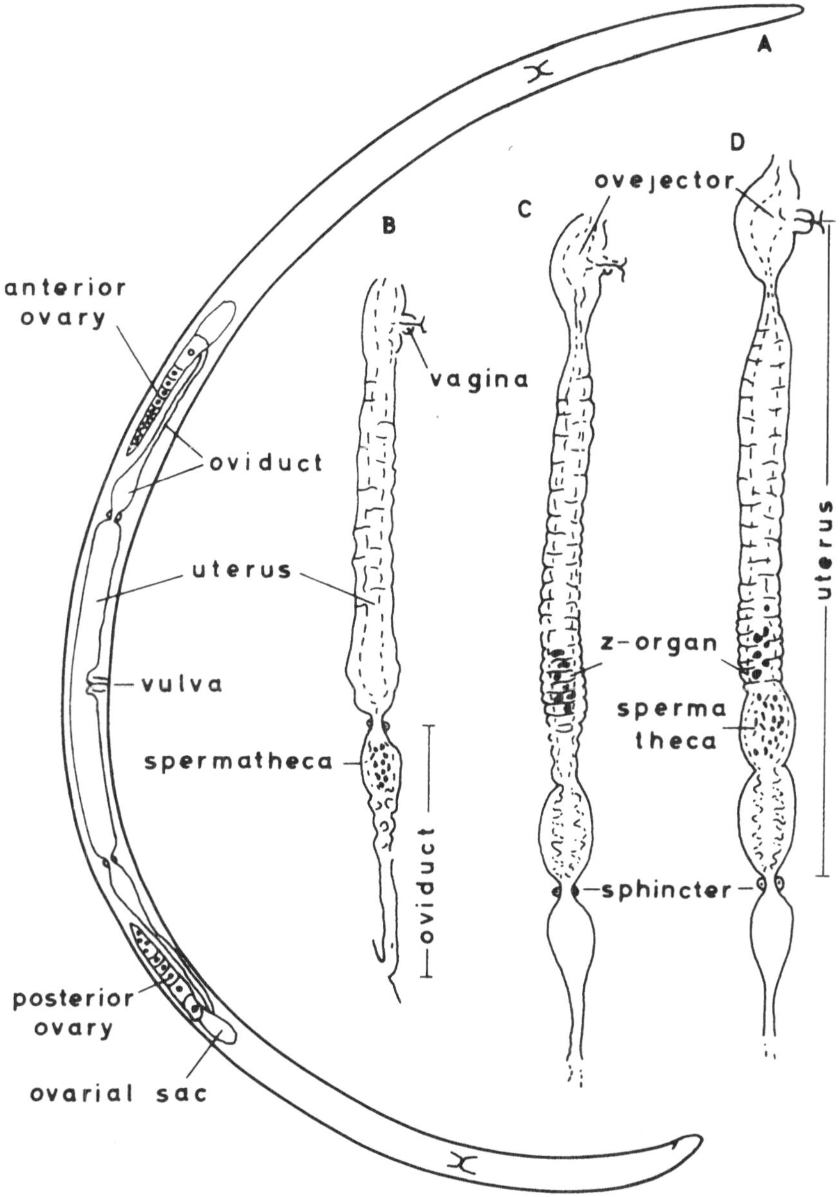

Fig. 5. Female reproductive system of Longidoridae. A: Position of the entire unspecialized, amphidelphic, didelphic system. B–D: Part of the posterior genital branch showing increasing complexity in the uterus.

spermatheca; (2) a long tubular portion with muscular wall that may show local differentiations such as the Z-organ, containing a variable number of irregular sclerotised or globular bodies of unknown nature; (3) a proximal enlarged part usually separated from the remainder of the uterus by a constriction and forming together with the proximal part of the opposite uterus a more or less kidney-shaped, well muscularized ovijector. This structure leads to the vagina, which is a small tube at a right angle to the body axis, with first a cross-shaped and then a slit-like lumen connecting with the transverse slit of the vulva. To the vagina attach different sets of muscles, some of which serve dilation, others constriction of the vagina, while still others suspend the vagina in the body thus preventing it from bulging out of the body during egg laying. Egg laying is also mediated by the vulval dilators which connect the vulva with the lateral body wall. Eggs are elongate oval in shape; the shell apparently is formed in the uterus and becomes rigid only after the eggs have been laid.

Males. In males the reproductive system consists of two testes and a single vas deferens whose terminal part is often called an ejaculatory duct (1). The posterior testis being reflexed, both testes are described as opposed and join each other where they open in the vas deferens. In each testis the gametes are surrounded by an epithelio-muscular sac. The oblique muscle bands almost completely surround the testis; they are weakly developed near the blind end of the sac, but well developed in the remaining part. Upon contraction they apparently inject sperm into the vas deferens. The junction between both testes and the vas deferens is not guarded by a special sphincter or a valve, but the oblique muscle bands continue over the vas deferens and are strongly developed here. The epithelial cells of the vas deferens are highly granular and apparently glandular; they probably produce a secretion that is mixed with the spermatozoa upon ejaculation. It seems logical to postulate that the obliquely arranged muscles of the whole vas deferens provoke the ejaculation of the sperm, and hence the term "ejaculatory duct" for the terminal part of the vas deferens seems rather inappropriate here. Caudally the vas deferens joins the very short rectum, forming a cloaca. The ejaculatory glands open into the system near this junction; the glands consist of a series of at least four uninucleate glands with long ducts lying at each side of the vas deferens. There are also three pairs of rectal glands connected with the dorsal wall of the cloaca. The function of both these glands is not properly understood, but they may perhaps act as cement glands or produce a lubricating material for the copulatory apparatus. Copulation is mediated through a pair of sclerotised structures, the spicules, enclosed in spicular sacs on which protractor and retractor muscles are inserted. The spicules of longidorids are arcuate and well developed; along their outer distal sides occur small lateral guiding pieces (= crura), representing the only more or less sclerotised parts of a protecting

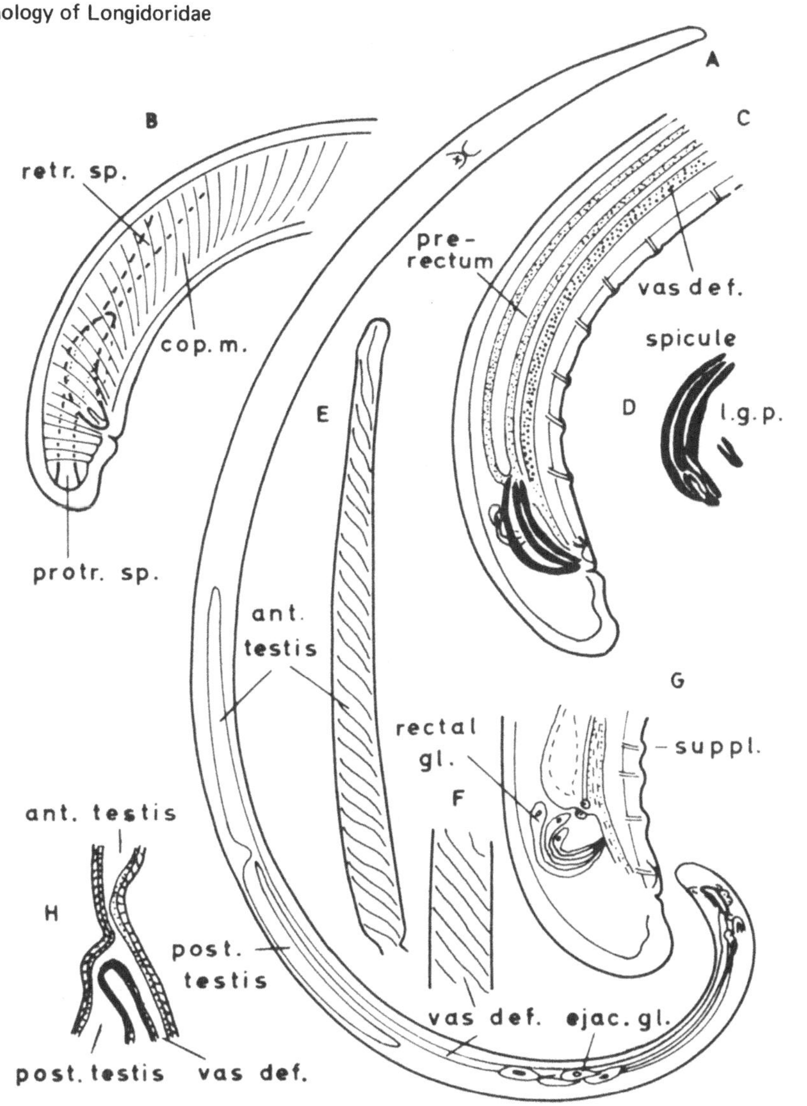

Fig.6. Male reproductive system of Longidoridae. A: Position of the entire, diorchic system. B: Muscles in the posterior body region. C: Posterior body region. D: Spicule and lateral guiding piece. E: Musculature of testis. F: Musculature of vas deferens. G: Junction of vas deferens and rectum. H: Junction of both testes with the vas deferens. ant = anterior, cop.m. = copulatory muscles, ejac.gl. = ejaculatory glands, l.g.p. = lateral guiding piece, post = posterior, protr.sp. = protractor spiculae, rectal.gl. = rectal glands, retr. sp. = retractor spiculae, suppl. = supplement, vas def. = vas deferens.

and guiding structure, the gubernaculum. During copulation the spicules are inserted in the vagina, forming a tube or trough through which the sperm is introduced in the female. The protraction is brought about by a large protractor muscle that runs from the proximal end of the spicule to the tail (terminal or subdorsal wall). Each spicule can be retracted by a well developed retractor muscle, also attached to the proximal end of the spicule but extending anteriad to the subdorsal body wall. Since longidorids do not possess caudal alae, the male's posterior end curves around the female during copulation. In relation to this a large number of obliquely arranged copulatory muscles occur in the posterior body region of the male. These muscles insert on the laterodorsal body wall and run posteriad to the subventral body wall, thus provoking not only curvation but also dorso-ventral flattening of the body when they contract. In this region the supplements or copulatory papillae are found, consisting typically of an adanal pair and a medioventral series (from 1 to 20), but in some Longidorus and Paralongidorus species the ventral series is arranged partly in a staggered, double row.

MORPHOLOGY OF JUVENILE FORMS

Sexual dimorphism seems to be rare in Longidoridae. In only two species of Xiphinema males and females have markedly different tail shapes (11). Two intersexes have been reported, both are essentially females with rudimentary (2) or rather well developed (3) but apparently non-functional male sexual characters.

So far only the adult morphology has been dealt with. The adult is, however, the last stage of the life cycle and it is also interesting to know how the other stages look like, particularly because juvenile forms may constitute a large part of a population.

During postembryonic development there is of course a considerable increase in body size, but often there is also a change in body shape and tail shape. Whereas body growth seems possible throughout the whole juvenile period, drastic changes, as for example in tail shape, occur during the four moulting processes that mark the transitions from one juvenile stage to the next and finally to the adult. When the different juvenile stages are compared one obtains information about the growth pattern of different organs or body regions. So, for example, it has been found that the oesophagus and the tail show a negative allometric growth. The reproductive system is elaborated mainly during moulting and especially during the last moulting period. From the third stage onward the spicular primordia can be recognised. The juvenile stages also provide information about the origin and composition of certain structures such as the anterior stomodaeum. Because of its importance for phylogenetic considerations it is worth having a clcser

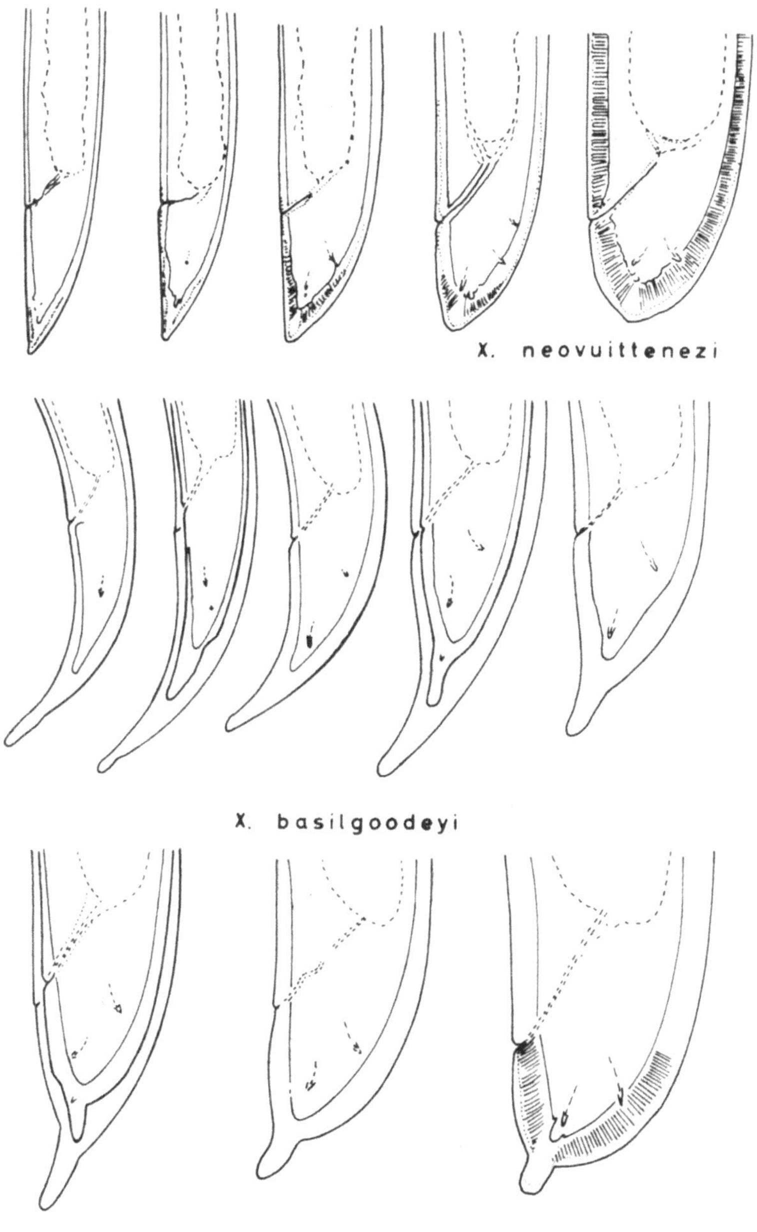

Fig. 7. Body shape of juvenile stages (1–4 and adults in one Longidorus and three Xiphinema species.

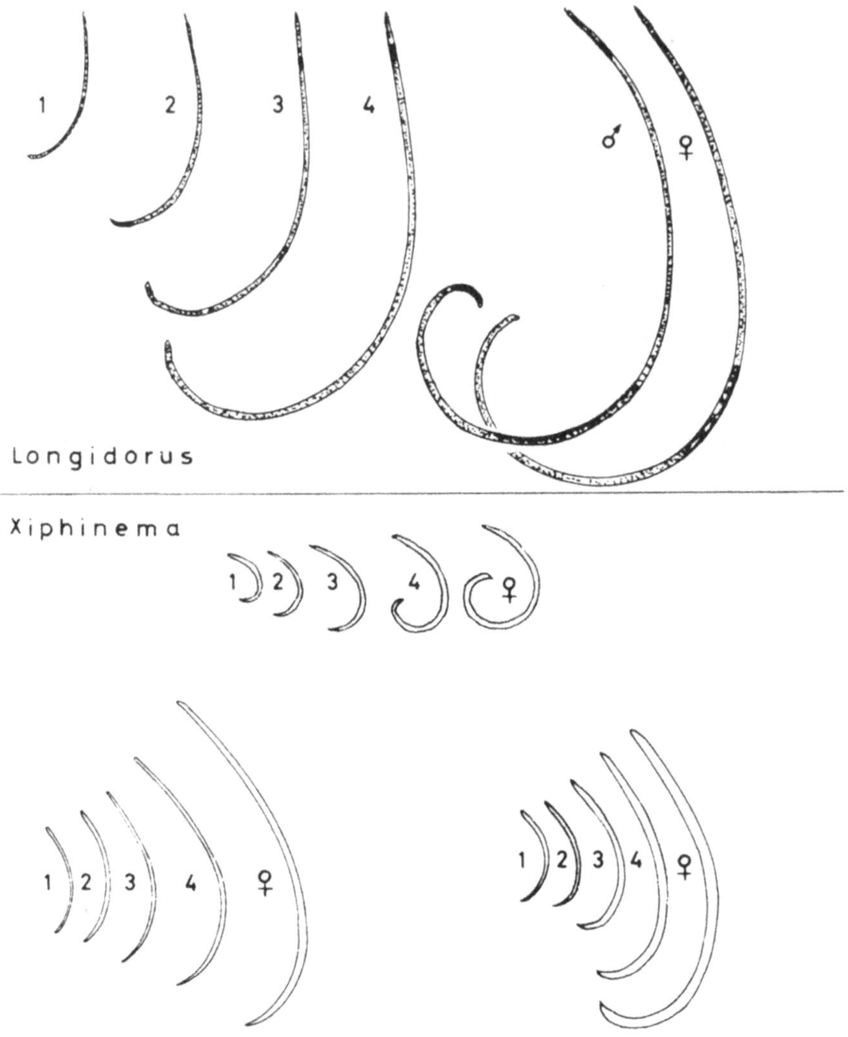

Fig. 8. Tail shape in the four juvenile stages and the adult in <u>X. neovuittenezi</u> and <u>X. basilgoodeyi</u>. For the latter species also the moulting stages are illustrated.

look at this.

The development of the anterior stomodaeum of dorylaims in general and of longidorids in particular starts with the elaboration of a tooth or odontostyle in the embryo, soon followed by the formation of a second one just behind it (9). When the first stage juvenile hatches it possesses a functional tooth or odontostyle on top of an odontophore and a replacement tooth or odontostyle within the wall of the odontophore region. Because of the length of the odontostyle in longidorids, only the anterior half of the replacement odontostyle is embedded within the wall of the odontophore. During the first moult the functional odontostyle of the first juvenile stage (J_1) is shed together with the cuticle of the body, cheilostome and inner lining of the anterior stomodaeum; the replacement odontostyle is shifted forward and will become the functional odontostyle of the next stage (J_2). In the meantime a new replacement odontostyle is secreted in the region of the odontophore (which at that time is also rebuilt and therefore not recognisable as such) by a large long necked cell whose cell body is situated in the narrow part of the oesophagus. Once it is completely formed the replacement odontostyle, however, is shifted backward so that it becomes enclosed within the wall of the slender portion of the oesophagus. The same phenomenon can be observed during the second (J_2- J_3) and third (J_3 - J_4) moults (6). In the last moult (J_4 - adult) a replacement odontostyle is no longer formed although occasionally a tip (mucro or vestigium) is formed and exceptionally even a complete odontostyle. The peculiar difference between the position of the replacement odontostyle in J_1 and J_2 - J_4 can only be interpreted after comparison with the situation in other Dorylaimida. As mentioned before, the odontostyle can be derived from a grooved tooth as found in bathyodontids and nygolaims. In the bathyodontids the replacement tooth is formed just behind the functional one and remains in that position in the intermoult. In nygolaims the replacement tooth is formed a short distance behind the functional tooth, within the region of the odontophore, and also remains in that position during the intermoult. In dorylaims with an odontostyle this situation only exists in the first juvenile stage; in the other stages there is a secondary posterior shift of the replacement odontostyle (8). This may be explained in terms of a different functioning of the anterior stomodaeum and the considerable elongation of it. This elongation is extreme in longidorids so that the replacement odontostyle is extending beyond the region of the odontophore in J_1. Protective caps prevent the needle-like odontostyle from piercing the wall of the oesophagus while the latter is moving forwards or backwards.

This example illustrates the importance of the inclusion of juvenile stages in morphological studies in order to understand the adult structures and to evaluate the relationships between the different groups of dorylaims.

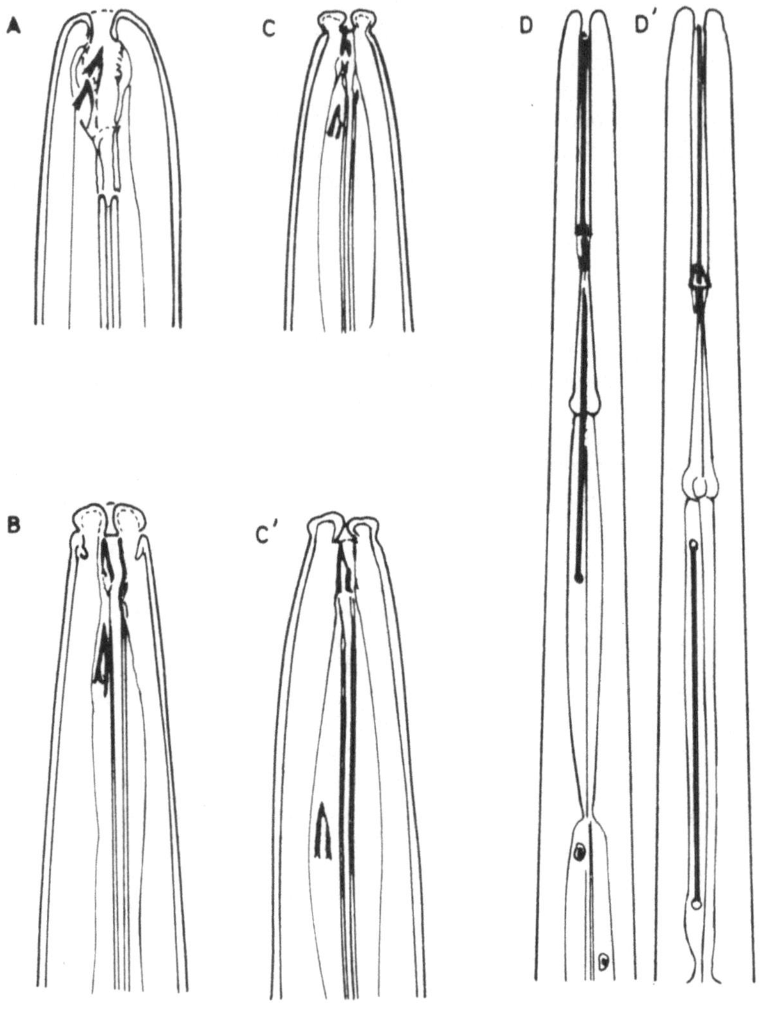

Fig. 9. Position of the replacement tooth or odontostyle in A: Mononchuloidea (Bathyodontina), all juvenile stages; B: Nygolaimoidea, all juvenile stages; C–C': Dorylaimoidea, first juvenile stage (C) and other juvenile stages (C'); D–D': Longidoridae, first juvenile stage (D) and other juvenile stages (D').

It is evident that morphological structures should be interpreted in relation to their function, but also morphogenesis provides a means of making comparative morphology more meaningful.

REFERENCES

1. Aboul-Eid, H.Z. (1969). Histological anatomy of the excretory and reproductive systems of Longidorus macrosoma. Nematologica 15, 437-450.

2. Aboul-Eid, H.Z. & Coomans, A. (1966). Intersexuality in Longidorus macrosoma. Nematologica 12, 343-344.

3. Cohn, E. & Mordechai, M. (1968). A case of intersexuality and occurrence of males in Longidorus africanus. Nematologica 14, 591.

4. Coomans, A. (1964). Stoma structure in members of the Dorylaimina. Nematologica 9 (1963), 587-601.

5. Coomans, A. (1965). Structure of the female gonads in members of the Dorylaimina. Nematologica 10 (1964), 601-622.

6. Coomans, A. & de Coninck, L. (1963). Observations on spear-formation in Xiphinema. Nematologica 3, 85-96.

7. Coomans, A. & Loof, P.A.A. (1970). Morphology and taxonomy of Bathyodontina (Dorylaimida). Nematologica 16, 180-196.

8. Coomans, A. & van der Heiden, A. (1971). Structure and formation of the feeding apparatus in Aporcelaimus and Aporcelaimellus (Nematoda: Dorylaimida). Z. Morph. Tiere 70, 103-118.

9. Flegg, J.J.M. (1968). Embryogenic studies of some Xiphinema and Longidorus species. Nematologica 14, 137-145.

10. Harris, J.E. & Crofton, H.D. (1957). Structure and function in nematodes: internal pressure and cuticular structure in Ascaris. J. exp. Biol. 34, 116-130.

11. Heyns, J. (1966). Studies on South African Xiphinema species, with descriptions of two new species displaying sexual dimorphism of the tail (Nematode: Dorylaimoidea). Nematologica 12, 369-384.

12. Inglis, W.G. (1966). The origin and function of the cheilostomal complex in the nematode Falcaustra stewarti. Proc. Linn. Soc. Lond. 177, 55-62.

13. Loof, P.A.A. & Coomans, A. (1972). The oesophageal gland nuclei of Longidoridae (Dorylaimida). Nematologica 18,

14. Luc, M. (1961). Structure de la gonade femelle chez quelques especes du genre Xiphinema Cobb, 1913 (Nematoda: Dorylaimoidea). Nematologica 6, 144-154.

15. Taylor, C.E. & Robertson, W.M. (1971). Ultrastructure of the guide ring and guiding sheath in Xiphinema and Longidorus. Nematologica 17, 303-307.

16. Taylor, C.E. & Robertson, W.M. (1973). The structure and musculature of the feeding apparatus in Longidorus and Xiphinema. In: The Longidoridae - Association of Applied Biologists Workshop Manual June 14, 1973. Harpenden: Rothamsted exp. Stn., 65-70.

DISCUSSION

Photographic evidence presented by Lamberti and Mrs. Grimaldi of a membranous envelope with scattered nuclei surrounding the gonads in Longidorus led to the suggestion that the space between membrane and gonads might represent a coelom. Coomans said a similar membrane existed in Priapulus but there was so far no embryological evidence of a coelom in nematodes. If the space referred to was regarded as a coelom then it would originate late in ontogenetic development and if it were a true coelom it could be seen in the embryo. Lippens asked about the difference between the structure as shown and a pseudocoelomic cavity, to which Coomans replied that the origin of the pseudocoelom was not known; perhaps it originated from a primitive cavity of the schizocoel type (with a few holes) or it may be a degenerate coelom. He said opinions of Anglosaxon and German authorities varied on this subject and various classifications and phytogenetic trees had consequently been erected.

The presence of a second stylet in adults of Longidorus sp. in a population from the Ardeche was referred to by Seinhorst; Coomans said this phenomenon appeared to be more common in some populations than others and referred to the explanation given by Dalmasso (A. Dalmasso, Mem.Mus.hist.nat. 61, 33-82 (1969) Loof pointed out that double stylets had been observed in other Dorylaims but was probably most common in Xiphinema; he considered the character of no taxonomic value, and neither is the presence of a mucro.

Lamberti observed that he had found several intersex forms to add to the "Coomans list". Lippens observed that the most common intersex was said to be female with male attachments and that some populations of marine nematodes commonly showed this condition, and some of which were fertilized suggesting that every individual was potentially female. Loof questioned whether the condition in nematodes should preferably be referred to as gynandromorphy. Hackney raised the possibility of nematode intersex forms and secondary sexual characters resulting from metabolic blockage of male or female hormones. Coomans replied that there was a danger in applying vertebrate definitions to nematodes, particularly as nothing was known of their endocrinology. Sex is not well fixed in invertebrates and he referred to examples of sex inversion in mermithids and in Meloidogyne.

Replying to a question about the nature of guide rings, Coomans stated that in Xiphinema there was a sclerotized annulus at the junction of the cheilostome and oesophostome (the true guide ring) which was used in taxonomy, and a second ring anterior to this formed an optical illusion by the folding of the guiding sheath, which should not be used in taxonomy. Robertson pointed out that electron microscopy had shown that the guide ring in Xiphinema and Longidorus is fused into the base of the cheilostome (see Robertson and Taylor in these Proceedings), and also explained that the true guide ring in Xiphinema moved anteriad when the stylet was fully protracted.

Pitcher asked for information about micro-organisms which may obscure the reproductive tract in X. americanum; he referred to published literature in which bacteria were said to be present and wondered whether the presence of these organisms had something to do with the difficulties encountered in culturing X. americanum. Coomans referred to the finding of microsporidia in Xiphinema (M.R. Morone de Lucia & S. Grimaldi de Zio. Nematol. Mediterranea 1, 66-68 (1973)), which was an important observation since as long as information is lacking one could assume that bacteria are symbiotic, but microsporidia are undoubtedly pathogens, and this may account for the difficulty in culturing Xiphinema species in which they occur. Should microsporidia also be present to some extent in natural populations, it has to be postulated that either the parasitism is mild (otherwise the parasite exterminates itself in the long run) or infested nematodes can still reproduce a sufficient number of eggs at the beginning of the parasitism to maintain the populations.

TAXONOMY OF *XIPHINEMA* COBB, 1913

MICHEL LUC

Laboratoire de Nematologie, O.R.S.T.O.M.

Dakar, Senegal.

HISTORICAL

The first species of *Xiphinema* described as such was *X. americanum*, the type species, by Cobb in 1913 (2). But twenty years before (1), Cobb himself had described a nematode from the Fiji Islands under the name of *Tylencholaimus ensiculiferus*, which was attributed later to *Xiphinema* by Thorne (13) and of which the accurate redescription, on topotypes, was published only last year by Southey and Luc (12). Somewhat earlier this species was determined by Cohn and Sher (3) from Hawaii and the Philippines and by Yeates (14) from the New Hebrides. Thus, nearly eighty years separate the first and the second record of the oldest established valid species of the genus. This example is given here only to consider with some care the doubts that are given on the validity of some species because they were "never found since their description".

CHARACTERIZATION OF THE GENUS

The genus *Xiphinema* belongs to the family Longidoridae Meyl, 1961 (11); this family is characterized among Dorylaimoidea by:

- a long attenuated odontostyle supported by a rather long basal extension (or odontophore)

- the posterior enlarged part of the oesophagus reduced in length, and

- similar tails of females and males (although two species of

Xiphinema described by Heyns (5) from South Africa, X. dimorphicaudatum and X. variabile, exhibit a distinct sexual dimorphism).

The family Longidoridae includes only three genera: Xiphinema, Longidorus and Paralongidorus.

The characters of the genus Xiphinema are rather constant and firmly established; they permit its easy separation from Longidorus and Paralongidorus, the differences between these last two genera being less strictly established. So it was justified, in our opinion, for Dalmasso (4) to establish the subfamily Xiphineminae with Xiphinema as the type and only genus as opposed to Longidorinae (Thorne, 1935) Dalmasso, 1969 emend., with the two genera Longidorus and Paralongidorus.

The diagnoses of these two subfamilies are as follows:

Xiphineminae:

Chromosome number (n) : 5 or 10. Length of adult not exceeding 6 mm. Coefficient "a" generally lower than 80. Stirrup-shaped amphids. Strong odontostyle, forked at the junction with odontophore : base of odontophore with well developed flanges. Guiding apparatus of the odontostyle apparently tubular and posteriorly located (see also Robertson and Taylor in these Proceedings). Squat muscular oesophageal bulb. Uterus often showing strong and peculiar differentiations. Shape and length of the tail very variable (coefficient "c'" varying from 0.6 to 15). Males with at most 7 supplements.

Longidorinae:

Chromosome number (n) : 7 (only three species of Longidorus observed). Length of adult variable but exceeding 6 mm in some cases. Coefficient "a" variable, over 80 for the largest species. Amphids pouch-like or stirrup-shaped. Weak odontostyle, not forked at the junction with odontophore : odontophore without or with very weakly developed basal flanges. Guiding apparatus of the odontostyle appearing as a single ring, anteriorly situated. Muscular oesophageal bulb elongated. Uterus without or with very weak differentiations. Tail generally conoid - rounded (coefficient "c" rarely exceeding 2.5). Males with more than 7 supplements (sometimes with 6 or 7, but rarely).

Moreover, Loof and Coomans (6) observed recently that the genera Longidorus and Paralongidorus on the one hand, and Xiphinema on the other show a distinct difference with regard to location, shape and size of the nucleus of the dorsal oesophageal gland. These observations support the subdivision of the Longidoridae into two subfamilies.

In fact Xiphinema species are among the most easily recognizable phytoparasitic nematodes under a dissecting microscope, mainly because of their long spear (odontostyle + odontophore) with refractive basal flanges. The only genus with which the Xiphinema species may be confused is the genus Xiphinemella Loos, 1950 (8), family Tylencholaimidae. In Xiphinemella the spear also possesses basal flanges, but it is much shorter than in Xiphinema. There is a terminal disc in front of the lip region which is always absent in Xiphinema, and the body is generally shorter and a peculiar cuticular ornamentation exists which is always absent in Xiphinema. Xiphinema species are readily differentiated from those of Longidorus and Paralongidorus on the structure and position of the guiding apparatus of the odontostyle (see above).

The genus Xiphinema appears as homogeneous and distinctly separated from the nearest genera. Only two species, X. longidoroides Luc, 1961 and X. sandellum Heyns, 1966, present a problem. Although included by Cohn and Sher (3) in their list of a valid species, they are not included in the subgenera proposed by these authors because the species present characters intermediate between Xiphinema and Longidorus. In fact close re-examination of paratypes of the two species show that their structure and position are very different (see Luc and Dalmasso in these Proceedings): X. longidoroides is a true Xiphinema, whereas X. sandellum represents an intermediate form between Xiphinema and Longidorus but closer to the latter genus.

Thus, apart from this questionable species, all the species belonging to the genus Xiphinema show clearly all and only the typical characters of the genus. This statement could appear to be a truism, but it is not the situation in all the genera of nematodes particularly the phytoparasitic ones.

THE SPECIES OF XIPHINEMA

At the present time, the genus Xiphinema comprises 68 valid species, 11 species inquirendae and one species incertae sedis, according to the paper on identification of Xiphinema by Luc and Dalmasso in these Proceedings.

What are the characters used to separate the species?

As in each case where the characters of the genus are strictly defined the separation of species is not very easy, as opposed to genera where characters are more loosely defined.

The main characters used for the definition and the separation of species are those of females, essentially because the males are rather common in only sixteen of the 68 species and juveniles were accurately described for 25 species only. So characters of males and juveniles can only be used supplementally to verify identification.

These main, female, characters listed by Luc and Tarjan (39) are:

1. length of the body (L).
2. position of vulva (V).
3. length of the spear.
4. type of genital tract : this point referring mainly to the possible reduction of the anterior branch.
5. presence of a Z differentiation in the uterus.
6. shape and length of the tail.

The accessory characters used are:

7. habitus.
8. outline of the fore-part of body.
9. presence or absence of males.
10. tail-shape of the different larval stages.

These last four characters have, however, to be used with some prudence.

Illustrations for the following characters accompany the paper by Luc and Dalmasso (in these Proceedings).

Length of the body.

There is little to say concerning this character. The body size varies in the genus from a mean value of 1.6 mm in some populations of X. turcicum to nearly 6 mm for X. ingens and some populations of X. turcicum. This character used alone is not very useful as the greatest number of species have rather similar body lengths. Moreover, in the same species this value may vary according to strains or geographical origin; for example the mean value of body length for several populations of X. elongatum from Rhodesia is about 2.5 mm whereas populations from Madagascar have a mean L value of 2.0.

Position of vulva (coefficient V).

The V value may vary in the genus from a mean of 27 (X. orthotenum) to a mean of 58 (X. opistohysterum). Logically, of course, the species with no or a greatly reduced anterior genital branch exhibit the lowest V values : in these cases the mean value is under 35. Some species with a little reduced anterior genital branch may have a higher V value (X. filicaudatum : V = 41) whereas species with two branches normally and equally developed may have a low V value (X. insigne : V = 32 - 35); in the latter case the genital tract is short, plain, without uterine differentiation. The value of V is rather constant within the species; however, a species recently described (10) from West Africa, X. bergeri, exhibits different V values according to populations : those found only in flooded rice fields have a mean V value of 32.4; other populations found in the rhizosphere of coffee and other trees have a mean V value of 38. This observation could perhaps be interpreted as the beginning of a splitting in the species, the first stage of a diverging speciation.

Length of the spear.

The term "spear" is used here for the whole structure comprising the odontostyle and the basal extension or odontophore. Despite the fact that these two parts have different origins it seems more logical to apply one term only for a structure acting as a whole. It must be noted that in descriptions of populations of known species and, especially in those of new species, the "total length of spear" should be given in addition to the lengths of the two parts. This is because it is easier to measure the whole spear and also the error in measuring is proportionately reduced.

As in the cases of L and V values, the length of spear is rather variable in the genus : it varies from about 122 µm (X. parvistilus) to about 325 µm (X. filicaudatum). The length may also vary within the species; for example a Nigerian population of X. radicicola has a mean spear length of 222 µm whereas an Australian population has a mean value of 170 µm only.

Thus, as in the case of L or V value, the length of the spear is not itself sufficient to characterize a species and the data so obtained should be used with some care and in close association with other characters.

Type of female genital tract

The tract is of primary importance for the determination of Xiphinema species. (Details on the anatomy are given by Dalmasso elsewhere in these Proceedings). Two points are of importance : first, the possible reduction of the anterior branch and in this case the degree of reduction; second, the possible peculiar differentiation of the uterus called "Z differentiation".

The reduction of the anterior genital branch may vary from species in which only the ovary is strongly reduced and apparently not functional (X. orbum) to those in which there is no trace of an anterior branch (X. radicicola). All intermediate situations were observed in the sixteen species that show such a characteristic in the female genital tract. Although this reduction was found to be slightly variable in two species (X. longicaudatum and X. simillimum), this character is one of the best and, combined with the tail length and shape, permits the determination of nearly each of the 16 species involved.

Peculiar uterine differentiations. Species having two equally, or nearly equally developed genital branches, may present variations in the development and size of this apparatus : the branches can be thin and short, such as generally observed in species with a short body (X. americanum and related species) or with a longer body but vulva rather anterior (X. bergeri, X. insigne); the branches can be longer and thick; and all intermediate stages are observed. Thus this character is not very useful for systematic purposes, particularly as the appearance of the genital tract may vary rather largely in the same species due to the age of the females, the stress of fixation, etc.

However, among those species with two genital branches equally developed, some exhibit a remarkable differentiation of the uterus referred to as the "Z differentiation". This term is derived from the "Z organ" which was described for the first time in the species X. ebriense Luc (9). This differentiated area of the uterus is always situated at the junction of the uterine pouch and the elongated distal part of the uterus; in fact, it constitutes an area of this last part of the tract. Since its first description, structures similar or identical to the Z organ have been described in 18 species. Although in some cases the descriptions and the drawings of different authors are not very well detailed, the examination of these drawings and, mainly, the observation of numerous paratypes lead me to differentiate two groups based on the structure commonly referred to as "Z organ" (9):

- first, a structure generally quite distinct from the other parts of the uterus, with strong circular muscles and an internal sclerotized wall finely striated longitudinally having in the central lumen sclerotized, refringent apophyses, well developed and often four in number. This structure is referred to as the "typical Z organ" (10). It is found in the following species: X. ebriense, X. hallei, X. ifacolum, X. imambaksi and X. rotundatum.

- second, a structure more variable from one species to another, not so distinct from the adjacent parts of the uterus, with weakly developed muscles; the sclerotization of the internal wall is weak or absent as are the longitudinal striations; there are no sclerotized apophyses but granules of various size, number and form according to species. This structure is called "Z pseudo-organ" and has been recorded in the following species: X. basiri, X. coxi, X. diversicaudatum, X. imitator, X. ingens, X. malagasi, X. marsupilami, X. mediterraneum, X. parvistilus, X. pini, X. turcicum,

X. variabile and X. zulu.

This division of the Z differentiation into two types was described by Luc (10) as tentative. It is not always possible to make a clearcut distinction : for example in X. pini although the wall is thin and of the Z pseudo-organ type, the internal bodies resemble apophyses more than globules. This division into two types was given essentially to focus the attention of Xiphinemologists and to motivate them, during description of new or redescription of known species, to make fine and accurate observations on this point. This structure appears remarkably constant in the same species; in only one case, as yet undescribed species from West Africa, I observed in some specimens a typical Z organ (muscular with four apophyses) and in other specimens an organ also muscular, but instead of the four apophyses there were four refringent globules.

Thus, Z differentiation is very useful for specific determination, but requires more detailed studies and careful observation.

Two species recently described from Madagascar (10), X.malagasi and X. spinuterus, exhibit a peculiar and different structure of the uterus. In X. spinuterus the cylindrical part of the uterus has a rather thin wall with "spines" on the inner wall; these "spines" pointed away from the vulva but do not seem sclerotized and each appears to be implanted in a cell of the wall; they are more numerous in the vicinity of the uterine pouch, which in this species is differentiated in a spermatheca. In X. malagasi where a Z pseudo-organ is present there are also similar spines but they are present only in the basal part of the Z pseudo-organ and on part of the length of the distal cylindrical part.

Shape and length of the tail.

The Xiphinema species exhibit one of the largest intra-generic variations in regard to the length and the shape of the tail. This part of the body may vary from long attenuated tails with a c' coefficient of 10 to 15 in X. filicaudatum and even 19 in some specimens of X. marsupilami, to nearly hemispherical tails with a coefficient c' of 0.6 in X. hygrophilum.

I prefer to use the c' coefficient rather than the true length of the tail, because it is less variable in the same species; but the true length of the tail is in some cases an important character and should always be given in the descriptions. The c' coefficient is not sufficient to characterize the tail of a species; species having similar c' values may have tails of different shape; for example, for the medium values of c', the tail can be regularly conical and straight, regularly conical but more or less ventrally curved, conical but with club-shaped or digitated tip and so on.

Thus, c' and tail shape are two very useful criteria for separating species and are the most accurate characters for this purpose combined with the structure of the genital tract.

Of course c' is variable in the same species, but not as much as for example body or spear length. The 68 *Xiphinema* species may be classified in the following arbitrary c' groups :

A : c' more than 5

B : c' from 2.5 to 5

C : c' from 1.5 to 2.5

D : c' from 1 to 1.5

E : c' less than 1

On this basis 10 species are grouped in A; 7 in B; 8 in C; 6 in D; and 14 in E. This accounts for 44 species, the 23 remaining species overlapping two of the groups as follows:

3 species on A and B

3 " B – C

9 " C – D

8 " D – E

No species overlap three categories, and thus using these groupings of coefficient c' for determination, an unknown population need only be compared with relatively few species.

The tail shape is difficult to describe accurately in words and therefore drawings are absolutely necessary. Nevertheless, arbitrary categories can be made and the following are proposed:

A : tail long-attenuated and pointed (c' over 7.5)

B : tail long conical (c' from 2.5 to 7.5)

C : tail regularly short conical (c' 2.5 or less)

D : tail short conical digitate (c' 2.5 or less)

E : tail short conical-rounded to hemispherical with a terminal peg, mucro or bulging

F : tail more or less regularly hemispherical

G : tail rounded – spatulate

If the 68 species are classified in these categories the result is as follows:

5	species fall in the category	A
12	- - - - - - - - - - - - - - -	B
9	- - - - - - - - - - - - - - -	C
9	- - - - - - - - - - - - - - -	D
8	- - - - - - - - - - - - - - -	E
7	- - - - - - - - - - - - - - -	F
1	- - - - - - - - - - - - - - -	G

The seventeen remaining species overlap two categories as follows:

3 in	A	and	B
3 "	B	"	C
1 "	B	"	D
2 "	C	"	D
1 "	C	"	F
7 "	E	"	F

These categories can be used with the value of coefficient c' to reduce drastically the number of known species to compare during the process of determination of an unknown population.

However, these data apply only to the "gross morphology" of the tail and in some cases fine details of its structure will be very useful. Some examples are:- X. attorodorum is a species rather frequent in West Africa that is very close to X. insigne; the two species can be separated by various measurements but examination of the tail-tip is in fact sufficient : in X. insigne the tail tip structure is of the normal type with the inner surface of the cuticle regular, whereas in X. attorodorum this inner surface forms a fine canal with a slight vesicle at its end; this structure is characteristic of the species and provides easy recognition.

- X. basiri and X. ifacolum are two species very close according to their various coefficients, length of body and of spear, gross morphology of the tail, etc. The Z differentiation is very different in the two species (see Luc and Dalmasso in these Proceedings), but identification of each is more readily made by looking at the structure of the tail-tip : in X. ifacolum the tail-tip shows a fine canal surrounded proximately by a fine muff, whereas in X. basiri the tail tip has no peculiarly differentiated structure and shows only a wide terminal canal, rather difficult to observe.

- X. bergeri is easily recognizable, even under a dissecting microscope, by its rather long tail, slightly ventrally curved and club-shaped at its end.

The length of the non-protoplasmic (= hyaline) part of the tail and/or its ratio to the total length of the tail can be useful in some cases. For instance, these measurements permit the easy differentiation of X. vulgare from X. setariae and of X. douceti from X. nigeriense.

There are other examples, but the aim of this discussion is to draw the attention of systematicians working on Xiphinema to the importance of observing and describing, with as much detail as possible, the structure of tail-tip and especially the internal details.

Many species possess a rounded tail, or a rounded tail but with a slight bulging or peg. These can be differentiated from species with a "true" hemispherical tail or from those in which the bulging is often so weakly developed that it could be easily overlooked, by means of a structure called by Southey (12) the "blind terminal canal". This is seen as a clear zone of the cuticle more or less free from the fine radial striations that surround the rest of the tail. The structure is never found in those species having a perfectly rounded tail such as X. hygrophilum, X. ensiculiferum, X. yapoense, X. macrostylum or X. turcicum; it is found in species where the hemispherical exhibits a more or less developed bulging, such as X. loosi, X. imitator or X. ingens.

It should be noted that some species with a normally hemispherical pegged tail such as X. index, X. vuittenezi or X. index may have exceptional specimens with rounded tails.

7-10 Accessory characters

Habitus. In heat relaxed specimens habitus may vary from nearly straight (X. orthotenum) or only slightly curved (X. hygrophilum, X. loosi) to rather closed spiral shape (X. americanum, X. rivesi); but the greatest number of species have a habitus varying from C-shaped, J-shaped or loose spiral and in these cases differences are difficult to appreciate if account is taken of the variability due to the nematode itself and to the more or less drastic action of the fixation.

Outline of fore-body. The outline of the labial region varies from rounded profile perfectly continuous with the body contour in some species (X. hygrophilum, X. cavenessi) to conspicuously set-off in some others (X. monohysterum, X. italiae). But in the great majority of species, the lip-region is more or less marked by a slight constriction or only by a minor alteration in the profile. Variations of such profiles are difficult to appreciate by themselves and often drawings provided by authors lack precision. Thus, this is a difficult character to use other than as a supplementary one, or in cases where there is a very marked shape.

Males. The presence or absence of males is a character to use with care, mainly because for many species of Xiphinema only a few representative individuals are known from the populations. Moreover the presence of males may be fortuitous : for example in numerous populations of the two West African species X. bergeri and X. ifacolum only one male has been found in each. It is possible too that in some species males are present in some strains and not in others : Loof and Maas (7) described two populations of the Surinamense species X. filicaudatum, in one of which males represented 20% of the population whereas they were absent in the other.

Larval tail shape. The tail shape of the larval stages may be very useful, but the complete set of such stages is known only for 25 species. It would be desirable to describe these stages but it is often difficult especially in tropical soil samples where it is not infrequent to find two, three or more Xiphinema species, and these most often imperfectly known at the adult stage. To ascribe the larval stages to the correspondent adults is in such cases very difficult.

SUMMARY

It must be emphazised that two characters have a real value in the specific identification of Xiphinema. These are

- the structure of the female genital tract and
- the length, shape and structure of the tail.

Carefully detailed observations and descriptions of these points would be of more value than a great number of measurements concerning various other structures, such as position of the guiding apparatus of the odontostyle, of the hemizonid, size of the basal oesophageal bulb.

Also it must be emphasized that large numbers of observations on various species of a genus like Xiphinema must normally lead to a clear cut distinction between the characters that have a real taxonomic and/or phylogenic value and those I prefer to call "key characters" that can be very useful for the determination of given species but have no signification at the generic level. It would appear that one of the aims of the taxonomy of nematodes should now be to differentiate more clearly, as far as possible, these two types of characters.

REFERENCES

1. Cobb, M.A. (1873). Nematodes mostly Australian and Fijian. MacLeay Mem. Vol., Linn.Soc. N.S.W. 252-308.

2. Cobb, M.A. (1913). New nematode genera found inhabiting fresh water and non-brackish soils. J. Wash. Acad. Sci. 3, 432-444.

3. Cohn, E. & Sher, S.A. (1972). A contribution to the taxonomy of the genus Xiphinema Cobb, 1913. J. Nematol. 4, 36-65.

4. Dalmasso, A. (1969). Etude anatomique et taxonomique des genres Xiphinema, Longidorus and Paralongidorus (Nematoda: Dorylaimidea). Mem. Mus. natn. Hist. nat. Paris, Ser. A. Zool. 61, 33-82.

5. Heyns, J. (1966). Studies on South African Xiphinema species, with description of two new species displaying sexual dimorphism of the tail. Nematologica 12, 369-384.

6. Loof, P.A.A. & Coomans, A. (1972). The oesophageal gland nuclei of Longidoridae (Dorylaimida). Nematologica 18, 213-233.

7. Loof, P.A.A. & Maas, P.W.Th. (1972). The genus Xiphinema (Dorylaimida) in Surinam. Nematologica 18, 92-119.

8. Loos, C.A. (1949). Notes on free-living and plant-parasitic nematodes from Ceylon. Nos 5 and 6. J. zool. Soc. India 1, 23-29, 30-36.

9. Luc, M. (1958). Xiphinema de l'ouest africain: description de cinq nouvelles espèces (Nematoda: Dorylaimidae). Nematologica 3, 57-72.

10. Luc, M. (1973). Redescription de Xiphinema hallei Luc, 1958 et description de six nouvelles especes de Xiphinema Cobb, 1913 (Nematoda: Dorylaimoidea). Cah. ORSTOM Ser. Biol. 21, 45-65.

11. Meyl, A.H. (1961). Die frielebende Erd-und Süsswasser-Nematoden (Fadenwürmer) in: Brohmer, P., Ehrman, P. and Ulmer, G.: Die Tierwelt Mitterleuropas 1 (5a) 164 pp.

12. Southey, J.F. & Luc, M. (1973). Redescription of Xiphinema ensiculiferum (Cobb, 1893) Thorne, 1937 and description of Xiphinema loosi n.sp. and Xiphinema hygrophilum n.sp. (Nematoda: Dorylaimoidea). Nematologica 19, 293-307.

13. Thorne, G. (1937). Notes on free-living and plant-parasitic nematodes. III. Proc. helm. Soc. Wash. 4, 16-18.

14. Yeates, G.W. (1973). Taxonomy of some soil nematodes from the New Hebrides. N.Z. J. Sci. 15, 673-697.

DISCUSSION

In reply to a question Luc said that the Z-organ was a constant feature of *Xiphinema* species although it was somewhat nondescript in one undescribed species and was not easily seen in all specimens. As a point of technique it is more easily seen in starved nematodes. Loof asked if there was any correlation between the Z-organ and tail shape and Luc replied there was not, going on to say that the presence of the Z-organ appears erratic in the genus and no correlation has been found with any other character, except that a Z-organ was never observed in species having no or reduced anterior genital branch. The function of the Z-organ was unknown. Coomans pointed out that there was also a pseudo Z-organ which was globular and less distinct than the Z-organ proper, which appeared sclerotized. Since these are fundamentally similar he suggested dropping the term 'pseudo' as it infers a fundamental difference between the two types of organ. Luc suggested Z-organ A and B as possible alter -natives. In discussing the reproductive apparatus Lamberti observed that the position of the vulva was possibly influenced by the host plant and by the soil type.

Discussing the creation by Cohn and Sher (E. Cohn and S.A. Sher J. Nematol. 4, 36-65 (1972)) of subgenera groupings within *Xiphinema*, Luc felt that this was premature and criticised the groupings because they were supported by only two characters *viz.* the genital tract and the tail. No correlation had been found between these two characters and to give predominance to one over another is purely arbitrary. Thus splitting other than the subgenera created by Cohn and Sher could be possible and with an equivalent value; "groups" without taxonomic value would be preferable. In response Cohn pointed out that the use of the subgenus category was optional, and it does not mean a change in the species name; changing a species from one subgenus to another is easy and it could form the basis for the creation of new groups. He admitted that there were flaws in his system but nevertheless felt that it provided the basis for taxonomic improvements : the genus *Xiphinema* was so variable that subgenera were useful to the taxonomist. Coomans added that it was not possible to use a single character for the creation of subgenera and one had to decide whether the criteria used for the groupings were valid, but he agreed that subgenera were useful. Hackney suggested that chromosome numbers in each species might be used as the basis for forming the groups. Cohn considered that the proposed grouping of *X. americanum* into five species was not justified solely on the basis of gonad differences found by Lamberti.

It was pointed out that taxonomy was a means of communicating information on relationships between species. It was suggested that it would be helpful to include biological data in the description of a species, but taxonomists present who did not favour this suggestion pointed out that other animal taxonomists were not asked to include biological information as a basis for species identification.

A "LATTICE" FOR THE IDENTIFICATION OF SPECIES OF *XIPHINEMA* COBB, 1913

MICHEL LUC AND A. DALMASSO

Laboratoire de Nématologie, O.R.S.T.O.M., Dakar, Sénégal and Station de Recherche sur les Nématodes, I.N.R.A., Antibes, France, respectively.

The following paper was not presented at the Institute but is included in these Proceedings as a supplement to the paper on the taxonomy of *Xiphinema* by M. Luc. It is part of a paper on the genus *Xiphinema* to be published in *Nematologia Mediterranea*, and only the "lattice" is included here. The figures accompanying this paper are also applicable to the previous paper by M. Luc.

THE LATTICE

The following lattice has been developed from one given by Stegarescu (8), in accordance with the principle of polytomous determination. It has some advantages over keys because it presents without any hierarchization all of the characters of each species. Thus it permits an easier diagnosis as there is no need to examine the different characters in a pre-established order, which is inevitably arbitrary and too restrictive. The "lattice" facilitates species determination even if the material to be examined is not perfect and this is not always possible with keys. Also it facilitates the establishment of the diagnosis of new species because it is easy to note for each character concerned which species are closest to the new one.

In order to afford the most simple and useful possible instrument of determination, only twelve characters were chosen, of which the last three can be considered as accessory ones. Each character was divided into at most five categories (with the exception of character D concerning the shape of the tail where seven categories were created) in order to avoid too many frequent and important overlappings. But such restricted subdivisions have a disadvantage

"LATTICE" FOR THE IDENTIFICATION OF XIPHINEMA SPECIES

(Explanations in the text)

	A	B	C	D	E	F	G	H	I	J
orthotenum	1	4	1	1	1	2	2 3	1	1	
chambersi	1	4	1	2	2	2 3	2	2	2 3	
monohysterum	1	4	1 2	2	2	2 3	2	4	2	
radicicola *	1	4	1 2	4	3	2 3	2 3	2	3	
australiae *	1	4	1 2	4	3	3	2 3	3	2 3	4
brasiliense	1	4	1 2	5	4	2	3	3	3	4
loosi	1	4	1 2	5 6a	5	2	2	3	2	
ensiculiferum *	1	4	1 2	6b	5	2	3 4	2	2	
denoudeni	1	4	2	5 6	4	2	2 3	4	3	
costaricense *	1	4	2	6b	5	2	3	2	2	
simillimum	1 2(1)	4	1 2	2	2	2	2	2 3	3	
longicaudatum	2	4	2	1 2	1	3	3	1	3	
krugi	2	4	2	3	5	2	2 3	1 2	2	3 4
filicaudatum	2	4	2 3	1	1	4	4	1	1 2	1
surinamense	2	4	2 3	6b	5	2 3 4	3 4	1	2	
orbum	3	4	1 2	2 3	2 3	3	1	2 3	4	2 3
hygrophilum	3	4	2 3	6b	5	1 2	3 4	1	2	
imambaksi	4	1	2 3	5	3 4	3	2 3	2	3	
ebriense	4	1	3	4	4	2	2	1	3	
hallei	4	1	3 4	2	1	3 4	2 3	3	3	
ifacolum	4	1	3 4	3	3	3 4	2	2 3	3 4	
rotundatum	4	1	3 4	3 6	4 5	2 3	2 3	1	3	
imitator	4		3 4	5 6a	4 5	2 3	1 2	2	3	3 4
pini	4		3 4	5 6a	5	3 4	2	2	4	3 4
zulu	4	2(3)	3	2	2	3	2	1	4	2
malagasi	4	2+3	3 4	2	2	3	2	3 4	3	
diversicaudatum	4	2	2 3	5(6)	4	4 5	3	1 2	3	5
marsupilami	4	2	3	1	1	4	4	2 3	1 2	
coxi	4	2	3	4	3 4	3 4	2 3	2 3	3	4
meridianum	4	2	3 4	2 4	3 4	3 4	2	4	3	3
parvistilus	4	2	3 4	4	3 4	2 3	1	3 4	3	4
basiri	4	2	3 4	4	3 4	3	2	2 3	3 4	3
ingens	4	2	3 4	5 6a	5	5	3 4	2	3	
turcicum	4	2	3 4	6b	5	4 5	3 4	1 2	3	
variabile	4	2	4	2 3	2 3	2 3	1	4	3 4	3
spinuterus	4	3	3	1	1	3	3	2 3	1 2	
bergeri *	4	4	1 2	2	1 2	2 3	1 2	2	2	
bakeri	4	4	1 2	4	3 4	4	3	4	3	
insigne *	4	4	2	2	2	2	1 2	3 4	2	2
arcum	(?)3 4(2)	4	2	6b(?)	4 5	2 3	2	2	2	
attorodorum *	4	4	2 3	2	2	2 3	2	3	2	
elongatum	4	4	2 3	3	3	2	1 2	3	3	
vulgare *	4	4	2 3	3 4	3	2 3	2	3 4	3	3
setariae *	4	4	2 3	3 4	3	2 3	2 3	3 4	3	
mammillatum *	4	4	2 3	5	5	2 3	2 3	1 2	3	
index *	4	4	2 3	5(6)	4 5	3	2 3	1 2	3	
macrostylum	4	4	2 3	6b	5	2	4	1 2	2	
flagellicaudatum	4	4	3	1	1	3	2 3	3	1 3	1
sahelense	4	4	3	4	3 4	4 5	2 3	2 3	2	3
paulistanum	4	4	3	4	4	2	2	1		
neovuittenezi	4	4	3	5(6)	4 5	3	2 3.	2 3	2 3	3 4
yapoense	4	4	3	6b	5	3	3	2 3	2 3	
cavenessi	4	4	3 4	1 2	1	2 3	1 2	1	1 2	
dimorphicaudatum	4	4	3 4	1 2	1 2	4 5	1 2	3	3	2
douceti *	4	4	3 4	2	1	2 3	2	1 2	3	
vanderlindei	4	4	3 4	2	1	3 4	1	4	1 2	
nigeriense *	4	4	3 4	2	1 2	2	2	1 2	2 3	
italiae	4	4	3 4	2 3	2 3	2 3 4	1 2	4	2	2
americanum	4	4	3 4	3	3 4	2	1 2	3 4	4	3
basilgoodeyi	4	4	3 4	5	4 5	3	1 2	2	3	
vuittenezi	4	4	3 4	5(6)	4 5	3 4	3 4	2 3	3	
pyrenaicum	4	4	3 4	5 6	5	3 4 5	3	1 2	3	
clavatum	4	4	3 4	7b	4 5	3	2 3	2	2 3	
opistohysterum *	4	4	4	3	3	2	1	4	4	
longidoroides	4	4	4	3	3	3 4	2 3	1	3	
mediterraneum *	4	4	4	3	3 4	2	1 2	4	4	3
rivesi	4	4	4	3	4	2	1 2	2	4	3
brevicolle	4	4	4	3	5	2	2	2 3	3 4	3

in that a few different species may have the same numerical values. In such cases, which are not frequent, notes are given below the lattice where complementary differential characters are indicated (such species are noted * in the lattice).

The lattice should be considered as a working instrument which is open to improvement; comments and suggestions from nematologists having used it would be greatly appreciated.

CODE OF THE LATTICE

(without any other indication, characters refer to the females)

A - Type of female genital tract:

- No anterior branch or anterior branch reduced to unorganized structure shorter than 2 vulval diameter (Fig. 1 E, F, G, H, I) 1

No anterior ovary but rest of anterior branch showing differentiation (uterus, sphincter and, in some cases, oviduct) and more than 2.5 vulvar diameter long. (Fig. 1B, D) 2

- Anterior ovary present but reduced and not functional; rest of anterior branch normal or reduced but differentiated (Fig. 1A) 3

- Two branches having approximately the same development 4

B - Type of uterine differentiation, if present:

- Z organ present (Fig. 2A, B) 1
- Z pseudoorgan present (Fig. 2 C, D, E, F, G & Fig.3A, B) 2
- Differentiation other than a Z organ or a Z pseudoorgan ("spines") (Fig. 3B, C) 3
- No differentiation in the uterus 4

C - Position of vulva (coefficient V):

- $V < 30$ 1
- $30 < V \leqslant 40$ 2
- $40 < V \leqslant 50$ 3
- $V > 50$ 4

D - Shape of the tail:

- Tail long attenuated and pointed ($c' > 7.5$) (Fig. 4A) ... 1

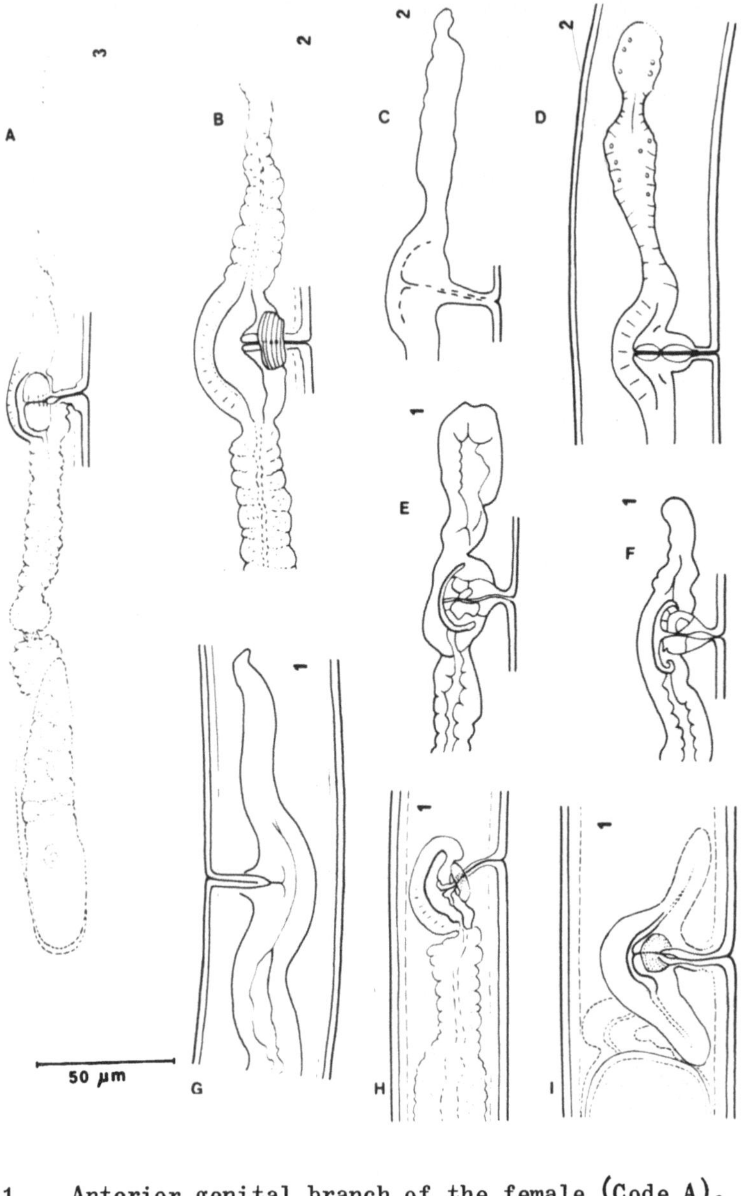

Fig. 1. Anterior genital branch of the female (Code A).
A: X. hygrophilum (Code A3, paratype, from Southey & Luc 1973). B: X. longicaudatum (Code A2, pop. Coffea sp./Ivory Coast, orig.) C: X. krugi (?) (pop. Florida, from Tarjan 1974). D: X. krugi (Code A2, pop. type, from Lordello 1955). E, F: X. costaricense (Code A1), pop. type, from Lamberti & Tarjan 1974). G: X. denoudeni (Code A1, pop. type, from Loof & Maas 1972). H: X. radicicola (Code A1, pop. forest/Ivory Coast, orig.). I: X. loosi (Code A1, paratype, from Southey & Luc 1973).

- Tail long conical ($2.5 < c' \leqslant 7.5$) (Fig. 4B,C, D, E, F).. 2
- Tail regularly short conical ($c' \leqslant 2.5$) (Fig.5A,B,C,D) . 3
- Tail short conical-digitate ($c' \leqslant 2.5$) (Fig. 5 I,J) 4
- Tail conical-rounded to hemispherical with a terminal peg, mucro or bulging (1) (Fig. 6 A,B,C,D)............. 5
- Tail more or less regularly hemispherical (2) (Fig.6F,G, J,K).. 6
- Tail rounded spatulate (Fig. 6 L) 7

*(1) Some species having normally a tail of the hemispherical type with a mucro, peg or bulge can have occasionally a "pegless" tail. Concerning these species, the normal figure 5 is followed by (6).

(2) For each species having a hemispherical tail, whenever possible, the presence (a) or the absence (b) of a "blind" terminal canal in the cuticle of the tail-tip is noted.

E - Ratio tail-length divided by anal diameter (coefficient c'):

- $c' > 5$ 1
- $2.5 < c' \leqslant 5$ 2
- $1.5 < c' \leqslant 2.5$ 3
- $1 < c' \leqslant 1.5$ 4
- $c' \leqslant 1$ 5

F - Body length:

- $L \leqslant 1.5$ mm 1
- $1.5 < L \leqslant 2.5$ 2
- $2.5 < L \leqslant 3.5$ mm 3
- $3.5 < L \leqslant 4.5$ mm 4
- $L > 4.5$ mm 5

G - Total spear length:

- spear $\leqslant$ 150 μm 1
- 150 m < spear $\leqslant$ 200 μm 2
- 200 μm < spear $\leqslant$ 250 μm 3
- spear > 250 m 4

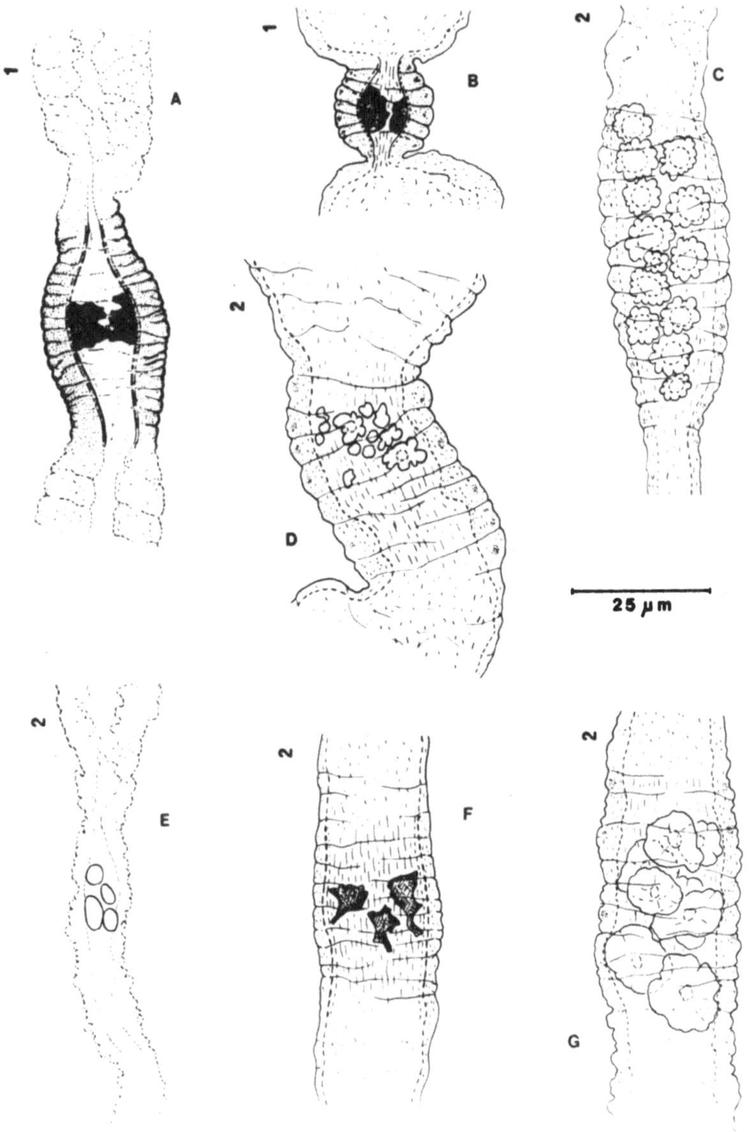

Fig. 2. Z differentiation (Code B). A: X. ifacolum (true Z organ, Code B1, pop. Ivory Coast, orig.). B: X. ebriense (true Z organ, Code B1, topotype, orig.). C: X. diversicaudatum (Z pseudoorgan, Code B2, pop. France, orig.). D: X. turcicum (Z pseudoorgan, Code B2, pop. Algeria, orig.). E: X. basiri (Z pseudoorgan, Code B2, paratype, orig.). F: X. pini (Z pseudoorgan, Code B2, paratype, orig.). G: X. ingens (Z pseudoorgan, Code B2, paratype, orig.).

H - Outline of the forepart of body:

- Without any constriction; outline progressively attenuated (Fig. 7A) 1
- Without any constriction but outline not progressively attenuated (Fig. 7C) 2
- With weak constriction and outline not progressively attenuated (Fig. 7D) 3
- With conspicuous constriction (Fig. 7E) 4

I - Habitus:

- Body straight or nearly so (Fig. 8A) 1
- Body weakly curved (Fig. 8 C, D)...................... 2
- Body hook-shaped or in C or J-shape (Fig. 8E) 3
- Body spiral-shaped (Fig. 8F, G) 4

J - Tail-shape of the fourth larval stage:

- Same divisions as for adult : see D.

K - Tail-shape of the first larval stage:

- Same divisions as for adult : see D.

L - Presence or absence of males:

- Males unknown or exceptional 1
- Males rather frequent 2

NOTES ON THE LATTICE

(1) X. simillimum: regarding the genital tract, the structure of anterior branch may be in question; in the original drawing by Loof and Yassin (7) (their Fig. 3B) a reduced and unorganized branch is clearly shown and their original description states: "anterior gonad very short, undifferentiated, except in the only egg-bearing specimen where it is rather long (110 µm) and somewhat differentiated".

(2) X. arcum: in the original description (3) it is stated : "anterior branch of reproductive system comparatively less developed than the posterior one"; the original illustration (Fig. 2A) (22) shows clearly a reduced anterior branch (length calculated on drawing : 240 µm, against 370 µm for the posterior one) and a very small anterior ovary (50 x 12 µm against 128 x 30 µm for the posterior one). Thus this species could appear as an intermediate between the hemispherical tailed

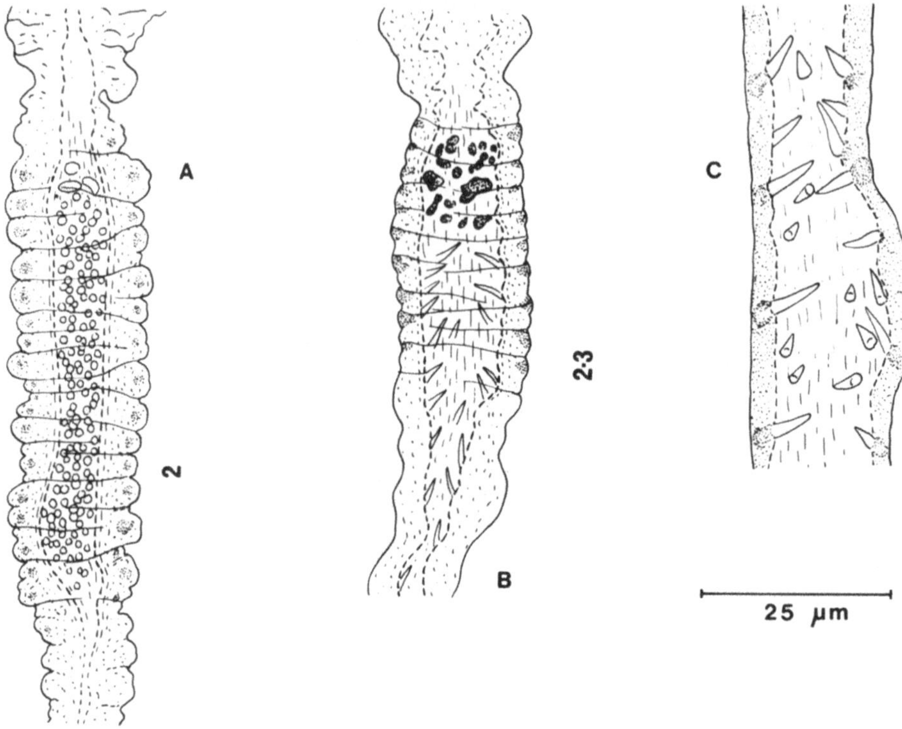

Fig. 3. Z differentiation (Code B). A: X.marsupilami (Z pseudoorgan, Code B2, paratype, from Luc 1973). B: X.malagasi (Z pseudoorgan and "spines", Code B2 + 3, paratype, from Luc 1973). C: X.spinuterus ("spines", Code B3, paratype, from Luc 1973).

Fig. 4. Shape of the female tail (Code D). A: X.spinuterus (Code D1, paratype, from Luc 1973). B: X.bergeri (Code D2, paratype, from Luc 1973). C: X.douceti (Code D2, paratype, from Luc 1973). D: X.nigeriense (Code D2, paratype, from Luc 1961). E: X.attorodum (Code D2, paratype, orig.). F: X.insigne (Code D2, paratype, orig.).

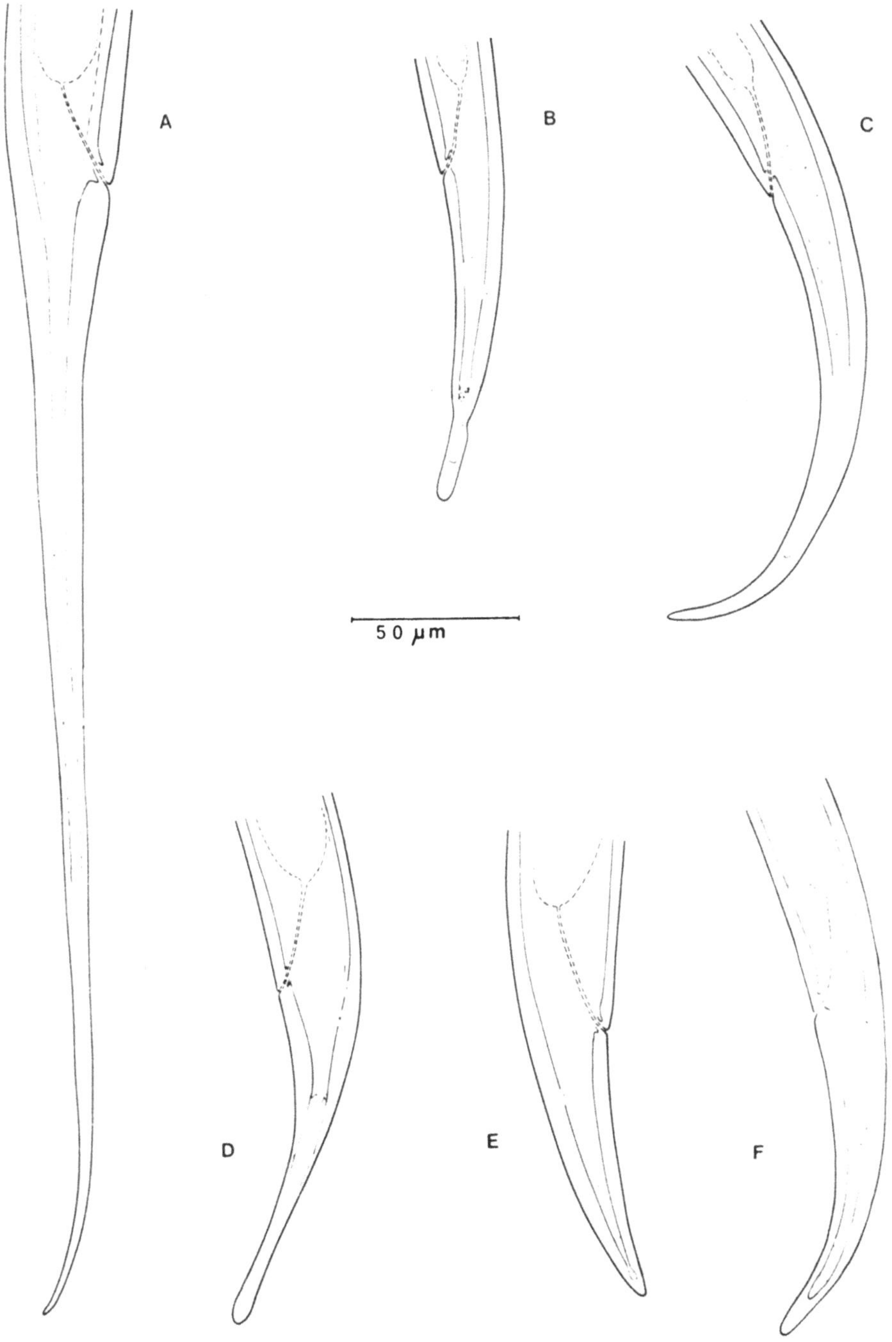
A
B
C
50 μm
D
E
F

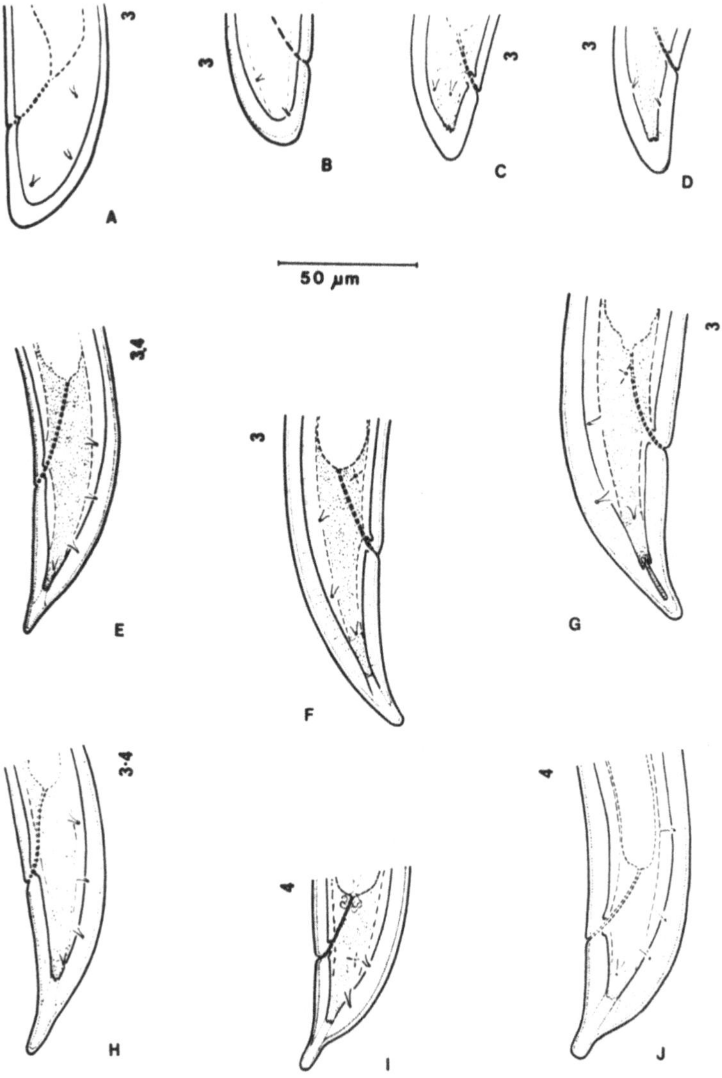

Fig. 5. Shape of the female tail (Code D). A: X.krugi (Code D3, type, from Lordello 1955). B: X.krugi (?) (Code D3, pop. Florida, from Tarjan 1974). C: X.brevicolle (Code D3, pop.Citrus/Ivory Coast, orig.). D: X.rivesi (Code D3, paratype, orig.). E: X.vulgare (Code D3/4, paratype, orig.). F: X.longidoroides (Code D3, paratype, orig.). G: X.ifacolum (Code D3, paratype, orig.). H: X.setariae (Code D3/4, paratype, orig.). I: X.ebriense (Code D4, topotype, orig.). J: X.basiri (Code D4, paratype, orig.)

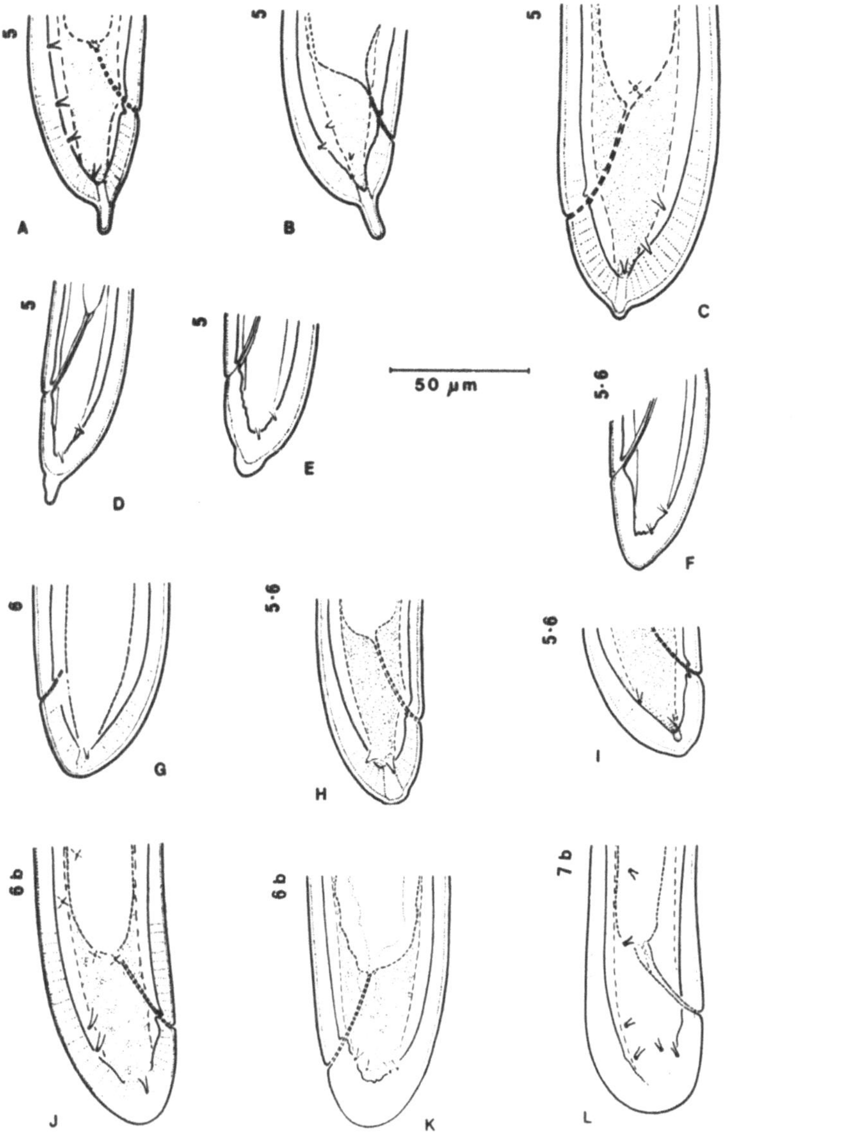

Fig. 6. Shape of the female tail (Code D). A: X.index (Code D5, pop.fig-tree, France, orig.). B: X.brasiliense (Code D5, syntype, Tarjan's orig. drawing). C: X.mammillatum (Code D5, lectotype, Luc & Tarjan 1963). D,E,F: X.denoudeni (respectively Code D5, D5/6, D6, Loof & Maas 1972). G: X.index (Code D6, pop.vine, France, Dalmasso's orig. drawing). H: X.loosi (Code D5/6a, type, Southey & Luc 1973). I: X.pini (Code D5/6a, paratype, orig.). J: X.yapoense (Code D6b, holotype, orig.). K: X.hygrophilum (Code D6b, type, Southey & Luc 1973). L: X.clavatum (Code D7b, type, Heyns 1965).

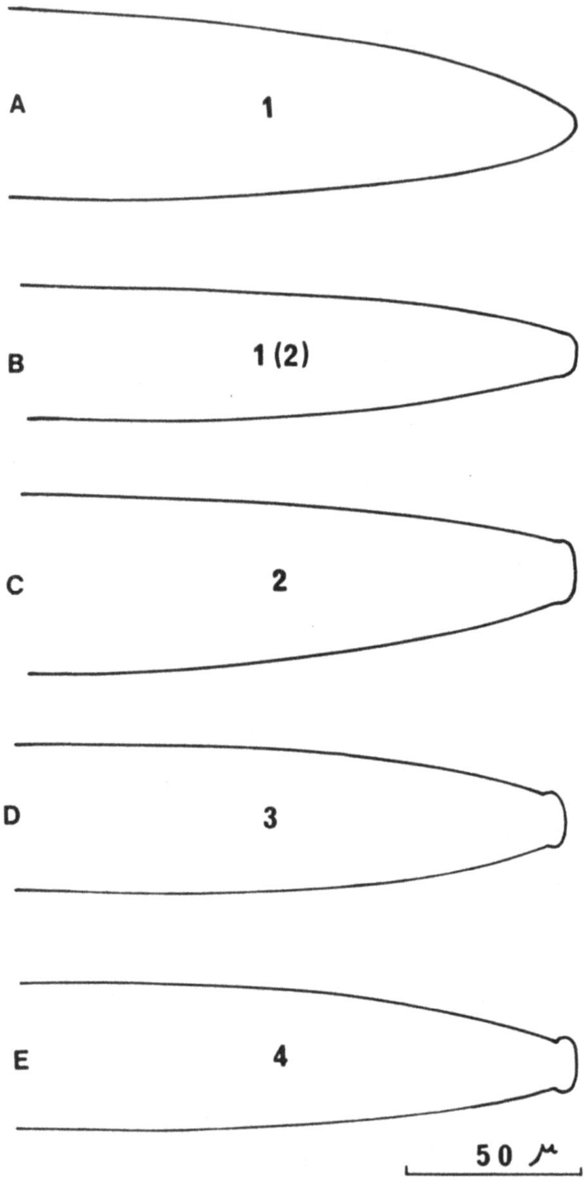

Fig. 7. Outline of the fore part of the body (Code H).
A: X.hygrophilum (Code H1). B: X.pyrenaicum (Code H1/2).
C: X.pini (Code H2). D: X.loosi (Code H3).
E: X.monohysterum (Code H4).

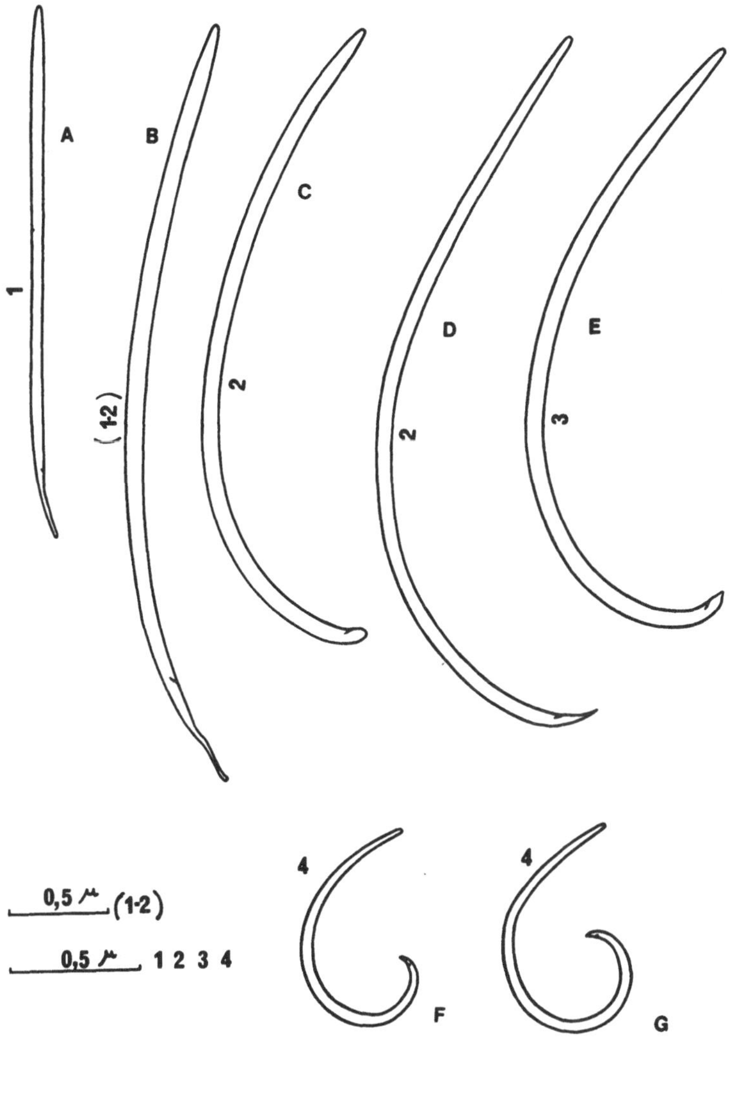

Fig. 8. "Habitus" of the adult female (Code I). A: X. orthotenum (Code I1). B: X. filicaudatum (Code I1/2). C: X. surinamense (Code I2). D: X. italiae (Code I2). E: X. index (Code I3). F,G: X. mediterraneum (Code I4).

species with two genital branches approximatively equal and X. hygrophilum in which the reduction of each part of the anterior branch is much more pronounced.

(3) X. zulu: the original description (2) does not give details on an eventual Z differentiation in the uterus, but figure 31 shows, at the place where this differentiation normally occurs, a different type of uterine wall and small bodies in the lumen, differing from spermatozoa also present in the spermatheca. This structure resembles a pseudoorgan Z of the X. marsupilami type, but less differentiated.

(4) X. filicaudatum: Loof and Maas (6) have found two populations of this species: one with rather abundant males (7 with 28 females), the other with no males in a population with 28 females.

HOW TO USE THE LATTICE

1 - The characteristics corresponding to each letter of the code are observed, measured or calculated; each of them is allocated the corresponding figure.

2 - These figures are put on a strip of transparent paper, so that each corresponds to the column of each letter, for example:

A	B	C	D	E	F	G	H	I	J	K	L
4	2	2/3	5	4	4/5	3	1/2	3	5	3	3

3 - This strip of paper is moved down from the top of the lattice until a horizontal line is found in which the figures correspond exactly to the figures put on the transparent strip; in the example given this is X. diversicaudatum. It is, however, necessary to continue to the last line, because of possible double correspondence. In this case, see notes following the lattice where supplementary differential characters are given.

4 - For each column, only one identical figure is needed; in the example chosen, for character C (V value) the figures corresponding are 2/3, because of overlapping on two divisions (V = 40-46). For this character all the species coded C2 or C3, and not only those coded C 2/3, are in correspondence. On the other hand species coded C1/2 or C 3/4 are to be discarded except in case of perfect identity of figures for all other letters.

5 - If no identity for the all letters is observed in any horizontal line, the specimens being identified are theoretically new, but it is more prudent to test, by careful comparison, the other species appearing close to them.

Differentiation of some species

As stated before, some pairs of species cannot be differentiated on the lattice because they have the same values for the main characters used. So for each of these cases supplementary details are given below which permit a practical differentiation of the species involved, or it is noted that a synonymization could be taken in consideration.

1: X. radicicola and X. australiae. In fact, as stated by Cohn and Sher (1) and as examination of types of the two species and of various other populations (of which the measurements were taken in consideration in the lattice) has shown, these two species are very closely related and it could seem valuable to synonymize them. A publication dealing with these species is in preparation by one of us (M.L.)

2: X. ensiculiferum and X. costaricense. As emphasized by the authors of the last species (5) these two species are very close. They can nevertheless easily be differentiated by the position of the vulva (V = 30.3 - 32.5 for X. ensiculiferum and 36.6 - 37.5 for X. costaricense). This difference is correlated with the weaker development of the vestigial anterior branch in X. ensiculiferum, another character to be taken in consideration.

3: X. insigne and X. bergeri; X. insigne and X. attorodorum. X. insigne is not well differentiated on the lattice from X. bergeri and from X. attorodorum although these two species are clearly separated from each other by the use of character "V" (position of vulva).

Independently of other characters, the best way to distinguish X. insigne from the two other species is to examine the structure of the tail-tip:

- in X. insigne (Fig. 4F) the tail tip is regularly conical rounded at the extremity and the inner surface of the cuticle is regular.

- in X. attorodorum (Fig. 4E) the tail tip is also regularly conical, rounded at the tip, but the inner surface of the cuticle forms a fine canal with a slight vesicle at its end; this structure is very characteristic of the species.

- in X. bergeri (Fig. 4B) the tail tip is clavate and the inner cuticular canal stops half-way to the inflated terminal part of the tail tip. This structure is characteristic of the species.

4: X. setariae and X. vulgare. Hyaline part of tail longer in X. setariae (23-29 μm) than in X. vulgare (13-20 μm).

5: X. mammillatum and X. index. The tail shape, especially the development and the position of the terminal mucro permit easy separation of these two species; in X. mammillatum (Fig. 6C), the peg is most frequently short and, above all, situated strictly in the long axis of the body; in X. index (Fig. 6A), the peg is more often rather long and well differentiated and in any case situated clearly on the ventral part of the tail. Moreover X. mammillatum is a very rare tropical species, never found again since its descrip-tion, in 1938, from Zaire; X. index is, on the contrary, a rather common species of temperate areas, namely in Europe and U.S.A.

6: X. nigeriense and X. douceti. These two species are close to each other but their differentiation is easy on the basis of morphology of the tail: in X. nigeriense (Fig. 4D), the tail is only slightly ventrally curved, not regularly conical-elongated; the tail-tip is rounded and the more often slightly clavate; the hyaline terminal portion is very large (58-64% of tail-length). In X. douceti (Fig. 4C), the tail is strongly ventrally arcuate, regularly conical-elongated; the tail-tip is more or less pointed but never clavate and the hyaline terminal position shorter (18-29% of tail-length).

7: X. opistohysterum and X. mediterraneum. In the diagnosis of the latter species, Lamberti and Martelli (4) wrote: "the differentiation between X. mediterraneum and X. opistohysterum seems difficult owing the strikingly similar overall appearance of the two species". Nevertheless, they give differential characters:

	X. opistohysterum	X. mediterraneum
length of the odontostyle.............	67 µm	87 µm
distance from the base of the spear uide to the fore-part	60 µm	78 µm
coefficient "c"	54	62

These figures apply to the mean values and overlaps may exist. Moreover, these authors stated that the shape of the tail is slightly different in the two species: the tip of the tail in X. mediterraneum is slightly "indented" on both the ventral and dorsal sides, whereas in X. opistohysterum this "indentation" does not exist or is present only on the ventral side. In fact, these differences are very weakly established, and it will not be surprising if, in the future, the study of numerous populations leads to the discovery of linking specimens which will question the validity of separating these two species.

Addendum

Shortly after the Institute, Dr. J.F. Southey drew our attention to two recently described species of Xiphinema not listed in our paper.

X. neoamericanum Saxena, Chhabra & Joshii, 1973 (described in: Zool.Anz. 191, 130-132, 1973) is quite close to X. americanum from which it differs essentially by the lip region being continuous with the rest of the body. It should be quoted in the following manner in the lattice:

A4 - B4 - C 3/4 - D3 - E 4/5 - F2 - G1 - H1 - I 3/4 - J? - K? - L1.

X. cubensis Razjivin in Razjivin, O'Reilly & Milian, 1973 (described in: Poeyama 108, 1-12 1973). This species, which should be named X. cubense, Xiphinema being neuter, resembles in some way X. hygrophilum by the continuous lip region, the rounded tail and the structure of the genital tract, but in X. cubense it is the posterior branch which shows the reduction of each of its different parts. This character makes this species unique in the genus and our lattice needs to be amended for this character (A5 : posterior branch reduced). The author describes the genital tract as having a Z organ, but examination of drawings and text lead us to state that what was described as a Z organ is in fact a differentiated muscular area of the uterus situated close to the ovijector; such a structure is not infrequent in the genus, but it bears no relation to a Z organ. This very interesting species certainly needs to be studied in the greatest detail. It could be tentatively quoted in the lattice as follows:

A5 - B4 - C3 - D6 - E4 - F 3/4 - G 2/3 - H1 - I? - J? - K? - L1.

REFERENCES

1. Cohn, E. & Sher, S.A. (1972). A contribution to the taxonomy of the genus Xiphinema Cobb. J. Nematol. 4, 36-65.

2. Heyns, J. (1965). Four new species of the genus Xiphinema (Nematoda : Dorylaimoidea) from South Africa. Nematologica 11, 87-99.

3. Khan, E. (1964). Longidorus afzali n.sp. and Xiphinema arcum n.sp. (Nematoda : Longidoridae) from India. Nematologica 10, 313-318.

4. Lamberti, F. & Martelli, G.P. (1971). Notes on Xiphinema mediterraneum. (Nematoda : Longidoridae). Nematologica 17, 75-81.

5. Lamberti, F. & Tarjan, A.C. (1974). Xiphinema costaricense n.sp. (Longidoridae, Nematoda), a new species of dagger nematode from Costa Rica. Nematologica mediter. 3, 1-11.

6. Loof, P.A.A. & Maas, P.W.Th. (1972). The genus Xiphinema (Dorylaimida) in Surinam. Nematologica 18, 92-119.

7. Loof, P.A.A. & Yassin, A.M. (1970). Three new plant-parasitic nematodes from Sudan, with notes on Xiphinema basiri Siddiqi, 1959. Nematologica 16, 537-546.

8. Stegarescu, O. (1966). (Application of the polytomous principal to determination of nematodes of the genus Xiphinema Cobb) in "Politomiceskij princip opredelenija zivotnyh i rastenij", 55-58.

TAXONOMY OF LONGIDORUS (MICOLETZKY) FILIPJEV AND PARALONGIDORUS SIDDIQI, HOOPER AND KHAN

FRANCO LAMBERTI

Laboratorio di Nematologica agraria del C.N.R.

Bari, Italy

HISTORICAL REVIEW AND DEFINITIONS

Bütschli in 1874 (5) and de Man in 1876 (10) described respectively *Dorylaimus maximus* and *D. elongatus* in the genus *Dorylaimus* Dujardin 1845.

Micoletzki (32) in his monograph on soil nematodes published in 1922 had noticed that although most of the *Dorylaimus* species were onthogenetically related (at that time about 150 species had been included in this genus), they could be divided in groups on the basis of peculiar morphological characters which he did not consider sufficiently important to justify establishment of different genera. Therefore, he divided this large genus into five subgenera. In the subgenus *Dorylaimus* (Longidorus) he included the species with long odontostyles, *D.* (*L.*) *elongatus*, which he considered to be the type species; *D.* (*L.*) *maximus*, now considered as a *Paralongidorus* species; and *D.* (*L.*) *pygmeus* (Steiner, 1914) transferred by Thorne (48) in 1939 into the new genus *Longidorella* Thorne, 1939.

In 1934 Filipjev (13) and Thorne (47) published separate notes on the classification of nematodes, both agreeing that *Longidorus* should be considered as a genus of the family Dorylaimidae de Man 1876. Filipjev's paper has the publication date 23 March, whereas that of Thorne's 7 April. Therefore the authority for the genus, which at the present time includes 32 valid species, must be attributed to the former.

The first list of species within the genus *Longidorus* is that given by Thorne and Swanger (50) in 1936. In this list are included:

L. diversicaudatus (Micoletzki, 1927) transferred by Thorne to the genus Xiphinema Cobb 1913;

L. elongatus (de Man, 1876), the type species redescribed and well defined on topotypes by Hooper in 1961 (19);

L. maximus (Bütschli, 1874) transferred by Siddiqi in 1964 (40) to the genus Paralongidorus Siddiqi, Hooper and Khan, 1963;

L. pygmeus (Steiner, 1914) transferred by Thorne in 1939 to Longidorella Thorne, 1939.

The most recent definition of the genus, however, is that of Siddiqi (38);

"Longidorinae: body elongate, over 2 mm long; body cuticle thick smooth, marked with very fine transverse striations. Lateral body pores usually in two rows, leading into clear pouch-like areas in hypodermal chords; in addition, a series of dorsal and ventral body pores are sometimes seen, especially on the extremities. Lip region smoothly rounded, continuous with or set off from the body, with six amalgamated lips bearing sixteen distinct papillae arranged in two circlets, six in inner and ten in outer. Amphids abnormally large, pouch-like, with very fine pore- or slit-like apertures situated at the base of the lateral lips. Stoma tubular, without sclerotization, less than five cephalic-widths in length.

Buccal spear elongate, cylindrical, much attenuated, not bearing pointed processes at base for attachment with its extension. Spear extension elongate, with or without basal swellings. Basal swellings on spear extension, if present, never forming distinct knobbed flanges. Spear guiding ring single, usually appearing cup-shaped in optical view, not extending into a sheath anteriorly, located around the anterior half of the spear usually at 2 to 4 labial widths from anterior end. Oesophageal bulb elongate-cylindroid, not distinctly set off from anterior slender portion of oesophagus, with inner walls forming an elongate valvular apparatus. Nuclei of the dorsal and the anterior pair of the sub-ventral oesophageal glands usually distinct, located near the anterior end and the middle of the oesophageal bulb respectively. Cardia present.

Vulva transverse, near middle of body. Ovaries usually paired, reflexed. No sexual dimorphism. Males, when present, with paired heavily reinforced spicules and supplements in the form of an ad-anal pair and a ventro-median series. Gubernaculum absent. Lateral guiding pieces for spicules present. Tails of both sexes similar, hemi-spheroidal to elongate-conoid."

This description can be considered still valid with the only alteration, according to the Aboul-Eid's studies (1) on the genera

Longidorus and Paralongidorus, "amphidial aperture pore-like, never slit-like", (Plate I, A, B).

Referring to the systematic position of Longidorus in the higher taxa, most authors agree at present to place the genus in the subfamily Longidorinae Thorne, 1935, within the family Longidoridae (Thorne, 1935) Meyl, 1960. However, it seems useful to mention that it was placed by Filipjev (13) in the subfamily Dorylaiminae (de Man, 1876), Filipjev, 1918 and by Chitwood (6) in the subfamily Tylencholaiminae Filipjev, 1934, both in the family Dorylaimidae de Man, 1876.

MORPHOLOGICAL AND BIOMETRICAL CHARACTERS OF LONGIDORUS

The major morphological and biometrical characters used for classification and identification of Longidorus species are:

- body size: the genus can be divided in three groups of species; one of small size having a body length included between 2 and 4 mm, a group of medium sized species ranging from 5 to 8 mm and large species above 8 up to over 10 mm;

- shape of the lip region: there are three main types: expanded with respect to the rest of the body, cylindrical and continuous with respect to the rest of the body (slight prelabial constriction may occur) and of sub acute shape, with rounded or truncate terminus, resulting from a marked and abrupt tapering of the body diameter toward the anterior end (Fig. 1).

- shape of the tail: three main types also are distinguishable in this morphological feature; they are: bluntly rounded tail, conoid with rounded terminus and conoid with more or less pointed terminus (Fig. 1);

- length of the odontostyle: is a rather stable character presenting only variations of about 10% within the species;

- distance of the spear guiding ring from the anterior end: this is also a fairly constant character;

- lengths of the tail and of the spear extension present a wide degree of variability and are therefore of only partial and secondary importance;

- the shape of the amphidial pouch which can be sacciform or bilobed is also of secondary importance.

With reference to the de Man indices it can be said that the ratio 'a' (total length of the body/width of the body at vulva) is the most useful, giving an indication of the general appearance of

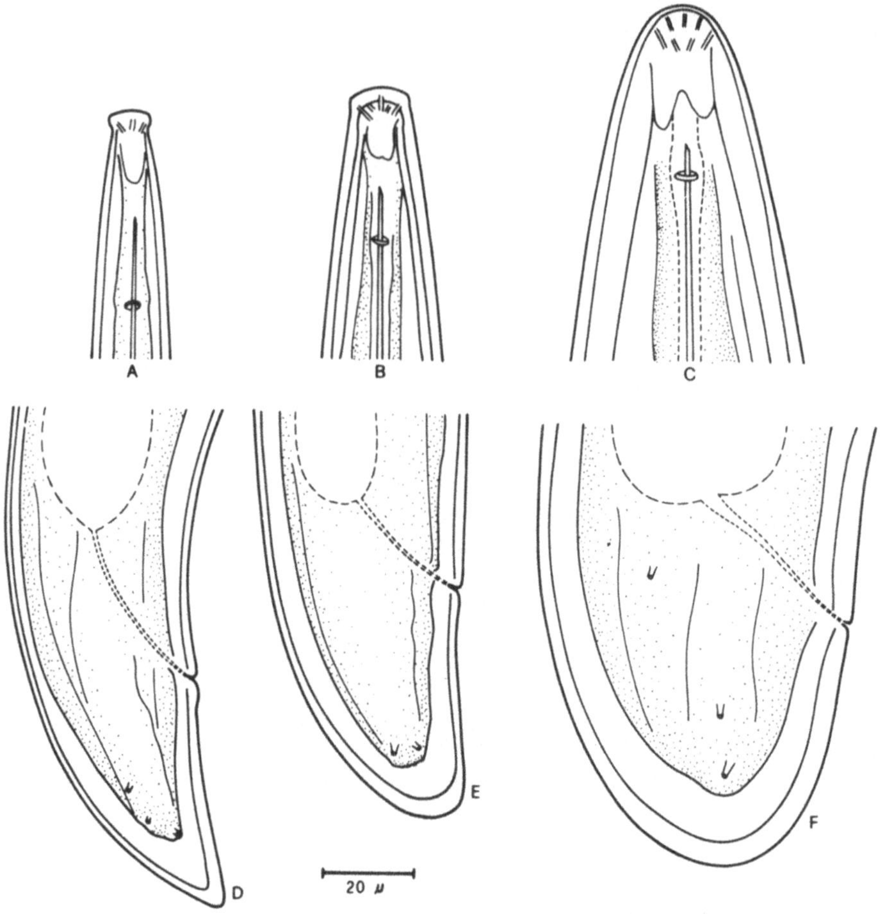

Fig. 1. Basic shapes of lip region and tail in Longidorus: A, lip region expanded e.g. L. siddiqii; B, lip region cylindrical separated from the rest of body by a slight prelabial constriction e.g. L. elongatus; C, lip region sub-acute with rounded terminus e.g. L. caespiticola; D, tail conoid with pointed terminus e.g. L. leptocephalus; E, tail conoid with rounded terminus e.g. L. elongatus; F, tail bluntly rounded e.g. L. macrosoma. Intermediate forms are common.

the nematode: more or less robust. Of minor importance is the 'b' ratio (total length of the body/length of the oesophagus) since the oesophagus length can be subjected to variations due either to the status of nutrition of the individuals or to the way of killing them (29). More useful is, without doubt, the ratio 'c' (total length of the body/length of the tail) although for the large species it comprises a rather wide range due to the variability of the tail length. The position of the vulva (V), very useful for

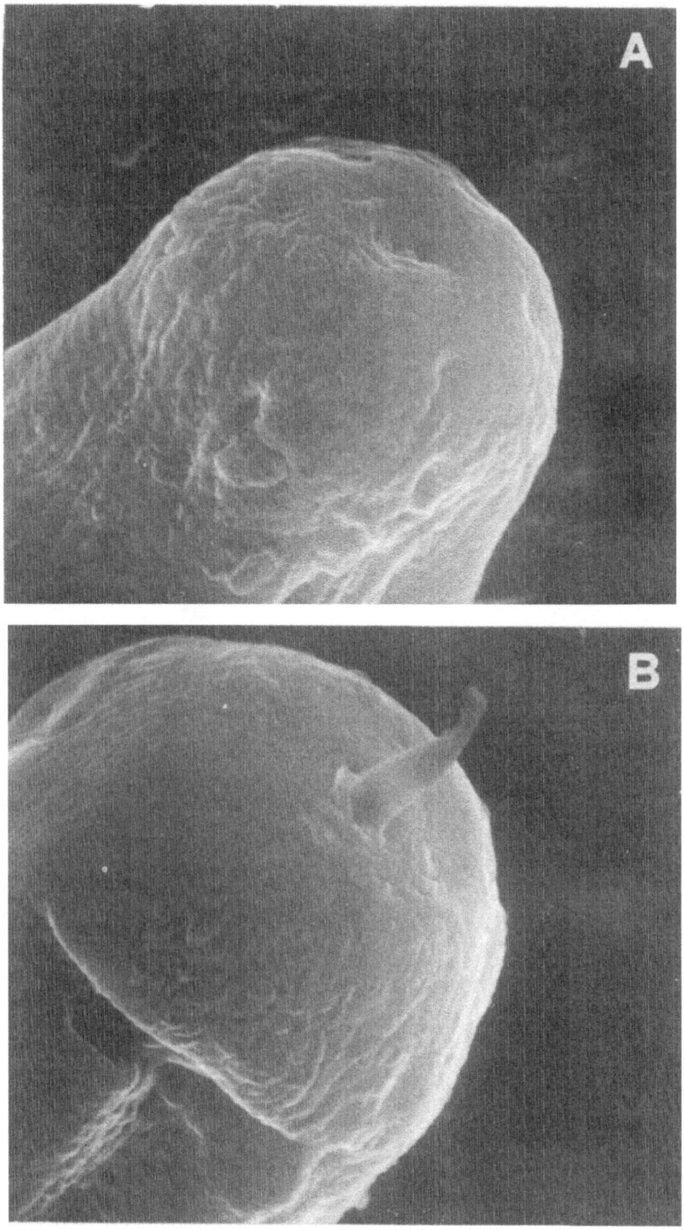

PLATE I. Scanning electronmicrograph of the anterior end of *Longidorus africanus* (A) and of *Xiphinema diversicaudatum* (B). The amphidial aperture is clearly evident in *Xiphinema* but not in *Longidorus* (X 3000).

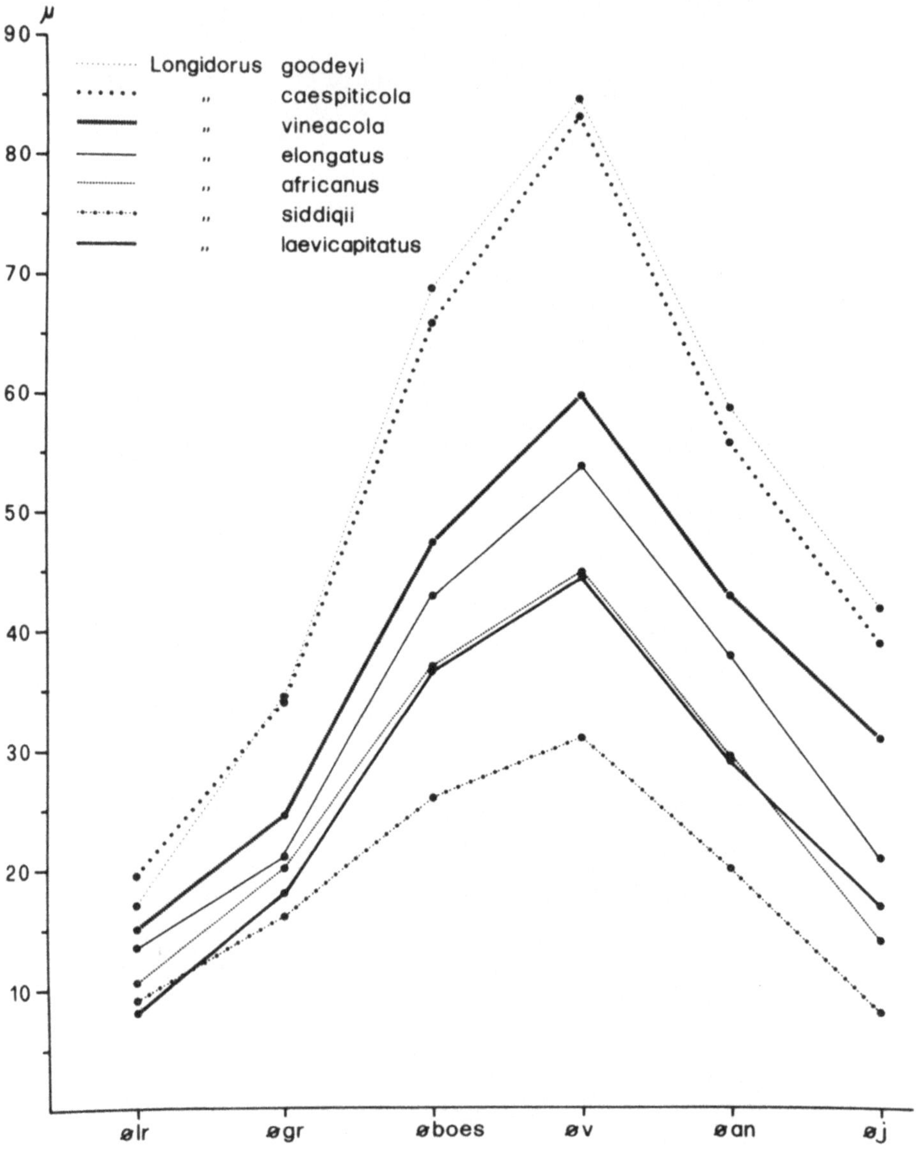

Fig. 2. Body profile of species of Longidorus: body diameter at lip region (lr), at the level of the guiding ring (gr), at the base of the oesophagus (b oes), at vulva (v), at anus (an), at the beginning of the jalin portion of the tail (j).

classification of species of Xiphinema, is of limited importance in Longidorus because in most species it is situated between 45 and 55% of the body length.

A good character also is the ratio 'c' (length of the tail/ width of the body at the anus); it gives a more precise figure of the proportion of the tail in the context of the whole animal and presents less variability than the ratio 'c'.

Other characters are the profile of the body width (27) (Fig. 2), the number of muscles in the oesophageal region (see Taylor and Robertson in these Proceedings) and the arrangements of the muscular region bands in the vaginal region (Seinhorst, personal communication) or the cuticle appearance as seen with the scanning electron microscope (Plate II, A-E) are suggested as useful features for species identification but their use is, at the present time, still very limited.

Juvenile stages can be used for species identification in this genus. The problem, in this case, is 'a lack' of information since not all authors describe and give measurements of juveniles. Very little information is also available on the structure of the female gonad of species of Longidorus. However, this character is unlikely to be of as great importance in Longidorus as it is in Xiphinema, since in the former genus only amphidial species are known.

VALID SPECIES OF LONGIDORUS

Type species: L. elongatus (de Man, 1876) Thorne and Swanger, 1936;
syn. Dorylaimus elongatus de Man, 1876;
D. (Longidorus) elongatus de Man, 1876 (Micol., 1922);
Trichodorus elongatus (de Man, 1876) Filipjev, 1921;
D. tenuis Linstow, 1879;
L. menthasolanus Konicek and Jensen, 1961 (25);
[synonomized by Siddiqi in 1962 (38)]
L. monohystera Altherr, 1953 (2);
[synonomized by Sturhan in 1963 (45)]

Other species: L. africanus Merny, 1966 (30 and 26);
L. attenuatus Hooper, 1961 (19 and 34);
syn. nec. L. elongatus in Thorne 1939 (48);
L. belondiroides Heyns, 1966 (17);

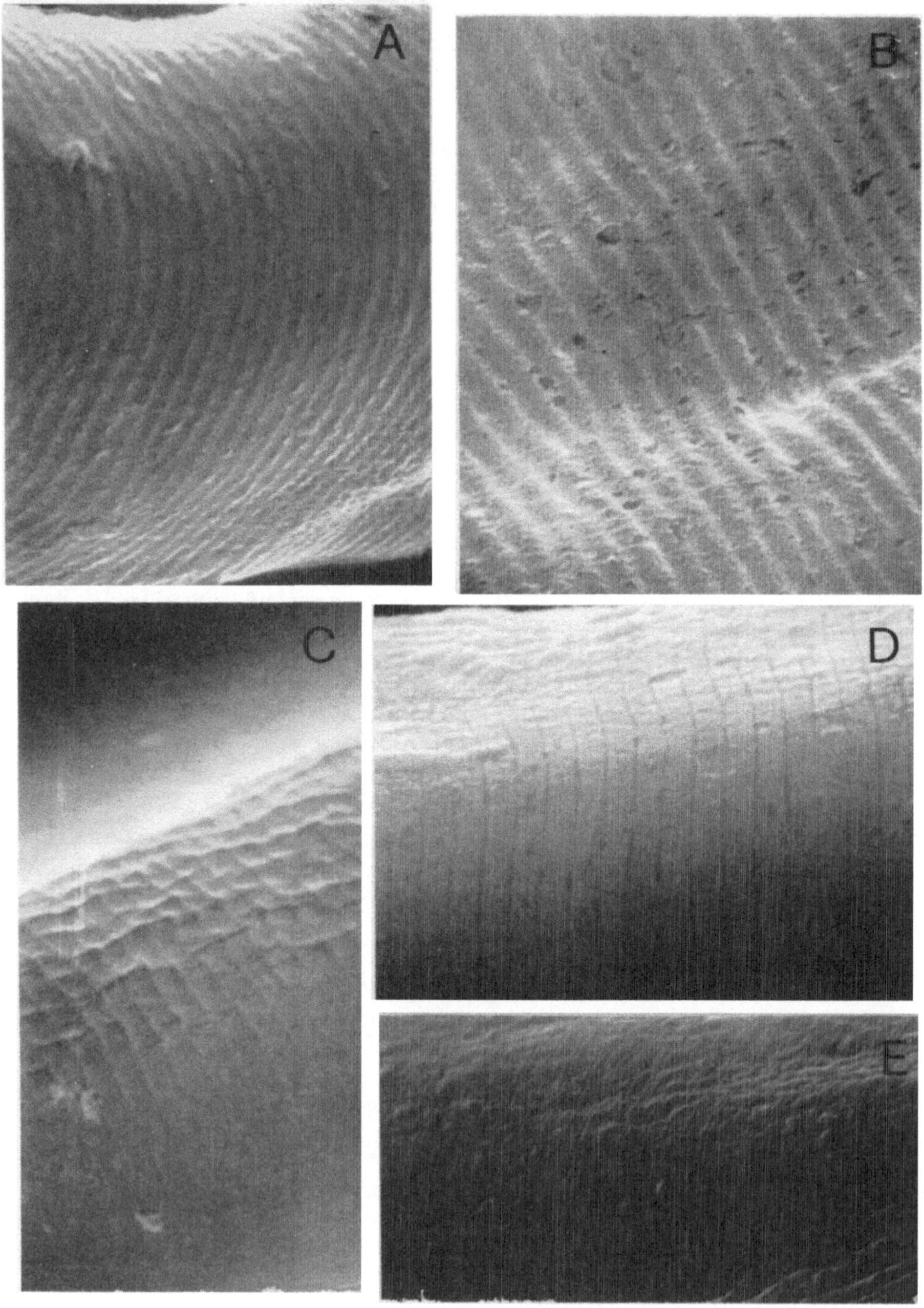

PLATE II. Scanning electronmicrograph of the tail region of L. africanus (A), L. caespiticola (B), L. elongatus (C), L. macrosoma (D) and Longidorus sp.(E) (X 3,000).

L. caespiticola Hooper, 1961 (19);

L. closelongatus Stoyanov, 1964 (44);

L. cohni Heyns, 1969 (18);

L. congoensis Aboul-Eid, 1970 (1);

L. globulicauda Dalmasso, 1969 (9);

L. goodeyi Hooper, 1961 (19);

L. indicus Prabha, 1973 (33);

L. jonesi Siddiqi, 1962 (38);

L. juvenilis Dalmasso, 1969 (9);

L. laevicapitatus Williams, 1959 (52);

L. leptocephalus Hooper, 1961 (19);

L. longicaudatus Siddiqi, 1962 (38);

L. macromucronatus Siddiqi, 1962 (38);

L. macrosoma Hooper, 1961 (19);

L. martini Merny, 1966 (30);

L. monile Heyns, 1966 (16);

L. moniloides Heyns, 1966 (16);

L. nirulai Siddiqi, 1965 (41);

L. pisi Edward, Misra and Singh, 1964 (11);

L. profundorum Hooper, 1965 (20);

L. reneyii Raina, 1966 (35);

L. saginus Khan, Seshadri, Weischer and Mathen, 1971 (23);

L. siddiqii Aboul-Eid, 1970 (1);

syn.: Xiphinema brevicaudatum Schuurmans Stekhoven, 1951 (36 and 37);

L. siddiqii (Siddiqi, 1959) Aboul-Eid, 1970 (1);

L. sylphus Thorne, 1939 (48);

syn.: L. striola Merzheevskaya, 1951 (31), new synonomy;

L. taniwha Clark, 1963 (7);

L. tardicauda Merzheevskaya, 1951 (31);

L. tarjani Siddiqi, 1962 (39);

L. vineacola Sturhan and Weischer, 1964 (46);

The following are considered as species inquirendae:

L. brevicaudatus (Sch. Stek., 1951 - Thorne, 1961 (36, 37 and 49) because it is described on the basis of a single juvenile stage;

L. heynsi Andrassy, 1970 (4), because it is described on the basis of a single male, new designation;

L. meyli Sturhan, 1963 (45), because it is inadequately described on the basis of material no longer existing.

GEOGRAPHICAL DISTRIBUTION OF LONGIDORUS

Concerning the geographical distribution of species of Longidorus, it can be said that the most widely distributed in temperate climates is L. elongatus. It has been reported, in fact, up to now, in all the countries in Northern and Central Europe, in Canada and in the northern states of the United States of America. The most common in subtropical and tropical climates is L. africanus, present in Northern and Eastern Africa, in California and Mexico and in the Middle East (26 and 28). Another species rather common in Northern and Central Europe as well as in the Mediterranean Basin is L. attenuatus. Other species more or less widespread are L. goodeyi, L. caespiticola, L. macrosoma, L. profundorum and L. vineacola in temperate climates (all Europe) and L. laevicapitatus, in tropical climates (in Africa, Antilles and Hawaii). All the other species are only of local importance.

PARALONGIDORUS

The genus Paralongidorus was established in 1963 by Siddiqi, Hooper and Khan (43) to include those species having characters as in Longidorus (position of the spear guiding ring) and as in Xiphinema (shape and aperture of the amphidial pouches). This was done on the basis of the species described at that time. But the finding of new species P. afzali (Khan, 1964) Siddiqi and Husain, 1965 and P. utriculoides (Corbett, 1964) Siddiqi and Husain, 1965 revealed the existence of intermediate forms between Longidorus and Paralongidorus. These two, in fact, have amphidial pouches sacciform as in Longidorus and amphidial apertures slit-like as in Paralongidorus. Siddiqi (41) stated that the type of amphidial aperture should be considered as the main differential character for the two genera (pore-like for Longidorus and slit-like for Paralongidorus) and this viewpoint is supported by Aboul-Eid (1), in contrast to Heyns (16 and 17) who considers the shape of the amphid pouch much more important than its aperture.

Aboul-Eid (1) also gives an amended diagnosis of the genus which, at the present, includes 24 species:

"Longidorinae: lip region may be offset by a constriction at the level of the amphidial opening. Amphid opening transverse slits, sublabial, often extending completely across the head in lateral view. Odontophore without flanges. Spear guiding ring located at not more than four labial widths from anterior end".

The morphological and biometrical features for species identification of Paralongidorus are the same used for Longidorus.

The genus Paralongidorus has been little studied. All the species appear to have a very limited geographical distribution. The only species found in several different countries in Europe and in North Africa is P. maximus.

VALID SPECIES OF PARALONGIDORUS

Type species: P. sali Siddiqi, Hooper and Khan, 1963 (43);

Other species: P. afzali (Khan, 1964) Siddiqi and Husain, 1965 (42);
syn.: L. afzali (Khan, 1964) (21);
P. beryllus Siddiqi and Husain, 1965 (42);
P. boshi Khan, Saha and Seshadri, 1972 (22);
P. capensis Heyns, 1966 (17);
P. citri (Siddiqi, 1959) Siddiqi, Hooper and Khan, 1963 (43);
syn.: Xiphinema citri Siddiqi, 1959 (37);
L. citri (Siddiqi, 1959) Thorne, 1961 (49);
P. epimikis Dalmasso, 1969 (9);
P. erriae Heyns, 1966 (16);
P. eucalypti Fischer, 1964 (14);
P. fici Edward, Misra and Singh, 1964 (12);
P. flexus Khan, Seshadri, Weischer and Mathen, 1971 (23);
P. georgensis (Tulaganov, 1937) Siddiqi, 1965 (41);
syn.: L. georgensis Tulaganov, 1937 (51);
P. hooperi Heyns, 1966 (16);

P. lutosus (Heyns, 1965) Aboul-Eid, 1970 (1);

syn.: L. lutosus Heyns, 1965 (15);

P. maximus (Butschli, 1874) Siddiqi, 1964 (40);

syn.: D. maximus Butschli, 1874 (5);

D. (L.) maximus (Butschli, 1874) Micol.1922 (32);

L. maximus (Butschli, 1874) Thorne and Swanger, 1936 (50);

P. microlaimus Siddiqi, 1964 (40);

P. nudus (Kirjaneva, 1951) n. comb.

syn.: L. nudus Kirjanova, 1951 (24);

P. paramaximus Heyns, 1965 (15);

P. remeyi (Altherr, 1963) Siddiqi and Husain, 1965 (42);

syn.: L. remeyi Altherr, 1963 (3);

P. sacchari Siddiqi, Hooper and Khan, 1963 (43);

P. spiralis Khan, Saha and Seshadri, 1972 (22);

P. strelitziae (Heyns, 1966) Aboul-Eid, 1970 (1);

syn.: L. strelitziae Heyns, 1966 (16);

P. utriculoides (Corbett, 1964) Siddiqi and Husain, 1965 (42);

syn.: L. utriculoides Corbett, 1964 (8);

P. xiphinemoides Heyns, 1965 (15);

KEY TO SPECIES OF LONGIDORUS (MICOLETZKY, 1922) FILIPJEV, 1934

1. Body length less than 5 mm 2
   Body length 5 mm or more 19

2. Body length less than 3 mm 3
   Body length between 3 and 4.9 mm 5

3. Lip region continuous with the rest of the body .... laevicapitatus Williams, 1959
   Lip region expanded 4

4. Tail conoid and short, ratio T/ABW 2 or less .... reneyii Raina, 1966
   Tail elongate, T/ABW 3 or more .... longicaudatus Siddiqi, 1962

5. Lip region continuous with the rest of the body ... 6
Lip region offset by a slight constriction or expanded .. 11

6. Tail conoid ... 7
Tail bluntly rounded ... 8

7. Tail with pointed terminus ... indicus Prabha, 1973
Tail subdigitate, ventrally arcuate ... nirulai Siddiqi, 1965

8. Amphidial pouch bilobed ... 9
Amphidial pouch not bilobed ... 10

9. Odontostyle 66–81 μ long ... congoensis Aboul-Eid, 1970
Odontostyle 107–125 μ long ... taniwha Clark, 1963

10. Odontostyle 87–99 μ long ... belondiroides Heyns, 1966
Odontostyle 107–120 μ long ... jonesi Siddiqi, 1962

11. Lip region separated from the rest of the body by a slight constriction ... 12
Lip region expanded ... 13

12. Tail conoid with rounded terminus ... africanus Merny, 1966
Tail conoid with pointed terminus ... sylphus Thorne, 1939

13. Odontostyle over 100 μ long ... macromucronatus Siddiqi, 1962
Odontostyle less than 100 μ long ... 14

14. Odontostyle 83 to 91 μ long ... martini Merny, 1966
Odontostyle less than 70 μ long ... 15

15. Tail conoid with pointed terminus ... 16
Tail conoid with rounded terminus ... 17

16. Body length 2.8–3.6 mm, distance of the spear guiding ring from the anterior end 20–24 μ ... juvenilis Dalmasso, 1969
Body length 3.5–4.5 mm, distance of the spear guiding ring from the anterior end 30–32 μ ... leptocephalus Hooper, 1961

17. Odontostyle 65–68 μ long, tail elongated with T/ABW around 2 ... siddiqii (Siddiqi, 1959) Aboul-Eid, 1970
Odontostyle less than 65 μ long, T/ABW less than 1.6 ... 18

18. Odontostyle 52–56 μ long, spear guiding ring 20–23 μ from the anterior end ... monile Heyns, 1966
Odontostyle 56–63 μ long, spear guiding ring 26–31 μ from the anterior end ... moniloides Heyns, 1966

19. Lip region continuous with the rest of the body ... 20
Lip region offset by a slight constriction or expanded 25

20. Body length 9 mm and over ... macrosoma Hooper, 1961
Body length less than 9 mm ... 21

21. c ratio more than 190 ... saginus Khan, Seshadri, Weischer & Mathen, 1971
c ratio less than 190 ... 21

22. Body length around 5 mm, always less than 6 mm .... <u>tardicauda</u> Merzheevskaya, 1951
Body length 6 mm and over 23

23. Amphidial pouch not bilobed .... <u>caespiticola</u> Hooper, 1961
Amphidial pouch bilobed 24

24. Lip region rounded ....<u>goodeyi</u> Hooper, 1961
Lip region truncate .... <u>profundorum</u> Hooper, 1965

25. Lip region offset by a slight constriction 26
Lip region clearly expanded 27

26. Body length 4.5–6 mm .... <u>elongatus</u> (de Man, 1876) Thorne and Swanger, 1936
Body length 6.9–9.2 mm .... <u>vineacola</u> Sturhan and Weischer, 1964

27. Odontostyle length more than 170 μ .....<u>tarjani</u> Siddiqi, 1962
Odontostyle length less than 170 μ 28

28. Odontostyle length between 100 and 125 μ 29
Odontostyle length less than 100 μ 30

29. Body slender, 'a' ratio around 200 .... <u>cohni</u> Heyns, 1969
Body more robust, 'a' ratio around 15 .... <u>closelongatus</u> Stoyanov, 1964

30. Odontostyle from 72 to 80 μ long .... <u>globulicauda</u> Dalmasso, 1969
Odontostyle longer than 80 μ .... <u>attenuatus</u> Hooper, 1961

N.B.: <u>L. pisi</u> Edward <u>et al</u>. not included for lack of information.

KEY TO SPECIES OF <u>PARALONGIDORUS</u> SIDDIQI, HOOPER AND KHAN, 1963

1. Body length 2 mm or less .... <u>eucalypti</u> Fisher, 1964
Body length more than 2 mm 2

2. Body length from 2.2 to 5.5 mm 3
Body length more than 5.5 mm 15

3. Lip region continuous with the rest of the body 4
Lip region clearly offset 7

4. Tail elongate with pointed terminus .... <u>flexus</u> Khan, Seshadri, Weischer and Mathen, 1971
Tail conoid or bluntly rounded 5

5. Odontostyle 70 to 93 μ long .... <u>spiralis</u> Khan, Saha and Seshadri, 1972
Odontostyle over 100 μ long 6

6. Tail bluntly rounded .... <u>boshi</u> Khan, Saha and Seshadri, 1972
Tail conoid with rounded terminus .... <u>sacchari</u> Siddiqi, Hooper and Khan, 1963

7. Body length 2.2 to 2.8 mm <u>sali</u> Siddiqi, Hooper and Khan, 1963
Body length from 3 mm to 5.5 mm 8

8. Tail bluntly rounded 9
Tail conoid with pointed or rounded terminus 10

9. Odontostyle more than 180 μ long ....
.... <u>remeyi</u> (Altherr, 1963) Siddiqi and Husain, 1965
Odontostyle about 100 μ long ....
.... <u>nudus</u> (Kirjanova, 1951) nov. comb.

10. Tail with pointed terminus 11
Tail with rounded terminus 12

11. Amphidial pouch stirrup shape ....
.... <u>beryllus</u> Siddiqi and Husain, 1965
Amphidial pouch bilobed ....
.... <u>afzali</u> (Khan, 1964) Siddiqi and Husain, 1965

12. Amphidial pouch sacciform slightly lobed ....
.... <u>utriculoides</u> (Corbett, 1964) Siddiqi and Husain, 1965
Amphidial pouch stirrup shape 13

13. Odontostyle about 100 μ long ..... <u>xiphinemoides</u> Heyns, 1965
Odontostyle from 60 to 80 μ long 14

14. Lip region with rounded terminus...<u>microlaimus</u> Siddiqi, 1964
Lip region with flat terminus ....<u>erriae</u> Heyns, 1965

15. Lip region continuous with the rest of the body ....
.... <u>strelitziae</u> (Heyns, 1966) Aboul-Eid, 1970
Lip region clearly offset 16

16. Body length around 6 mm ....
.... <u>georgensis</u> (Tulaganov, 1937) Siddiqi, 1965
Body length more than 6.5 mm 17

17. Tail bluntly rounded...<u>maximus</u> (Butschli, 1874) Siddiqi, 1964
Tail conoid 18

18. Tail with pointed terminus ....
.... <u>lutosus</u> (Heyns, 1965) Aboul-Eid, 1970
Tail with rounded terminus 19

19. Odontostyle over 150 μ long 20
Odontostyle less than 150 μ long 21

20. Odontostyle length 205 to 216 μ ...... <u>epimikis</u> Dalmasso, 1969
Odontostyle length 170 to 181 μ .... <u>hooperi</u> Heyns, 1966

21. V 43-45% <u>citri</u> (Siddiqi, 1959) Siddiqi, Hooper and Khan, 1963
V 47-51% 22

22. T/ABW less than 1 .... <u>paramaximus</u> Heyns, 1965
T/ABW more than 1 ....<u>capensis</u> Heyns, 1966

N.B.: <u>P. fici</u> Edward <u>et</u> <u>al</u> (11) is not included because of lack of information.

REFERENCES

1. Aboul-Eid, H.Z. (1970). Systematic notes on Longidorus and Paralongidorus. Nematologica 16, 159-179.

2. Altherr, E. (1953). Nematodes du sol du Jura vandois et Francais (1). Bull.Soc.Vand. Sci. Nat. 65, 429-460.

3. Altherr, E. (1963). Contribution a la connaissance de la faune des sables submerges en Lorraine. Nematodes. Ann. Speleologie 18, 53-98.

4. Andrassy, I. (1970). Nematoden aus einigen Fluss-Systemen Südafrikas. Opusc. Zool. Budapest 10, 179-219

5. Bütschli, O. (1874), Zue Kenntnis der freilebenden Nematoden insbesondere der des kieler Hafens. Abh. Seskemb. naturf. Ges. 9, 237-292.

6. Chitwood, B.G. (1957), A new species of Xiphinemella Loos, 1950 (Nematoda), from Florida. Proc. helm. Soc. Wash. 24, 53-56.

7. Clark, W.C. (1963). A new species of Longidorus (Micol.) Dorylaimida, Nematoda). N.Z. J. Sci. 6, 607-611.

8. Corbett, D.C.M. (1964). Longidorus utriculoides n.sp. (Nematoda :Dorylaimidae) from Nyasaland. Nematologica 10, 496-499.

9. Dalmasso, A. (1969). Etude anatomique et taxonomique des genres Xiphinema, Longidorus et Paralongidorus (Nematoda: Dorylaimidae). Mem. Mus. natn. Hist. nat., Paris, Serie A. Zoologie 61, 33-82.

10. de Man, J.G. (1876). Onderzoekingen over vrij in de aarde levende Nematoden. Tijdschr. Ned. Dierk. Ver. 2, 78-196.

11. Edward, J.C., Misra, S.L. & Singh, G.R. (1964). Longidorus pisi n.sp. (Nematoda: Dorylaimoidea) associated with the rhizosphere of Pisum sativum from Uttar Pradesh, India. Jap. J. appl. Ent. Zool. 8, 310-312.

12. Edward, J.C., Misra, S.L. & Singh, G.R. (1964). A new species of Paralongidorus (Nematoda: Dorylaimoidea) from Allahabad, Uttar Pradesh, India. Jap. J. appl. Ent. Zool. 8, 313-316.

13. Filipjev, I.N. (1934). The classification of the free-living nematodes and their relation to parasitic nematodes. Smithson. misc. Coll. 89, 1-63.

14. Fisher, J.M. (1964). Dolichodorus adelaidensis n.sp. and Paralongidorus eucalypti n.sp. from S. Australia. Nematologica 10, 464-470.

15. Heyns, J. (1965). New species of the genera Paralongidorus and Longidorus (Nematoda: Dorylaimoidea) from South Africa. S. Afr. J. agric. Sci. 8, 863-874.

16. Heyns, J. (1966). Further studies on South African Longidoridae (Nematoda). S. Afr. J. agric. Sci. 9, 927-944.

17. Heyns, J. (1966). Paralongidorus capensis n.sp. and Longidorus belondiroides n.sp. with a note on L. taniwha Clark, 1963 (Nematoda: Longidoridae). Nematologica 12, 568-574.

18. Heyns, J. (1969). Longidorus cohni n.sp. a nematode parasite of Alfalfa and Rhodes grass in Israel. Israel J. agric. Res. 19, 179-183.

19. Hooper, D.J. (1961). A redescription of Longidorus elongatus (de Man, 1876) Thorne and Swanger, 1936 (Nematoda: Dorylaimoidea), and description of five new species of Longidorus from Great Britain. Nematologica 6, 237-257.

20. Hooper, D.J. (1965). Longidorus profundorum n.sp. (Nematoda: Dorylaimidae). Nematologica 11, 489-495.

21. Khan, E. (1964). Longidorus afzali n.sp. and Xiphinema arcum n.sp. (Nematoda: Longidoridae) from India. Nematologica 10, 313-318.

22. Khan, E. Saha, M. & Seshadri, A.R. (1972). Plant parasitic nematodes from Kumaon Hills, India. I. Two new species of Paralongidorus (Nematoda: Longidoridae). Nematologica 18, 38-43.

23. Khan, E., Seshadri, A.R., Weischer, B. & Mathen, K. (1971). Five new nematode species associated with coconut in Kerala, India. Indian J. Nematol. 1, 116-127.

24. Kirjanova, E.S. (1951). Nematody pochvy Khlipkovogo polya i tseling v Golodnoi Stepi (Uzbekistan). Trav. Inst. Zool. Akad. Sci. SSSR 9, 525-557.

25. Konicek, D.E. & Jensen, H.J. (1961). Longidorus menthasolanus, a new plant parasite from Oregon (Nematoda: Dorylaimoidea). Proc. helm. Soc. Wash. 28, 216-218.

26. Lamberti, F. (1969). Morphological variations and geographical distribution of Longidorus africanus Merny (Nematoda: Longidoridae). Phytopath. medit. 8, 137-141.

27. Lamberti, F. (1970). A new character for species identification in the genus Longidorus (Micol.) Filipjev. Proc. Xth Int. Nemat. Symp., Pescara, 1970, 18-19.

28. Lamberti, F. (1970). Longidorus africanus Merny, un grave parassita di Barbabietola da zucchero e Lattuga e largamente diffuso nel Bacino del Mediterraneo. Phytopath. medit. 9, 176-179.

29. Lamberti, F. & Sher, S.A.(1969). A comparison of preparation techniques in taxonomic studies of Longidorus africanus Merny. J. Nematol. 1, 193-200.

30. Merny, G. (1966). Nematode d'Afrique tropicale: un nouveau Paratylenchus (Criconematidae), deux nouveaux Longidorus et observations sur Longidorus laevicapitatus Williams, 1959 (Dorylaimidae). Nematologica 12, 385-395.

31. Merzheevskaja, O.I. (1951). Novye vidy nematod. Sborn. Nauchn. Trudov Akad. Nauk. Belorussk. SSR, Inst. Biol. 2, 112-120.

32. Micoletzky, H. (1922). Die frielebenden Erdnematoden. Erch. Naturgesch. 87, 1-650.

33. Prabha, M.J. (1973). Two species of Longidorus from Marathwada Region. Nematologica 19, 62-68.

34. Prota, U., Lamberti, F., Bleve-Zacheo, T. & Martelli, G.P.(1971) I Longidoridae (Nematoda: Dorylaimoidea) dei vigneti sardi. Redia 52, 601-618.

35. Raina, R. (1966). Longidorus reneyii sp. nov. (Nematoda: Longidoridae) from Srinagar, Kashmir. Indian J. Ent. 28, 438-441.

36. Schuurmans Stekhoven, J.H. (1951). Nematodes saprooaires et libres du Congo Belge. Mem. Inst. Sci. nat. Belg. ser. 2. 39, 1-77.

37. Siddiqi, M.R. (1959). Studies on Xiphinema spp. (Nematoda: Dorylaimoidea) from Aligarh (North India), with comments on the genus Longidorus Micoletzky, 1922. Proc. helm. Soc. Wash. 26, 151-163.

38. Siddiqi, M.R. (1962). Studies on the genus Longidorus Micoletzky, 1922 (Nematoda: Dorylaimoidea), with descriptions of three new species. Proc. helm. Soc. Wash. 29, 177-188.

39. Siddiqi, M.R. (1962). Longidorus tarjani n.sp. found around oak roots in Florida. Nematologica 8, 152-156.

40. Siddiqi, M.R. (1964). Xiphinema conurum n.sp. and Paralongidorus microlaimus n.sp. with a key to the species of Paralongidorus (Nematoda: Longidoridae). Proc. helm. Soc. Wash. 31, 133-137.

41. Siddiqi, M.R. (1965). Longidorus nirulai n.sp. a parasite of Potato plants in Shillong, India, with a key to species of Longidorus (Nematoda: Dorylaimoidea). Proc. helm. Soc. Wash. 32, 95-99.

42. Siddiqi, M.R. & Husain, Z. (1965). Paralongidorus beryllus n.sp. (Nematoda: Dorylaimoidea) from India. Proc. helm. Soc. Wash. 32, 243-245.

43. Siddiqi, M.R., Hooper, D.J. & Khan, E. (1963). A new nematode genus Paralongidorus (Nematoda: Dorylaimoidea) with description of two new species and observations on Paralongidorus citri (Siddiqi, 1959) n.comb. Nematologica 9, 7-14.

44. Stoyanov, D. (1964). A contribution to the nematode fauna of the grapevine. Rastit. Zasht. 12, 16-24.

45. Sturhan, D. (1963). Beitrag zur Systematik der Gattung Longidorus. Nematologica 9, 131-142.

46. Sturhan, D. & Weischer, B. (1964). Longidorus vineacola n.sp. Nematologica 10, 335-341.

47. Thorne, G. (1934). The classification of the higher groups of Dorylaims. Proc. helm. Soc. Wash. 1, 19.

48. Thorne, G. (1939). A monograph of the nematodes of the superfamily Dorylaimoidea. Capita Zool. 8, 1-261.

49. Thorne, G. (1961). Principles of Nematology. London, Mc-Graw Hill Book Company, Inc. 553 pp.

50. Thorne, G. & Swanger, H.H. (1936). A monograph of the nematode genera Dorylaimus Dujardin, Aporcelaimus n.g., Dorylaimoides n.g. and Pungentus n.g. Capita Zool. 6, 1-223.

51. Tulaganov, A.T. (1937). Nematoden der Tomate und des sie umgebenden Bodens. Zool. Anz. 118, 283-285.

52. Williams, T.D. (1959). Studies on the nematode soil fauna of sugarcane fields in Mauritius. 3. Dorylaimidae. Occ. Pap. Maurit. Sug. Ind. Res. Inst. 3, 1-28.

DISCUSSION

Attention was drawn to the variability within Longidorus elongatus. Wyss had found a population with a male:female ratio of 1:1, whilst Seinhorst reported a ratio 1:3 in one population and another population with no males. Seinhorst said that he had been able to identify three groups within the species, the separation being made on the basis of odontostyle length, tail shape, the number of sphincter muscles associated with the vagina, and the shape of the sphincter. Lamberti reported that he had examined 35 different populations of L. elongatus from several countries and had found them to be similar.

There was considerable discussion and exchange of viewpoints on the identification and geographical distribution of Longidorus spp. Lamberti remarked that all reported populations of L. elongatus from North Africa were in fact L. africanus. Caveness observed that only three Longidorus spp. had been found in Nigeria, and Lamberti added that he had sent Longidorus specimens from Nigeria to S.A. Sher, University of California, and all had been identified as L. attenuatus. Cohn pointed out that what previously had been referred to as L. brevicaudatum should now be called L. siddiqi; it was an economically important nematode in Israel. Hooper pointed out the L. vineacola and L. closelongatus were similar, but quite distinct species, and L. euonymus was between L. closelongatus and L. elongatus; L. euonymus has been implicated as a vector of Eonymus virus (V.R. Mali and D.J. Hooper, Nematologica 19, 459-467 (1974)).

MORPHOLOGY OF TRICHODORID NEMATODES

DAVID J. HOOPER

Nematology Department, Rothamsted Experimental Station,

Harpenden, Herts.

GENERAL MORPHOLOGY

Nematodes of the genera Trichodorus and Paratrichodorus range from 0.5 to 1.5 mm long. They are relatively plump and the females have a very short usually blunt tail so that they often look cigar-shaped especially when moribund. See Figs.1-4 for basic structures.

The cuticle

This is loose and wrinkles as the body bends; in fixed specimens particularly in the genus Paratrichodorus, the cuticle has a swollen appearance; it looks smooth under a light microscope but electron micrographs show minute transverse striations. Hirumi et al. (3) note that the cuticle in P. (Nanidorus) christiei consists of two main layers, loosely connected in the cephalic region, and that these layers are subdivided further. Raski et al. (9) reported eight layers in the cuticle of P. (P.) allius. More details of the ultrastructure of the cuticle and of other structures are given by Robertson and Taylor in these Proceedings.

Somatic musculature

The body muscles of this group are not very conspicuous. They are apparently the platymyarian type in which the muscle cells flatly extend toward the cuticle and meromyarian where there are few cells in each transverse section quadrant of the body musculature. Hirumi, et al (5) note that the muscle cells in P. (N.) christiei may be considered as a very primitive type capable of performing only simple sluggish movements.

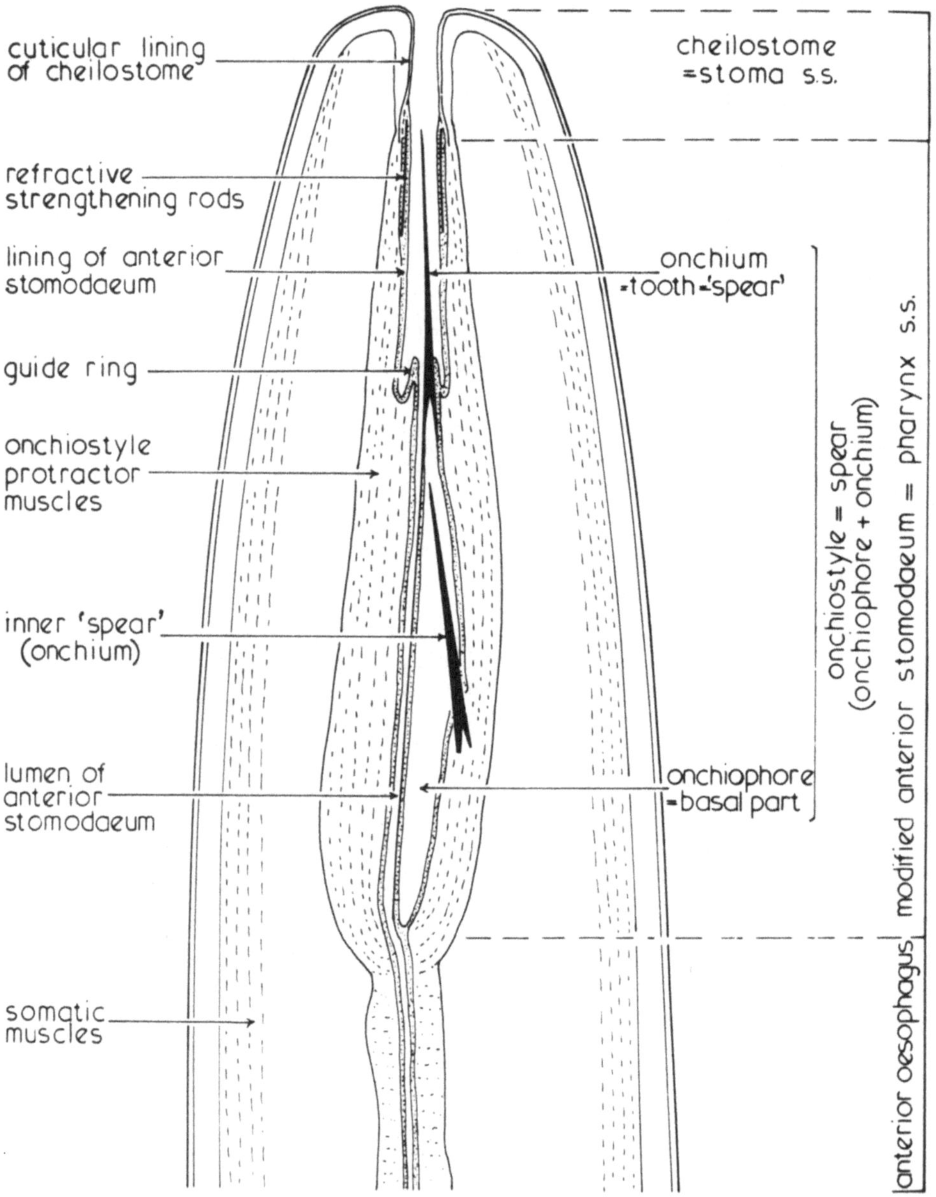

Fig. 1. Schematic drawing of head region of a trichodorid nematode (the inner 'spear' is usually visible in juveniles of all trichodorid spp. but only in adults of <u>Paratrichodorus</u> (<u>Nanidorus</u>) spp.).

Lip region

This is dome-shaped and set off from the neck contour by the slight elevations caused by an outer ring of 10 papillae formed by one on each lateral lip and two on the outer margins of the sub-lateral lips. There is an inner ring of 6 papillae around the stoma.

Amphids

These paired lateral sensory organs are sublabial; the opening is a transverse elipse leading to a cup-shaped pouch which is separated posteriorly by a constriction from the sensilla pouch.

Stoma sensu stricto = Cheilostome

A relatively short region, triradiate in section and lined with cuticle.

Pharynx sensu stricto = modified anterior stomadaeum = oesophastome

This region contains the onchiostyle which is surrounded by prominent protractor muscles. The onchiostyle varies in length between species from 20μ to 85μ; it is ventrally curved with a solid tip that projects into the stoma. About a third of the way along the onchiostyle there is a guide ring formed by a fold in the pharyngeal wall. The middle region of the onchiostyle connects on the dorsal side with the muscles in the dorsopharyngeal wall. Just behind this region of the onchiostyle at its widest part there is an oblique dorsal aperture through which is inserted the anterior half of an "inner spear". This inner spear is not readily seen with a light microscope in adults except those in the subgenus *P.* (*Nanidorus*); see Seinhorst (10). The inner spear presumably replaces the anterior end of the onchiostyle but formation or shedding of the spear has not been observed during moulting (8). A narrow pharyngeal lumen ventrally adjacent to the onchiostyle connects the stoma to the oesophagus.

Oesophagus and nerve ring

The narrow anterior half of the oesophagus gradually expands into a pear-to spatulate-shaped posterior oesophageal bulb. The nerve ring usually surrounds the oesophagus about half way along the narrow part. The oesophagus has a generally glandular appearance and Bird (2) reports that the basal bulb in *T. porosus* is non muscular. The oesophageal bulb may abut onto the intestine, overlap it ventro laterally or the intestine may extend forward dorsolaterally. Five oesophageal gland nuclei are present, one dorsal and two pairs subventral; their position may vary between species and Siddiqi (11) uses their position and the type of oesophageal junction with the intestine to separate genera and subgenera in this group.

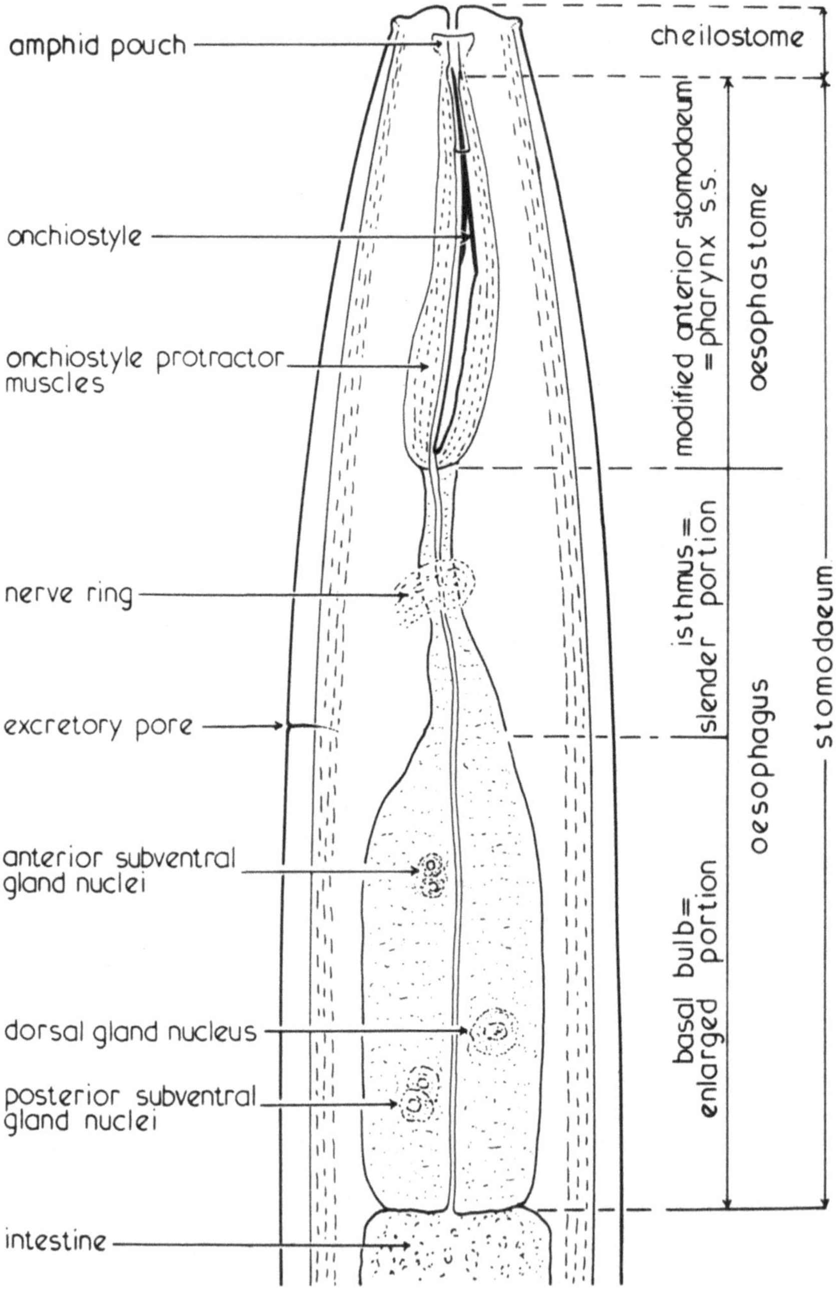

Fig. 2. Basic structure of anterior region of a trichodorid nematode.

Intestine

This long straight tube from the base of the oesophagus to the rectum in the female, or cloaca in the male, does not show any obvious division into various sections as do some other Dorylaims but Hirumi and Hung (4) noted that the cellular structure was divided into an anterior granular region and a posterior micro-villous region.

Excretory pore

This is located ventrally opposite the mid to posterior part of the oesophagus and exceptionally posterior to the oesophageal region.

FEMALES

Reproductive system

With one exception all species in this group have a median placed vulva and paired, opposed, reproductive systems with short reflexed ovaries as is typical in many Dorylaim nematodes. In bi-sexual species an oval spermatheca is present between the uterus and oviduct. *T. monohystera* has a single anterior outstretched ovary, the vulva is posterior and the posterior ovary is rudimentary. The vulval opening varies between species and may be pore-shaped or a transverse or a longitudinal slit. A refractive (sclerotized) ring is often present at the junction of the vulva with the vagina and in lateral view the refractive pieces are characteristically shaped, varying from inconspicuous dots to triangular blocks or transverse rods. Shape and size of the vagina is also characteristic having a well developed muscular wall in *Trichodorus* spp. and usually occupying about half a body width whereas in *Paratrichodorus* spp. it is generally smaller and less well developed (see also Loof in these Proceedings).

Body pores

Females in the subgenus *Nanidorus* do not have caudal or lateral body pores but all other species have a pair of subterminal caudal pores. Most species also have lateral body pores; in *Trichodorus* spp. one pair of lateral body pores are usually located within a body width behind the vulva, one or more other pairs of pores may also be present. In *Paratrichodorus* spp. the number and their position varies somewhat within and between species (6); Kuiper and Loof (6) report that *P. pachydermus* may have up to four pores on one side. However, the pores are not usually within a body width of the vulva in *Paratrichodorus* spp. Lateral body pores are not readily seen against the granular background of the intestine of specimens viewed laterally. They are much more easily

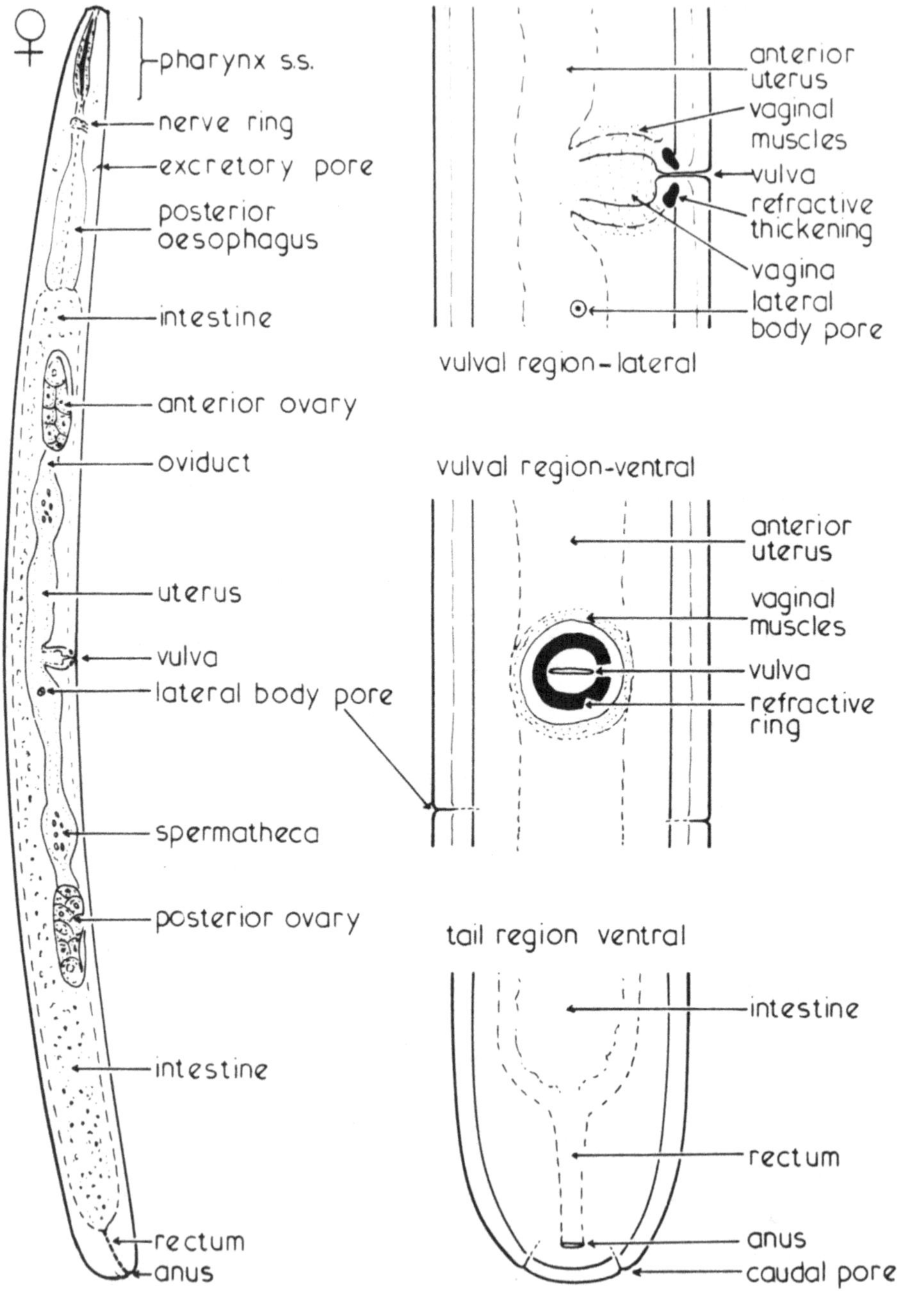

Fig. 3. Basic structure of a Trichodorus female

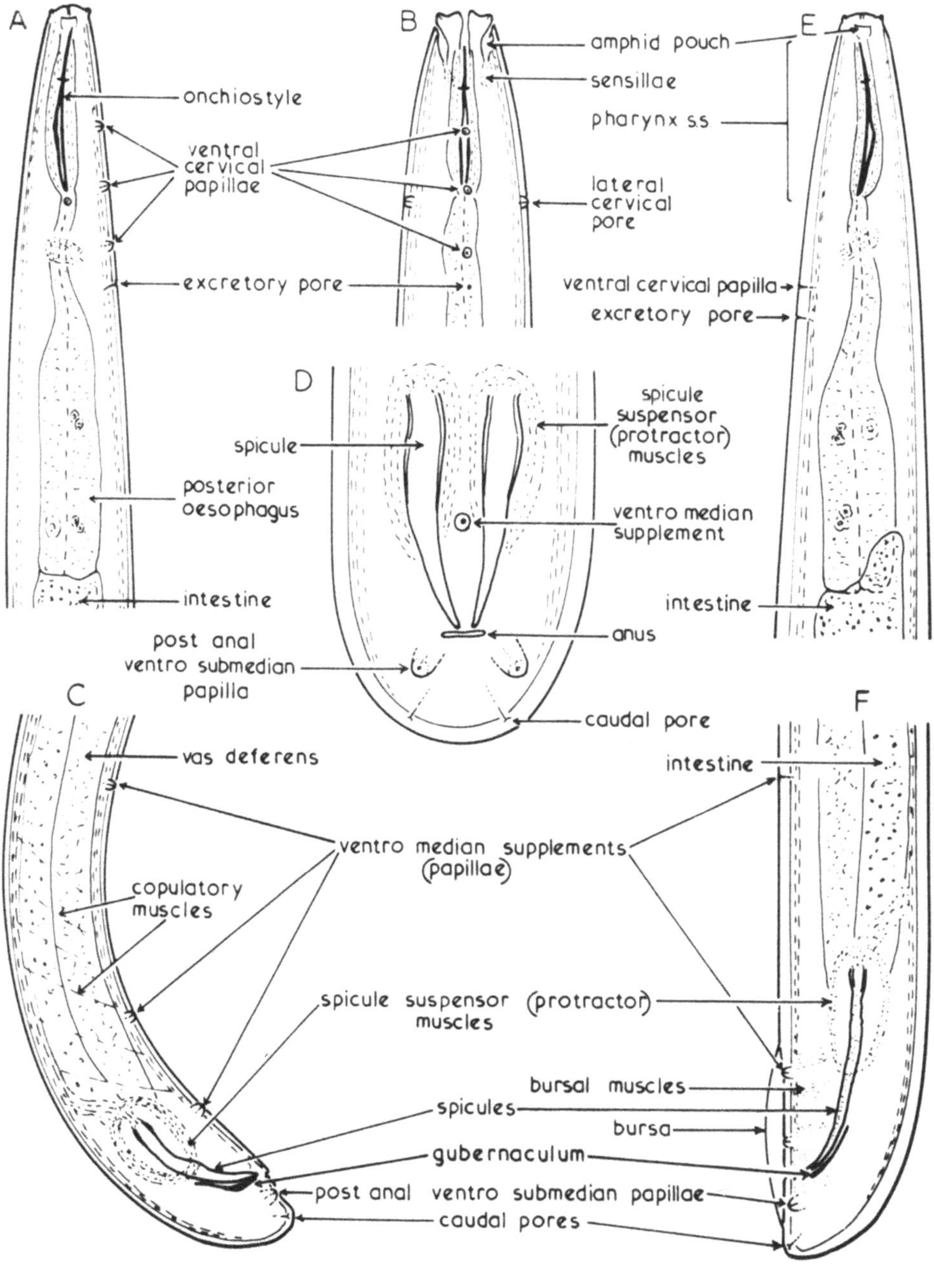

Fig.4. Basic structure of a trichodorid male: A–D, Trichodorus: A, anterior region – lateral; B, anterior end – ventral; C, posterior region – lateral; D, posterior end – ventral; E–F, Paratrichodorus:– E, anterior region – lateral; F, posterior region – lateral.

seen projecting out of the sides of specimens in dorsal or ventral view. *P. porosus* has a pair of ventromedian pores just anterior and posterior to the vulva and *P. atlanticus* has a pair of ventro-submedian pores just posterior to the vulva.

MALES

Reproductive system

There is a single outstretched testis. Anteriorly it begins with a relatively short geminal zone followed distally by a seminal vesical, containing oval to sausage shaped sperms, followed by the longer, glandular, vas deferens leading to the cloaca. The cloaca contains paired copulatory spicules. The proximal half of each spicule is enclosed by a capsule of suspensor muscles which seem to act as protractor muscles by a squeezing effect caused by their contraction; retractor muscles connect from the head of the spicules to the top of the capsule and further retractor muscles connect the capsule to the subdorsal body wall. Siddiqi (11) describes and illustrates the morphology and function of this region and notes that the spicule suspensor muscles are generally less well developed in *Paratrichodorus* spp. The spicules are elongate, ventrally arcuate tending to be more curved in *Trichodorus* than *Paratrichodorus* but in either genus spicules of some species have faint transverse striations. Spicules are usually of a characteristic shape for each species, the proximal end usually is the widest part and sometimes swollen and set off (cephalated); in most species the shaft gradually tapers to the narrow tip which may be slightly bifurcate. The spicules of some species, mainly in *Trichodorus*, have a stouter appearance the distal half being quite wide and it may even have a ventral flange as in *T. velatus*. Also, in some species the spicules are setose having bristles directed towards the spicule tip. The bristles are most easily seen on extruded spicules. They are present only on the middle region of spicules of some species whereas in *T. sparsus* the middle third or more is setose (7) (see also Loof in these Proceedings).

Copulatory muscles and bursa

Trichodorus spp. *sensu stricto* do not have a bursa (caudal alae) but the copulatory muscles are fairly well developed, running obliquely from the latero-dorsal to the subventral surface of the body wall and extending from the cloacal region to about two spicule lengths in front. These muscles cause the posterior part of the body to curl ventrally when killed by heat. In *Paratrichodorus* spp. a bursa is present but the copulatory muscles are not so well developed and are limited to the bursal region; males in this genus stay relatively straight when killed by heat.

Pores and papillae

Most males have a lateral cervical pore on each side opposite or just posterior to the base of the onchiostyle. Many males also have ventromedian cervical papillae varying from one to four in number; they are usually anterior to the excretory pore although in a few species one papilla may be posterior. In *Trichodorus* spp. these cervical papillae are usually relatively prominent compared with the excretory pore, whereas in *Paratrichodorus* they tend to be less conspicuous and somewhat similar in appearance to the excretory pore. Up to three preanal supplementary ventromedian papillae (supplements) are present, the most anterior often being less distinct than the others. At least one supplement is within the range of the spicules and in *Paratrichodorus* when there are two they are within the bursal region and well separated from the anterior third if it is present. All species, except those in the subgenus *Nanidorus*, have a pair of subterminal caudal pores. All species have one, and a few occasionally two, pairs of post anal ventrosubmedian papillae.

REFERENCES

1. Allen, M.W. (1957). A review of the nematode genus *Trichodorus* with descriptions of ten new species. *Nematologica* **2**, 32–62.

2. Bird, G.W. (1971). Digestive system of *Trichodorus porosus* *J. Nematol* **3**, 50–57.

3. Hirumi, H., Chen, T.A., Lee, K.J. & Maramorosch, K. (1968). Ultrastructure of the feeding apparatus of the nematode *Trichodorus christiei*. *J. Ultrastructure Res.* **24**, 434–453.

4. Hirumi, H. & Hung, C.L. (1969). Ultrastructure of intestinal epithelium in a plant parasitic nematode, *Trichodorus christiei*. *J. Nematol.* **1**, 292.

5. Hirumi, H., Raski, D.J. & Jones, N.A. (1971). Primitive muscle cells of Nematodes: Morphological aspects of platymyarian and shallow coelomyarian muscles in two plant parasitic nematodes, *Trichodorus christiei* and *Longidorus elongatus*. *J. Ultrastructure Res.* **34**, 517–543.

6. Kuiper, K. & Loof, P.A.A. (1962). *Trichodorus flevensis* n.sp. (Nematoda: Enoplida) a plant nematode from new polder soil. *Versl. Meded. plzietenk. Dienst Wageningen* **136**, 193–200.

7. Loof, P.A.A. (1973). Taxonomy of the Trichodorus aequalis complex (Diphtherophorina). Nematologica 19, 49–61.

8. Moreton, H.V. & Perry, V.G. (1968). Life cycle and reproductive potential of Trichodorus christiei. Nematologica 14, 11.

9. Raski, D.J., Jones, N.O. & Roggen, D.R. (1969). On the morphology and ultrastructure of the esophageal region of Trichodorus allius Jensen. Proc. helminth. Soc. Wash. 36, 106–118.

10. Seinhorst, J.W. (1970). Replacement stylet in adult Trichodorus nanus. Nematologica 16, 30.

11. Siddiqi, M.R. (1974). Systematics of the genus Trichodorus Cobb, 1913 (Nematoda: Dorylaimida) with descriptions of three new species. Nematologica 19, 259–278. (Year 1973).

DISCUSSION

The fate of the larval tooth after the moult aroused considerable interest and various views were expressed. Coomans suggested that it passed back into the intestine and was then defaecated. Support for this theory was given by Lippens, who suggested an analogy with the chromadorids; by Weischer who reported seeing the sclerotised cuticular lining of the oesophagus in the intestine; and by Coomans with a slide of Aporcelaimellus showing parts of the stomadeum wall in the intestine at a late stage of moulting.

Robertson reported that so-called 'bristles' on the male spicules appeared arcuate in transverse section when viewed by electron microscopy, resembling air-intakes on a car. Seinhorst suggested that they were fixation artifacts but Wyss disagreed, at least in respect of Trichodorus similis. No function was suggested for them.

Wyss reported his observations on extrusions from the amphids in living nematodes and posed the question of their function. The majority opinion favoured their function as chemoreceptors, perhaps working with phasmids in detecting concentration gradients. Coomans referred to work with one-amphid mutants from which it could be inferred that amphids were used as chemoreceptors and that a bilateral arrangement was necessary to direct the nematode towards a stimulus. Roggen suggested that they might be mechanoreceptors, sensitive to shearing stresses imposed by turgor pressure. Coomans remarked that the openings of the labial papillae suggested that they were mechanoreceptors rather than chemoreceptors. Lippens said that he had observed in Aporcelaimellus that its papillae could stick the nematode to glass, and thus make it act as suckers.

Lippens pointed out that the excretory pore is always posterior to the nerve ring and should not be confused with another pore-like structure, anterior to the nerve ring and without any terminal cellular apparatus. Loof observed that it is better to express its position relative to body length (as for the 'v' index) rather than to the oesophagus.

Referring to differences distinguishing *Trichodorus* from *Paratrichodorus*, Seinhorst said that *Paratrichodorus* skin is more liable than *Trichodorus* to swell in response to chemical stimuli, including fixatives, detergents and water. Lippens suggested that the presence of *lacunae* in the collagen layer between inner and outer cuticles could account for this.

TAXONOMY OF TRICHODORIDAE

P. A. A. LOOF

Landbouwhogeschool,

Wageningen, The Netherlands.

HISTORICAL REVIEW

Members of Diphtherophorina were first described by de Man in 1880/1884 (27,28). At that time it was customary to include all spear-bearing nematodes with a bottle-shaped oesophagus not containing a valvular apparatus, in the genus Dorylaimus Dujardin, 1845. De Man described a species Dorylaimus primitivus, which nowadays is considered a Trichodorus. In the same papers he described the new genus Diphtherophora with the single species D. communis. It was chiefly the peculiar characters of the cuticle and the complicated spear, as well as the shape of the oesophagus, which made him exclude this species from Dorylaimus.

In 1913 Cobb (8) described the genus Trichodorus with the single species T. obtusus. Micoletzky (29) transferred Dorylaimus primitivus de Man to Trichodorus and synonymized T. obtusus Cobb with it.

In 1935 Thorne, recognizing similarities between Diphtherophora and Trichodorus, erected, within the superfamily Dorylaimoidea, the family Diphtherophoridae, distinguishing two subfamilies, Diphtherophorinae with the genera Diphtherophora de Man, 1880 and Triplonchium Cobb, 1920, and Trichodorinae with Trichodorus Cobb, 1913 (38). In 1961 Clark (6) raised this family to superfamily rank, Diphtherophoroidea, with two families Diphtherophoridae and Trichodoridae. In 1970 Coomans & Loof (12) raised its rank to suborder Diphtherophorina with a single superfamily Diphtherophoroidea. In 1974 Siddiqi (35) gave superfamily rank to the Trichodoridae which now became Trichodoroidea. The suborder Diphtherophorina is now included, together with the suborders Bathyodontina,

Dorylaimina, Mononchina and Trichosyringina, in the Order Dorylaimida.

The superfamily Diphtherophoroidea contains the single family Diphtherophoridae with the genera *Diphtherophora* de Man, 1880, *Tylolaimophorus* de Man, 1880 (of which *Triplonchium* Cobb, 1920 probably is a synonym), *Longibulbophora* Yeates, 1967 and the imperfectly known *Brachynemella* Cobb, 1933. The superfamily Trichodoroidea contains the single family trichodoridae.

Thus the systematic placement of trichodorids is now:

Nematoda	
Subclass:	Adenophori von Linstow, 1905
Infraclass:	Enoplia Pearse, 1942
Order:	Dorylaimida Pearse, 1942
Suborder:	Diphtherophorina Coomans & Loof, 1970
Superfamily:	Trichodoroidea (Thorne, 1935) Siddiqi, 1974
Family:	Trichodoridae (Thorne, 1935) Clark, 1961.

Until recently the family Trichodoridae contained only the genus *Trichodorus* Cobb, 1913. This genus was reviewed by Micoletzky (29), who incorrectly synonymized *Leptonchus* Cobb, 1920 with it, by Thorne (39) and by Allen (1). Whereas in Thorne (39) the genus was still monotypic, Allen mentioned twelve species, ten of which were described by him as new. Since then the number of species has increased constantly; at present there are 45 specific names which at some time or another have been combined with the generic name *Trichodorus*. Esser (14) gave a compendium.

At the XIth International Symposium of Nematology, held at Reading in 1972, Siddiqi proposed to split the genus *Trichodorus* into three genera:- *Trichodorus*, *Paratrichodorus* and *Nanidorus*. As a result of the discussions held during the symposium, he sank *Nanidorus* to the rank of subgenus of *Paratrichodorus*. In the Abstracts of Papers, distributed among the participants at the opening of the symposium, all these generic names were mentioned with diagnoses and type species, but this volume was not a publication in the sense of the Rules. Siddiqi's paper was ultimately published in Nematologica in January, 1974 (35), so that the generic and subgeneric names proposed by him are available from that date. *Paratrichodorus* was now divided into three subgenera:- *Paratrichodorus*, *Atlantadorus* and *Nanidorus*.

MORPHOLOGICAL CHARACTERS OF *DIPHTHEROPHORINA*, *DIPHTHEROPHOROIDEA* AND *TRICHODOROIDEA*.

The suborder Diphtherophorina is distinguished from Dorylaimina by the following morphological characters:

1. The musculature is meromyarian (among the Dorylaimina, however, there is also one meromyarian group, the leptonchids).

2. The structure of the spear is quite different; it is referred to as an onchiostyle as opposed to the odontostyle of the Dorylaimina.

3. The basal part of the oesophagus is not cylindrical, but pyriform or elongate-bulboid.

4. The arrangement of the oesophageal gland nuclei is different: the dorsal nucleus lies at the same level as, or posterior to the anterior pair of ventrosublateral nuclei.

5. An excretory pore is present.

6. The posterior part of the mid-gut is not differentiated into a prerectum.

7. Rows of lateral body pores are lacking, and only a few pores are present.

8. The body cavity is filled with peculiar inclusions.

9. The males are monorchic.

10. The males lack an adanal pair of supplements.

11. The spicule musculature is of a unique character, quite different from the usual type found in dorylaims and many other groups.

12. The shape of the spicules is not dorylaimid.

13. The males possess usually additional cervical pores.

14. The males possess a gubernaculum and lack lateral guiding pieces.

15. The replacement tooth is formed in the immediate vicinity of the functioning spear.

The last two characters - which are primitive - are common to Diphtherophorina and the superfamily Nygolaimoidea of the suborder Dorylaimina.

The Diphtherophoroidea and the Trichodoroidea differ as follows. In the Diphtherophoroidea the onchiostyle is short, straight, with basal knobs and often with complicated sclerotizations near the apex. The male supplements are reduced. The female rectum is directed obliquely to the longitudinal body axis and the anus lies well anterior to the terminus. In the Trichodoroidea the onchiostyle is long, curved, without basal knobs or distinct apical sclerotizations; it is embedded in a muscular sheath. The male supplements are well developed. The female rectum runs almost parallel to the longitudinal body axis and the anus lies subterminally.

THE GENERA AND SUBGENERA OF *TRICHODORIDAE*

Allen (1) already had pointed out that there were two species groups in the old genus *Trichodorus*, distinguished by presence or absence of caudal alae in the males, but at that time he saw no reason for splitting the genus. Siddiqi (35) found more characters associated with the one mentioned and raised the two groups to generic rank. Because the type species *T. obtusus* Cobb, 1913 belongs to the group without caudal alae, the generic name *Trichodorus* is now restricted to that species group. The distinguishing characters between the genera *Trichodorus* and *Paratrichodorus* are listed in Table 1.

Table 1. Differentiating characters between *Trichodorus* and *Paratrichodorus*

	Trichodorus	*Paratrichodorus*
Caudal alae	absent	present
Supplements	more or less evenly spaced, S-1 and S-2 anterior to spicules	S-2 and S-3 close together, within reach of spicules; or only S-3 present
Tail (male) in death	curved ventrad	straight
Suspensor muscles	conspicuous, oval to circular	inconspicuous
Vaginal sclerotization	strong	weak
Cuticle after fixation	not swollen	swollen
Lateral pores near vulva	present	absent

Two remarks must be made. First, *T. cylindricus* Hooper, 1962 (16) is somewhat intermediate, the male tail being only slightly bent ventrad in death, caudal alae are indicated, the spicules are almost straight and show transverse striae. Second, *T. monohystera* Allen, 1957 (1) was not considered by Siddiqi under either genus; he remarked that possibly a third generic name would have to be erected for this species. For the moment I leave *T. monohystera* in *Trichodorus* s.str. because of the male characters.

Table 2 gives the differences between the three subgenera of *Paratrichodorus*.

Table 2. Differentiating characters between subgenera of Paratrichodorus

	Paratrichodorus	Atlantadorus	Nanidorus
Supplements	3	3	1
Lateral and caudal pores in females	present	present	absent
Lateral cervical pores in males	usually absent	present	absent
Excretory pore	near nerve ring, well anterior to oesophagus base	near nerve ring, well anterior to oesophagus base	near or behind oesophagus base
Ventrosublateral oesophageal gland nuclei	close to oesophago-intestinal junction, far behind dorsal nucleus	well anterior to oesophago-intestinal junction, close to dorsal nucleus	close to oesophago-intestinal junction, far behind dorsal nucleus

CHARACTERS USED FOR SPECIFIC DIFFERENTIATION

The five demanian indexes give but little aid in this family. Body length is about the same in the great majority of species, only those of the subgenus Nanidorus are consistently smaller than the others. The coefficient "a" is roughly 15-25, only P. teres has "a"=30-40. The indexes "b" and "V" do not give any help at all; V. is markedly aberrant only in T. monohystera. The index "c" may be used in a few cases, e.g. in distinguishing males of P. pachydermus from those of P. anemones and P. teres (21,23). Onchiostyle length and spiculum length have some value, but the variability should be taken into account.

We realize nowadays that we should rely more upon qualitative and meristic characters. The most important of these are in the group under consideration:

For males: 1) shape of spicules and gubernaculum; 2) distribution of supplements; 3) number and distribution of lateral and ventral cervical pores; 4) number and size of caudal papillae.

For females: 1) shape of vulva (longitudinal slit, transverse slit or pore); 2) shape of sclerotization of vagina; 3) mode of sperm storage in the uteri; 4) presence or absence of ventral and subventral advulvar pores; 5) number and location of lateral epidermal pores.

For males and females: 1) shape of the posterior part of the oesophagus; 2) location of excretory pore; 3) occasionally tail shape may be of value; e.g. P. acutus has a peculiar female tail shape; in males there are clear differences between P. pachydermus on one hand, P. teres and P. anemones on the other; and between T. cylindricus on one hand, T. primitivus and T. similis on the other.

INTRASPECIFIC VARIATION

It is important to know the reliability of the characters we use for distinguishing taxa. In trichodoridae, as in many other nematode groups, absolute and relative dimensions are rather variable. Onchiostyle length is often reliable, but in two species, T. aequalis and T. sparsus, long-speared and short-speared populations are known (24). The number of cervical papillae in the males is fairly constant, but the relative distances may vary somewhat, e.g. in T. sparsus. The number of lateral body pores in females is variable also, at least in the genus Paratrichodorus (21,23); the variation pattern, however, is certainly characteristic for several species (21,23). In Trichodorus s.str. the number seems more constant, but little is known about its variation in this genus. In T. monohystera Allen (1) found two pairs of lateral pores, Loof (22) only one. Supplement number is very constant except in the T. aequalis-group (24). Spiculum length, on the contrary, is surprisingly variable in several species: in T. monohystera the range is 51-59 μ, the maximum value being 116% of the minimum (n = 9); the figures for some other species are: T. pakistanensis (n = 25): 49-58 μ (118%); P. pachydermus (n = 7): 44-54 μ (123); T. obscurus (n = 7): 50-62 μ (124%); T. proximus (n = 5): 48-65 μ (135%); T. aequalis (n = 7): 35-50 μ (143%); T. cylindricus (n = 20): 33-49 μ (148%); T. sparsus (n = 177): 43-65 μ (151%); T. californicus (n = 11): 43-70 μ (163%).

THE GENUS TRICHODORUS SENSU RESTRICTO

Definition: The cuticle does not swell strongly upon fixation. The dorsal oesophageal gland nucleus lies rather posteriorly. The ventrosublateral oesophageal glands do not overlap the mid-intestine. Females with a pair of lateral body pores within one body-width from the vulva. Vaginal sclerotization generally strongly developed. Males numerous; the diagonal copulatory muscles extend well anterior to the spicular region, so that the male tail curves ventrad in death. No caudal alae. Spicules arcuate, with conspicuous suspensor muscles. Supplements three, more or less evenly spaced, the two anterior ones nearly always located anterior to the spicules. Number of ventromedian cervical pores most often 2 or 3, rarely 1 or 0.

Cobb (8) gave a combined description of the genus Trichodorus and the species T. obtusus, with the indication "n.gen.,n.sp.". Thus T. obtusus Cobb, 1913 is type species, by original designation, of the then monotypic genus. The status of this species has never been clear. Micoletzky (29) synonymized it with T. primitivus (de Man, 1880). Thorne (39) accepted this synonymy with a short discussion; Allen (1) with a more detailed one, his principal argument being that T. primitivus recently had been found to occur in the U.S.A. The type locality of T. obtusus was Arlington Farm, Virginia, U.S.A.

Siddiqi (35) followed Thorne and Allen, without discussion.

The identity of T. primitivus is unambiguous, the type specimens being preserved; they are poor, but give just enough information. In 1963 Seinhorst (31) showed that Allen's description (1) of T. primitivus referred to two species: the females were the real primitivus, the males represented an undescribed species now known as T. similis Seinhorst, 1963. These two species differ markedly by the shape of spicules and gubernaculum: in T. similis the spicules are thick, evenly curved, the gubernaculum is also regular and lies on the dorsal side of the spicules (Fig.1,L), whereas in T. primitivus the gubernaculum lies partly on the ventral side of the

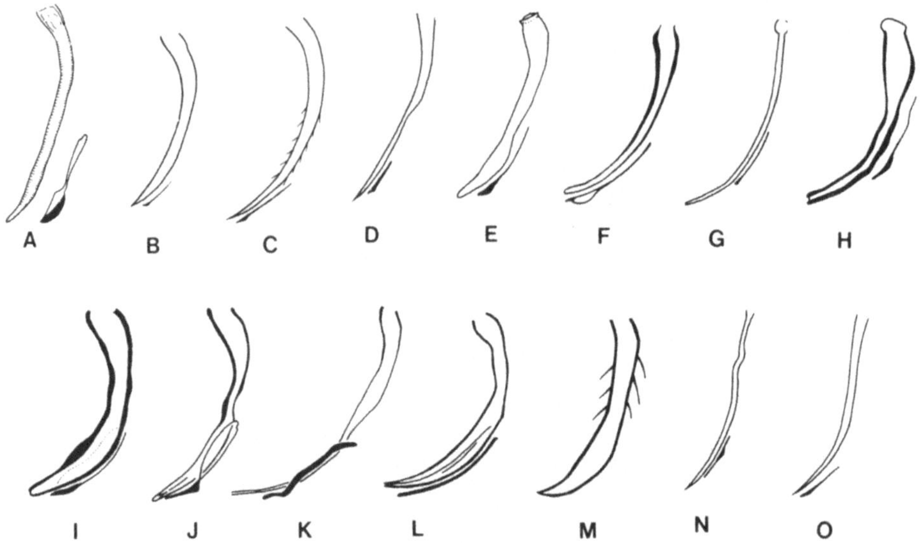

Fig. 1. Spicules of: A: Trichodorus borneoensis; B: T. aequalis; C: T. sparsus; D: T. hooperi; E: T. cottieri; F: T. californicus; G: T. monohystera; H: T. lusitanicus; I: T. velatus; J: T. viruliferus; K: T. primitivus; L: T. similis; M: T. cylindricus; N: Paratrichodorus anemones; O: P. tunisiensis.
Redrawn from: Allen (1): F,G; Clark (7): E; Hooper 16): A; Hooper (17): J; Hooper (18): I; Loof (2): N; Loof (24): B,C,D,M; Seinhorst (31): K,L; Siddiqi (34): O; Siddiqi (35): H.

spicules and the latter are irregular, very thin in the middle part (Fig.1,K). Cobb's illustration of T. obtusus shows a male tail end agreeing with similis, definitely not with primitivus. So, as Seinhorst already concluded, obtusus and primitivus are two different species. The question remains whether obtusus and similis are identical. Cobb's description unfortunately is very short; he does not give any indications about the vaginal sclerotization of the female, which in similis as well as in primitivus has a characteristic shape (Fig.2, C and D). We must go entirely by his two illustrations: head end and male tail. Onchiostyle length can be calculated as 85-90 μ from the illustration, about 75μ from the

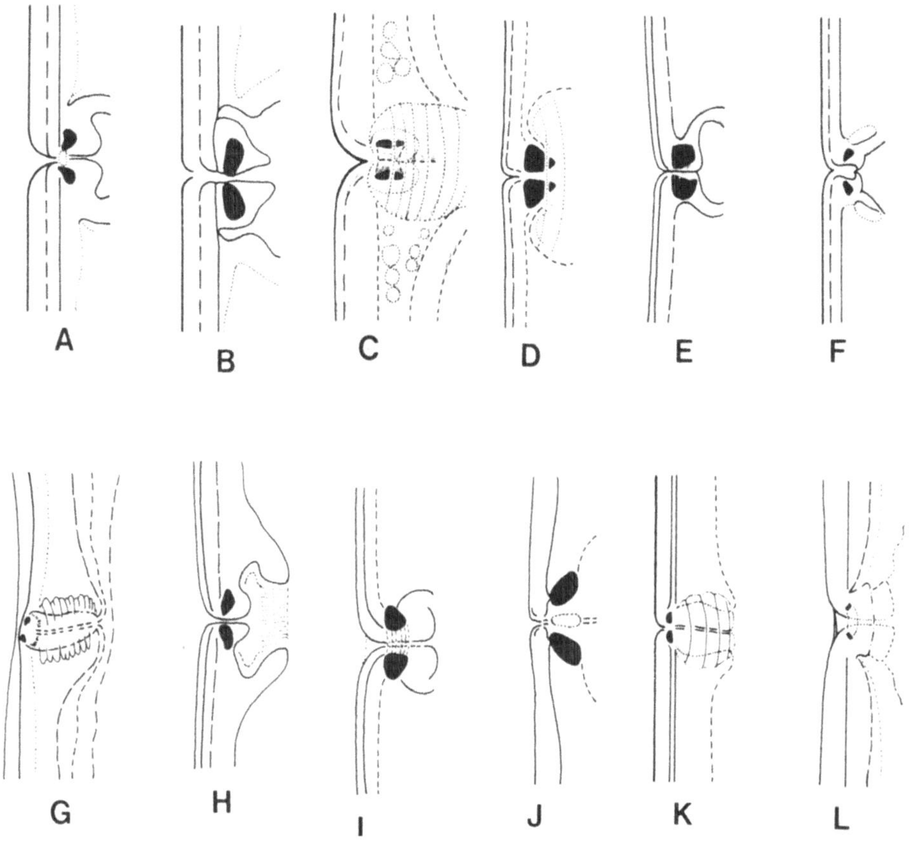

Fig. 2. Vagina of:
A: *Trichodorus borneoensis*; B: *T. cylindricus*;
C: *T. primitivus*; D: *T. similis*;
E: *T. variopapillatus*; F: *T. viruliferus*;
G: *T. pakistanensis*; H: *T. velatus*;
I: *T. lusitanicus*; J: *T. proximus*;
K: *Paratrichodorus rhodesiensis*; L: *P. tansaniensis*.

Redrawn from:
Allen (1): J; Hooper (16): A,B; Hooper (17): F;
Hooper (18): E,H; Seinhorst (31): C,D;
Siddiqi (33): G; Siddiqi (35): I,L;
Siddiqi & Brown (36): K.

formula on p.441. Spicule length can be calculated as 56 µ, body length is given as 1.0-1.1 mm. Body length is somewhat too great for similis, spicule length and especially onchiostyle length are much greater than in similis, in which Seinhorst gives spicule length as 36-40 µ and onchiostyle length as 38-43 µ. On the basis of these data I cannot do otherwise than consider similis and obtusus as two different species. Whether Allen's (1) report of T. primitivus occurring in the U.S.A. referred to primitivus or to similis could not be checked; the latter seems more probable in view of Cobb illustrating only the male.

In 1931 Thorne collected trichodorids on Arlington Farm. He described the species (39) as primitivus = obtusus, but Allen (1) redescribed it as a new species T. obscurus. Length of body, onchiostyle and spicules agree with obtusus, but the spicules are much less strongly curved.

Siddiqi (35) gave an up-to-date key to the species of Trichodorus, but he omitted four species: T. obtusus Cobb, 1913 which he considered identical with T. primitivus; T. monohystera Allen, 1957 which he tentatively excluded from the genus; T. elegans Allen, 1957 and T. hooperi Loof, 1973, which he overlooked. I therefore give new keys: one for males, modified from Siddiqi, and one for females. The latter, however, is tentative as it appears that it is not always possible to identify isolated females of this genus. Fig.1 gives some spicule shapes, which I have always determined along the axis. Fig.2 illustrates vaginal shape in a number of species.

Key to Trichodorus, males (modified from Siddiqi (35))

1. No ventromedian cervical papillae located in the onchiostyle region . . . . 2.
   One or two ventromedian cervical papillae in the onchiostyle region . . . . .14.

2. Ventromedian cervical papillae 2 or 3. . . . . . . 3.
   Ventromedian cervical papillae 1 or 0. . . . . . . 9.

3. Ventromedian cervical papillae 2 . . . . 4.
   Ventromedian cervical papillae 3 . . . . 7.

4. Tail contour evenly rounded, terminal cuticle not conspicuously thickened . . . . 5.
   Tail contour not evenly rounded, terminal cuticle conspicuously thickened . . . . 6.

5. Two supplements within reach of spicules; spicules striated (Fig.1,A). . . . . .borneoensis
   One supplement within reach of spicules; spicules not striated (Fig.1,B). . . . . .aequalis

6. Spicules setose, curved strongly proximally (Fig.1,C). . . . . .sparsus
Spicules not setose, almost straight (Fig.1,D) .hooperi

7. Onchiostyle length 72–75 μ . . . . .longistylus
Onchiostyle length 58–60 . . . . .cedarus
Onchiostyle length under 55 μ. . . . . .8.

8. Onchiostyle length 52–54 μ; body length under 0.65 mm. . . . . .kurumeensis
Onchiostyle length 39–48 μ; body length over 0.80 mm. . . . . .pakistanensis

9. No ventromedian cervical papillae. . . . . .obscurus
One ventromedian cervical papilla. . . . . .10.

10. Posterior supplement anterior to spicules. . . .proximus
Posterior supplement within reach of spicules. .11.

11. Onchiostyle length over 130 μ. . . . . .elegans
Onchiostyle length under 85 μ. . . . . .12.

12. Spicula expanded in distal third (Fig.1,E) . . .cottieri
Spicula not expanded in distal third, regular. .13.

13. Onchiostyle length 54–82 μ; spicules tapering, not conspicuously thin and slender (Fig.1,F) . .californicus
Onchiostyle length 43–48 μ; spicules cylindrical, conspicuously thin and slender (Fig.1,G). .monohystera

14. Spicules with transverse striae, curved only weakly (Fig.1,M); male tail almost straight . .cylindricus
Spicules not striated; male tail curved distinctly. . . . . .15.

15. Gubernaculum at level with, or ventral to spicules; convex anteriorly over most of its length (Fig.1,K) . . . . .primitivus
Gubernaculum not so. . . . . .16.

16. Spicules constricted near middle . . . . .17.
Spicules not constricted near middle (ig.1,L) .19.

17. Ventromedian cervical papillae 2; spicules with small rounded head (Fig.1,H) . . . .'. .lusitanicus
Ventromedian cervical papillae 3; spicules with broad head . . . . .18.

18. Spicules ventrally expanded in distal third (Fig.1,I). . . . .*velatus*
Spicules not ventrally expanded in distal third (Fig.1,J). . . . .*viruliferus*

19. Onchiostyle length 38–42 μ . . . .*similis*
Onchiostyle length 51–54 μ . . . .*variopapillatus*
Onchiostyle length over 70 μ . . . .*obtusus*

N.B. Males of *T. castellanensis* seem to differ from *T. primitivus* by the shape of the gubernaculum.

Tentative key to *Trichodorus*, females

1. Mono-prodelphic, vulva about 80% . . . .*monohystera*
Didelphic, vulva about 50–60%. . . . . .2.

2. Onchiostyle length over 130 μ. . . . .*elegans*
Onchiostyle length under 100 μ . . . . .3.

3. A pair of ventrosubmedian pores behind vulva . .*cedarus*
No ventromedian pores behind vulva . . . . .4.

4. Cuticular layers strongly demarcated, separated by an irregular or wavy line . . . . .5.
Cuticular layers not separated by a distinct, irregular or wavy line . . . . .6.

5. Two pairs of lateral pores . . . .*aequalis + sparsus*
One pair of lateral pores. . . . .*hooperi*

6. Sclerotization of vagina appearing as two thick straight lines parallel to the vaginal lumen (Fig.2,C); three pairs of lateral pores . . . .*primitivus*
Sclerotization of vagina different; one or two pairs of lateral pores . . . . .7.

7. Two pairs of lateral pores . . . . .8.
One pair of lateral pores. . . . . .9.

8. Sclerotization of vagina appearing as very large pieces (Fig.2,E); body length under 1 mm. . . .*variopapillatus*
Sclerotization of vagina appearing as smaller, oblique pieces (Fig.2,J); body length over 1 mm.*proximus*

9. Vulva a pore .10.
Vulva a transverse slit. . . . . .11.

10. Onchiostyle length 53–57 μ; scleroti ation of vagina reniform to pyriform (Fig.2,A). . . . . . borneoensis
Onchiostyle length 41–44 μ; sclerotization of vagina triangular with obtuse angles (Fig.2,I) .lusitanicus

11. Onchiostyle length 55–80 μ . . . . .12.
Onchiostyle length 35–53 μ . . . . .14.

12. Body length 0.57–0.64 mm . . . . .longistylus
Body length 0.70–1.25 mm . . . . .13.

13. Lateral body pores distinctly behind vulva . . .cottieri
Lateral body pores at level of vulva . .californicus + obscurus

14. Vaginal sclerotization dot-like (Fig.2,G). . . .pakistanensis
Vaginal sclerotization not dot-like. . . . . . .15.

15. Vaginal sclerotization very large, trapezoid, almost contiguous (Fig.2,D). . . . . .similis
Vaginal sclerotization smaller, not contiguous .16.

16. Vaginal sclerotization two oblique triangles (Fig.2,F). . . . . .viruliferus
Vaginal sclerotization more parallel to body axis, elongate-droplike to broadly triangular. .17.

17. Onchiostyle 35–45 μ; vagine deep, with small bulge near the middle (Fig.2,B). . . . . .cylindricus
Onchiostyle 42–50 μ; vagina less deep, with large lobes distally (Fig.2,H) . . . . .velatus

N.B. Not identifiable: T. castellanensis, T. kurumeensis and T. obtusus.

Remarks on some species

T. castellanensis. Described from few, poorly fixed specimens. I examined one male and one female. The female does not show any differences from T. primitivus. About the male I am less sure. The spicules are not very distinct, they might possibly have the same shape as those of primitivus. The gubernaculum, on the other hand, shows distally an angular dorsal thickening which I have not found in this shape in primitivus. So I must keep these species apart for the moment, but examination of fresh material is necessary for definitely settling their status.

T. cedarus. Described from poorly fixed specimens, which may account for the rather small body length (the same holds for T. kurumeensis and T. longistylus). Vaginal sclerotization said to be "comparatively large, conspicuous", but the illustration does not indicate this detail. Lateral body pores not mentioned.

T. boscurus. Syn. T. primitivus apud Thorne, 1939 (39) and Goodey, 1951 (15).

THE GENUS PARATRICHODORUS SIDDIQI, 1974

Definition: Cuticle swelling strongly upon fixation. Females without lateral body pores within one body-width from the vulva; vaginal sclerotization weakly developed. Males with caudal alae; the copulatory muscles do not extend anterior to bursa; male tail straight in death. Suspensor muscles inconspicuous. Two supplements within reach of caudal alae, well separated from the third which often is reduced; or only one supplement present, a short distance anterior to the cloacal aperture. Ventromedian cervical papillae rarely more than one.

Type species: P. tunisiensis (Siddiqi, 1963) Siddiqi, 1974, by original designation.

Subgenus Paratrichodorus Siddiqi, 1974. Definition: Posterior ventro-sublateral glands extending somewhat over the intestine, their nuclei located close to the oesophago-intestinal junction. Nucleus of dorsal gland near anterior end of enlarged oesophageal part. Mzles possessing one pair of large ventrosubmedian postcloacal papillae. Females with subterminal pores.

Type species: P. tunisiensis (Siddiqi, 1963) Siddiqi, 1974, by original designation.

Key to Paratrichodorus (Paratrichodorus) males (modified from Siddiqi (35))

1. Overlapping part of oesophagus about one body-width long. . . . . lobatus
   Overlapping part of oesophagus shorter than one body-width . . . . 2.

2. Onchiostyle length 75–87 µ . . . . acaudatus
   Onchiostyle length under 55 µ. . . . . 3.

3. Body length over 0.8 mm. . . . . 4.
   Body length under 0.8 mm . . . . 5.

4. Testis with well-developed sperm; one ventromedian cervical papilla present. . . . . tunisiensis
   Testis degenerate, without sperm; no cervical papillae . . . . teres

5. Three ventromedian cervical papillae; onchiostyle length 29–36 μ . . . . .mirzai
One or no ventromedian cervical papilla; onchiostyle length 37–44 μ . . . . .6.

6. Distance S-2* to anus about 45% of spiculum length . . . . .alleni
Distance S-2 to anus about 90% of spiculum length . . . . .7.

7. Spicule length 31 μ. . . . . .allius
Spicule length 42–44 μ . . . . .rhodesiensis

N.B. Males are unknown in P. tansaniensis.

Key to Paratrichodorus (Paratrichodorus) females (modified from Siddiqi (35))

1. Oesophageal overlap one body-width long. . . . .lobatus
Oesophageal overlap shorter than one body-width . . . . .2.

2. Onchiostyle length 75–83 μ . . . . .acaudatus
Onchiostyle length under 55 μ. . . . . .3.

3. Lateral body pores present . . . . .4.
Lateral body absent. . . . . .5.

4. Body stout (a = 23–29), uteri with spermatheca filled with sperm. . . . . .tunisiensis
Body slender (a = 27–41), uteri without sperm. .teres

5. Onchiostyle length 36 μ or less. . . . . .mirzai
Onchiostyle length 37 μ or more. . . . . .6.

6. Vaginal sclerotization appearing as two bold dots in lateral view (Fig.2,K) . . . . .rhodesiensis
Vaginal sclerotization appearing reniform or rod-like in lateral view . . . . .7.

7. Body length 0.64–0.78 mm; onchiostyle length 37–48 μ; vaginal sclerotization reniform. . . .allius
Body length 0.46–0.65 mm; onchiostyle length 35–40 μ; vaginal sclerotization in shape of two tiny rods (Fig.2,L). . . . . .tansaniensis

N.B. Females are unknown in P. alleni

---

*S-2: Second supplement

Remarks on some species

P. allius. The original description (29) mentioned only two supplements, but Siddiqi (35) found that there are three.

P. teres. Males are generally very rare; even in those populations in which they are numerous, they are non-functional (23).

P. tunisiensis. The female rectum is adpressed to the ventral body side. Males can be distinguished from those of P. anemones by the regular spicules (Fig.1, N and O).

Subgenus Atlantadorus Siddiqi, 1974. Definition: Posterior ventrosublateral glands do not overlap the intestine, but the intestine may have a short dorasl or dorsolateral overlap over the oesophageal bulb. Posterior ventrosublateral gland nuclei anterior to oesophago-intestinal junction, close to dorsal nucleus. Males usually with two pairs of large ventrosubmedian postcloacal papillae. Females with subterminal pores.

Type species: P. (A.) atlanticus (Allen, 1957) Siddiqi, 1974, by original designation.

Key to Paratrichodorus (Atlantadorus) males (modified from Siddiqi (35))

1. Onchiostyle length 64 μ or more. . . . . atlanticus
   Onchiostyle length 60 μ or less. . . . . 2.

2. Supplements 2, both within spicular range; spicule length 36–39 μ . . . . porosus
   Supplements, 3, S-1 well anterior to head of retracted spicules; spicules longer than 40 μ. . . . . 3.

3. Tail short, round; spicules irregular (Fig.1,N); distance S-2 to anus 38–64% of spiculum length. . . . . anemones
   Tail longer, trapezoid; spicules regular; distance S-2 to anus 75–94% of spiculum length . . . . pachydermus

Key to Paratrichodorus (Atlantadorus) females (modified from Siddiqi (35))

1. Subventral or ventromedian pores near vulva present. . . . . .2.
No subventral or ventromedian pores near vulva . . . . .3.

2. Onchiostyle length 64 μ or more; a pair of subventral pores behind the vulva. . . . . . . .atlanticus
Onchiostyle length 50 μ or less; two ventral pores on each side of the vulva. . . . . .porosus

3. Lateral pores behind vulva 0–2 on each body side; sperm contained within round spermatheca. . . . . .anemones
Lateral pores behind vulva 2–5 on each body side; sperm distributed all through the uteri. . . . . .pachydermus

Remarks on some species:

P. anemones. Females can be distinguished from those of P. tunisiensis by the more broadly rounded tail and the not adpressed rectum. Copulation obligatory.

Subgenus Nanidorus Siddiqi, 1974. Definition: Nuclei of posterior ventrosublateral oesophageal glands close to oesophage-intestinal junction; dorsal nucleus lies anteriorly in bulb. Excretory pore near, or behind, base of oesophagus. Lateral and subterminal pores absent. Males with only one supplement; spicules very long; one pair of large ventrosubmedian papillae near terminus. No cervical pores. A second onchiostyle tip usually inserted in dorsal slit of the functioning onchiostyle in adults.

Type species: P. (N.) nanus (Allen, 1957) Siddiqi, 1974, by original designation.

Key to Paratrichodorus (Nanidorus) males (modified from Siddiqi (35))

1. Basal region of oesophagus irregular, extending over intestine ventrally and subventrally; onchiostyle length 28–31 μ . . . . .minor
Basal region of oesophagus not overlapping; onchiostyle length 21–23 μ . . . . .nanus

N.B. Males are unknown in P. acutus and P. renifer.

Key to Paratrichodorus (Nanidorus) females
(modified from Siddiqi (35))

1. Tail acute to subdigitate. . . . . acutus
   Tail hemispherical . . . . 2.

2. Oesophagus base overlapping mid-intestine. . . .
   ventrally and subventrally . . . . minor
   Oesophagus base not overlapping. . . . . 3.

3. Excretory pore opposite oesophagus base;
   onchiostyle length 31-34 μ . . . . nanus

Remarks on some species:

P. minor and P. christiei. The description of P. minor appeared in June, 1956, that of P. christiei Allen, 1957 in February, 1957. From Allen's text it is clear that at that time he was not aware of the description of P. minor. Several authors (2,16,34) have noted that the two species agree in all essential characters with the exception of onchiostyle length; according to the original descriptions the onchiostyle length is 29-34 μ in minor, 33-47 μ in christiei. Siddiqi (34) added spiculum length as differentiating character: 50 μ in minor, 60-65 μ in christiei.

I examined females (n = 76) from several localities and hosts. Onchiostyle length was as follows:

Venezuela (n = 23), potato, strawberry, blackberry: 26-32 μ.
Nigeria (n = 2): 27 μ.
Sanford, Florida, U.S.A. (Type host and locality)(n = 4): 30-32 μ.
Java, Indonesia, potato (n = 5): 27-32 μ.
Wädenswil, Switzerland, Anemone spec. (n = 11): 29-33 μ.
Stockholm, Sweden, Gardenia spec. (n = 22): 26-32 μ.
Norrköping, Sweden, Saintpaulia ionantha (n = 1): 26 μ.
South Africa (n = 3): 29-32 μ.
Brasil, Bahia, Hevea and Ananas (n = 5): 26-33 μ.

Mr H. Brinkman, Plantenziektenkundige Dienst, informed me that he had measured at least 20 females from Rhododendron sp., Boskoop, The Netherlands: 31-34 μ.

These results made me doubt whether Allen's indication "onchiostyle length 33-47 μ in christiei" is correct, the more so as all species of the subgenus Nanidorus have very short onchiostyles (nanus 21-23 μ; renifer 31-34 μ; acutus 22-26 μ; minor 26-34 μ). Miss E. M. Noffsinger kindly lent me five paratypes of christiei (four females, one male); these were found to have onchiostyle length 27-33 μ. Vagina shape in the female paratypes agreed wholly with that of the populations mentioned above.

As to spiculum length, the difference seems big, but in the first place from Allen's illustration a value of 57 μ was calculated (the same value was found in the male examined), and secondly, if minor and christiei were conspecific, the longest spiculum known would measure 130% of the shortest, a variation width exceeded in several species of Trichodoridae.

I am, therefore, of opinion that these two species should be synonymized. The valid name is P. minor.

GEOGRAPHICAL DISTRIBUTION

Our knowledge about geographical distribution is still fragmentary; what is known reflects the countries which have been searched rather than the actual distribution of trichodorids. P. minor is cosmopolitan but may have been distributed by man. P. porosus also has a very wide area: California, Brasil, India, Ceylon and Japan.

VARIOUS REMARKS

It is common to find more than one species of Trichodoridae in one soil sample. On sandy soils in the Netherlands T. similis and P. pachydermus often occur together, occasionally T. primitivus, T. sparsus or P. teres are also present. In forest nurseries in Western Germany I often found T. sparsus and P. pachydermus together, now and then also T. similis and P. teres occurred. A marine clay sample from the North of the Netherlands contained a mixed population of T. primitivus and T. cylindricus. In Airolo, Switzerland, I found T. sparsus and T. cylindricus together. A sample from the bulb growing district of the Netherlands even contained five species: T. similis, T. primitivus, P. nanus, P. pachydermus and P. anemones (23). Many more examples could be given, but the above are sufficient to indicate that one should always identify at least 10-20 specimens from a sample before giving a specific name to the trichodorid population. P. nanus may have been frequently overlooked owing to its short stout body and inconspicuous vulva and vagina. It is known from Western Europe and Tunisia, but it may have a wider distribution.

Trichodoridae are extremely sensitive to all kinds of influences. When killed and/or fixed in an improper way, they may collapse strongly with the result, that important diagnostic characters, e.g. lateral body pores, may become invisible. Killing with a drop of hot fixative and processing by the Seinhorst method gives good results.

LIST OF DESCRIBED SPECIES OF TRICHODORUS WITH THEIR PRESENT STATUS

acaudatus Siddiqi, 1960 (32): Paratrichodorus (Paratrichodorus) acaudatus (Siddiqi, 1960) Siddiqi, 1974.
acutus Bird, 1967 (5): Paratrichodorus (Nanidorus) acutus (Bird, 1967) Siddiqi, 1974.
aequalis Allen, 1957 (1): Trichodorus aequalis Allen, 1957.
alleni Andrassy, 1968 (3): Paratrichodorus (Paratrichodorus) alleni (Andrassy, 1968) Siddiqi, 1974.
allius Jensen, 1963 (20): Paratrichodorus (Paratrichodorus) allius (Jensen, 1963) Siddiqi, 1974.
anemones Loof, 1965 (23): Paratrichodorus (Atlantadorus) anemones (Loof, 1965) Siddiqi, 1974.
atlanticus Allen, 1957 (1): Paratrichodorus (Atlantadorus) atlanticus (Allen, 1957) Siddiqi, 1974.
borneoensis Hooper, 1962 (16): Trichodorus borneoensis Hooper, 1962.
bucrius Lordello & Zamith, 1958 (25): syn. of Paratrichodorus porosus.
californicus Allen, 1957 (1): Trichodorus californicus Allen, 1957.
castellanensis Arias, Jimenez & Lopez, 1965 (4): Trichodorus castellanensis Arias, Jimenez & Lopez, 1965 (species inquirenda).
cedarus Yokoo, 1964 (41): Trichodorus cedarus Yokoo, 1964.
christiei Allen, 1957 (1): syn. of Paratrichodorus minor.
clarki Yeates, 1967 (40): syn. of Paratrichodorus lobatus.
cottieri Clark, 1963 (7): Trichodorus cottieri Clark, 1963.
cylindricus Hooper, 1962 (16): Trichodorus cylindricus Hooper, 1962.
elegans Allen, 1957 (1): Trichodorus elegans Allen, 1957.
flevensis Kuiper & Loof, 1962 (21): syn. of Paratrichodorus teres,(19).
granulosus (Cobb, 1920) Micoletzky, 1922 (9): Leptonchus granulosus Cobb, 1920.
hooperi Loof, 1973 (24): Trichodorus hooperi Loof, 1973.
kurumeensis Yokoo, 1966 (42): Trichodorus kurumeensis Yokoo, 1966.
litchi Edward & Misra, 1970 (13): syn. of Trichodorus pakistanensis.
lobatus Colbran, 1965 (11): Paratrichodorus (Paratrichodorus) lobatus (Colbran, 1965) Siddiqi, 1974.
longistylus Yokoo, 1964 (41): Trichodorus longistylus Yokoo, 1964.
lusitanicus Siddiqi, 1974 (35): Trichodorus lusitanicus Siddiqi, 1974.
minor Colbran, 1956 (10): Paratrichodorus (Nanidorus) minor (Colbran, 1956) Siddiqi, 1974.
mirzai Siddiqi, 1960 (32): Paratrichodorus (Paratrichodorus) mirzai (Siddiqi, 1960) Siddiqi, 1974.
monohystera Allen, 1957 (1): Trichodorus monohystera Allen, 1957.
musambi Edward & Misra, 1970 (13): syn. of Paratrichodorus mirzai.
nanus Allen, 1957 (1): Paratrichodorus (Nanidorus) nanus (Allen, 1957) Siddiqi, 1974.
obscurus Allen, 1957 (1): Trichodorus obscurus Allen, 1957.
obtusus Cobb, 1913 (8): Trichodorus obtusus Cobb, 1913.
pachydermus Seinhorst, 1954 (30): Paratrichodorus (Atlantadorus) pachydermus (Seinhorst, 1954) Siddiqi, 1974.

pakistanensis Siddiqi, 1962 (33): Trichodorus pakistanensis Siddiqi, 1962.
porosus Allen, 1957 (1): Paratrichodorus (Atlantadorus) porosus (Allen, 1957) Siddiqi, 1974.
primitivus (de Man, 1880) Micoletzky, 1922 (27): Trichodorus primitivus (de Man, 1880) Micoletzky, 1922.
proximus Allen, 1957 (1): Trichodorus proximus Allen, 1957.
rhodesiensis Siddiqi & Brown, 1965 (36): Paratrichodorus (Paratrichodorus) rhodesiensis (Siddiqi & Brown, 1965) Siddiqi, 1974.
similis Seinhorst, 1963 (31): Trichodorus similis Seinhorst, 1963.
sparsus Szczygie, 1968 (37): Trichodorus sparsus Szczgie, 1968.
teres Hooper, 1962 (16): Paratrichodorus (Paratrichodorus) teres (Hooper, 1962) Siddiqi, 1974.
tunisiensis Siddiqi, 1963 (34): Paratrichodorus (Paratrichodorus) tunisiensis (Siddiqi, 1963) Siddiqi, 1974.
variopapillatus Hooper, 1972 (18): Trichodorus variopapillatus Hooper, 1972.
velatus Hooper, 1972 (18): Trichodorus velatus Hooper, 1972.
viruliferus Hooper, 1963 (17): Trichodorus viruliferus Hooper, 1963.

N.B. Two Paratrichodorus species have been described directly under that generic name and hence are not included in the above list, viz. P. (Nanidorus) renifer Siddiqi, 1974 and P. (Paratrichodorus) tansaniensis Siddiqi, 1974. The present status is unchanged.

REFERENCES

1. Allen, M.W. (1957). A review of the nematode genus Trichodorus with descriptions of ten new species. Nematologica 2, 32-62.

2. Allen, M.W. (1957). Supplement: Dorylaimoidea described since the monographs on Dorylaims and Dorylaimoidea were published in 1936 and 1939. In: G. Thorne, A monograph of the nematodes of the superfamily Dorylaimoidea, 2nd edition. Capita zoologica 8 (5), 262-263.

3. Andrassy, I. (1968). The scientific results of the Hungarian soil zoological expedition to the Brazzaville-Congo. 31. Nematoden aus Grundwasser. Ann.Univ.Sci.Budap. 9-10, 3-26.

4. Arias Delgado, M., Jiménez Millán, F. & López Pedregal, J.M. (1965). Tres nuevas especies de nematodos posibles fitoparásitos en suelos españoles. Publ.Inst.Biol.Apl. 38, 47-58.

5. Bird, G.W. (1967). Trichodorus acutus n.sp. (Nematodea: Diphtherophoroidea) and a discussion of allometry. Can.J.Zool. 45, 1201-1204.

6. Clark, W.C. (1961). A revised classification of the Order Enoplida (Nematoda). N.Z.J.Sci. 4, 123-150.

7. Clark, W.C. (1963). A new species of Trichodorus (Nematoda: Enoplida) from Westland, New Zealand. N.Z.J.Sci. 6, 414-417.

8. Cobb, N.A. (1913). New nematode genera found inhabiting fresh water and non-brackish soils. J.Wash.Acad.Sci. 3, 432-444.

9. Cobb, N.A. (1920). One hundred new nemas. Contrib.Sci. Nematol. 9, 217-343.

10. Colbran, R.C. (1956). Studies of plant and soil nematodes 1. Two new species from Queensland. Qd.J.agric.Sci. 13, 123-126.

11. Colbran, R.C. (1965). Studies of plant and soil nematodes 9. Trichodorus lobatus n.sp. (Nematoda: Trichodoridae), a stubby-root nematode associated with citrus and peach trees. Qd.J.agric.anim.Sci. 22, 273-276.

12. Coomans, A. & Loof, P.A.A. (1970). Morphology and taxonomy of Bathyodontina (Dorylaimida). Nematologica 16, 180-196.

13. Edward, J.C. & Misra, S.L. (1970). Two new species of Trichodorus from Uttar Pradesh, India. Allahabad Farmer 44, 167-171.

14. Esser, R.P. (1971). A compendium of the genus Trichodorus (Dorylaimoidea: Diphtherophoridae). Proc. Soil and Crop Sci.Soc.Fla. 31, 244-253.

15. Goodey, T. (1951). Soil and freshwater nematodes. London, Methuen, 390pp.

16. Hooper, D.J. (1962). Three new species of Trichodorus (Nematoda: Dorylaimoidea) and observations on T. minor Colbran, 1956. Nematologica 7, 273-280.

17. Hooper, D.J. (1963). Trichodorus viruliferus n.sp. (Nematoda: Dorylaimida). Nematologica 9, 200-204.

18. Hooper, D.J. (1972). Two new species of Trichodorus (Nematoda: Dorylaimida) from England. Nematologica 18, 59-65.

19. Hooper, D.J., Kuiper, K. & Loof, P.A.A. (1964). Observations on the identity of Trichodorus teres Hooper, 1962 and T. flevensis Kuiper & Loof, 1962. Nematologica 9, 646.

20. Jensen, H.J. (1963). Trichodorus allius, a new species of stubby-root nematode from Oregon (Nemata: Dorylaimoidea). Proc.helminth.Soc.Wash. 30, 157-159.

21. Kuiper, K. & Loof, P.A.A. (1962). Trichodorus flevensis n.sp. (Nematoda: Enoplida), a plant nematode from new polder soil. Versl.PlZiekt.Dienst Wageningen 136, 193-200.

22. Loof, P.A.A. (1964). Free-living and plant parasitic nematodes from Venezuela. Nematologica 10, 201-300.

23. Loof, P.A.A. (1965). Trichodorus anemones n.sp. with a note on T. teres Hooper, 1962 (Nematoda: Enoplida). Versl. PlZiekt.Dienst. Wageningen 142, 132-136.

24. Loof, P.A.A. (1973). Taxonomy of the Trichodorus-aequalis-complex (Diphtherophorina). Nematologica 19, 49-62.

25. Lordello, L.G.E. & Zamith, A.P.L. (1958). Nota sôbre o gênero Trichodorus Cobb, 1913 com descricao de Trichodorus bucrius n.sp. (Nematoda, Dorylaimoidea). Ann.Acad.bras.Sci.30, 103-105.

26. Mamiya, Y. (1967). Descriptive notes on three species of Trichodorus (Dorylaimida: Trichodoridae) from forest nurseries in Japan. Appl.Ent.Zool. 2, 61-68.

27. de Man, J.G. (1880). Die einheimischen, frei in der reinen Erde und im süssen Wasser lebenden Nematoden. Tijdschr. ned.dierk.Ver. 5, 1-104.

28. de Man, J.G. (1884). Die frei in der reinen Erde und im süssen Wasser lebenden Nematoden der niederländischen Fauna. Leiden, Brill, 206 pp.

29. Micoletzky, H. (1922). Die freilebenden Erd-Nematoden. Arch. Naturgesch. A 87, 1-650.

30. Seinhorst, J.W. (1954). On Trichodorus pachydermus n.sp. (Nematoda: Enoplida). J.Helminth. 28, 111-114.

31. Seinhorst, J.W. (1963). A redescription of the male of Trichodorus primitivus (de Man), and the description of a new species T. similis. Nematologica 9, 125-130.

32. Siddiqi, M.R. (1960). Two new species of the genus Trichodorus (Nematoda: Dorylaimoidea) from India. Proc.helminth.Soc. Wash. 27, 22-27.

33. Siddiqi, M.R. (1962). Trichodorus pakistanensis n.sp. (Nematoda: Trichodoridae) with observations on T. porosus Allen, 1957, T. mirzai Siddiqi, 1960 and T. minor Colbran, 1956, from India. Nematologica 8, 193-200.

34. Siddiqi, M.R. (1963). Trichodorus spp. (Nematoda: Trichodoridae) from Tunisia and Nicaragua. Nematologica 9, 69-75.

35. Siddiqi, M.R. (1974). Systematics of the genus Trichodorus Cobb, 1913 (Nematoda: Dorylaimida), with descriptions of three new species. Nematologica 19, 259-278.

36. Siddiqi, M.R. & Brown, K.F. (1965). Trichodorus rhodesiensis and Amphidelus trichurus, two new nematode species from cultivated soils of Africa. Proc.helminth.Soc.Wash. 32, 239-242.

37. Szczygiel, A. (1968). Trichodorus sparsus sp.n. (Nematoda, Trichodoridae). Bull.Ac.Pol.Sci. 16, 695-698.

38. Thorne, G. (1935). Notes on free-living and plant-parasitic nematodes II. Proc.helminth.Soc.Wash. 2, 96-98.

39. Thorne, G. (1939). A monograph of the nematodes of the superfamily Dorylaimoidea. Capita zoologica 8 (5), 1-261.

40. Yeates, G.W. (1967). Studies on nematodes from dune sands 4. Diphtherophoroidea. N.Z.J.Sci. 10, 322-329.

41. Yokoo, T. (1964). On the stubby root nematodes from the Western Japan. Agric.Bull.Saga Univ. 20, 57-62.

42. Yokoo, T. (1966). On a new stubby root nematode (Trichodorus kurumeensis n.sp.) from Kyushu, Japan. Agric.Bull.Saga Univ. 23, 1-6.

DISCUSSION

In a general debate on describing new species it was generally agreed that a single specimen formed an inadequate basis for designating a new species and that any such proposals should be rejected. However, it is hard to establish the minimum number admissible; much depends on the quality of the diagnosis. Ten to fifteen individuals form an acceptable basis for establishing reliable ranges and means of quantitative data, while more than 30 would be needed to express confidence limits. Morphometric characters used in diagnosis must be reliable and understandable, in some (e.g. the 'b' index) the reliability may vary between species, growth stages etc.

Discussion of the relative merits of polytomous, dichotomous and trichotomous keys tended to favour the former, but Loof felt that keys alone were insufficient; Hooper emphasised that good descriptions were also essential and it is better not to rely on two or three characters only. Coomans suggested that illustrated keys would be very helpful and Loof found punch card keys useful. For practical reasons he also recommended breaking up keys for large genera arbitrarily, even if subgenera have not been designated.

The problem of determining overlapping stylet length ranges (especially in females) is facilitated by having enough specimens, but, because of their slight curvature, errors in stylet measurements are small, whereas male spicules are best measured along their median line. Alphey observed that TAF will shorten the onchiostyle and is therefore best avoided for *Trichodorus* spp. Pre-anal supplements and cervical papillae were considered reliable characters, except for the latter in *T. variopapillatus*.

Various queries concerning the status of several species were discussed e.g. the apparent localisation of *elegans* (seen only once); *anemones* (not found in Germany); *sparsus* (not found in UK); and *hooperi* which is similar to *sparsus* (found in UK but so far not elsewhere); these cases were not considered abnormal. Loof suggested that such problems might be resolved as experience widened, and Seinhorst agreed that *Trichodorus* spp. are characterised by their limited distribution, even within relatively small areas.

Paratrichodorus was thought, because its cuticle tended to swell like that of *Diphtherophora*, to be more primitive than *Trichodorus*, but Loof pointed out that the analogy did not hold good for *Triplonchium*.

FUNCTIONAL MORPHOLOGY OF DORYLAIMIDA

D. ROGGEN

Eenheid voor Dierkunde Fakulteit Wetenschappen Urije

Universiteit Brussel, Triomflaan, Brussels (Belgium)

INTRODUCTION

If morphology is to be a complete science it should do more than merely describe, it should explain its observations. In morphology this has been done, and still is done, by comparative techniques. This has led tothe viewpoint that morphology is explained by phylogeny.

In my opinion, however, no explanation has been given unless the phylogeny itself has been explained. Furthermore, I find it hard to believe that historical reasons per se can be responsible for the occurrence of morphological features in an organism. For instance, the fact that all nematodes have a triradiate pharynx is not explained by invoking an hypothetical ancestor which had a triradiate pharynx. Indeed, it is not clear why the ancestor had a triradiate pharynx to start with. Secondly, it is also not clear why the triradiate pharynx was retained while other features of the ancestor changed and were lost to the point that it is now not possible to recognize that ancestor unequivocally.

The functional morphologist tries to explain a feature by stressing its adaptive value. Much of my thinking about this subject has been influenced by Rudwick (4), to whom I shall not refer any more, but whose ideas will be present throughout this paper.

Firstly it should be stressed that the study of the function of a morphological feature is completely independent of the study of its evolution: the detection of adaptation does not depend upon the study of its origin. Hence functional arguments are a real addition to morphology and their relationship to phylogenetical reasoning is not circular, as is the case with homology. Indeed because the recognition of homology and of phylogeny is based upon the study of the same similarities between organisms, both concepts are not independent.

GENERAL METHODOLOGY

A functional morphological analysis is an exercise in model construction, or in Rudwick's (4) terminology "a paradigm construction".

One starts from the function and tries to describe the structure(s) that would be capable of fulfilling the function with the maximal efficiency attainable within the limits imposed by the material. Success in this approach does not depend upon the range of adaptations that happen to be possessed by organisms, but by the range of our understanding of the engineering problems. This functional approach can be combined with phylogenetical studies in several ways, upon which I will only give some brief comments.

a. In my view proof that a structure is functionally determined is connected with the possibility of using the structure's features in phylogenetical reasoning. Indeed, if a certain function can be performed by one structure only, the occurrence of this structure in two taxa has no phylogenetical implications. On the other hand, if a function can be performed by n structures, the occurrence of the same structure in two taxa has phylogenetical implications as the probability of the common occurrence being due to chance alone is $1/n^2$.

b. Every feature of an organism, except perhaps vestigial organs, should be considered adaptive. Failure to see the adaptive value of a structure is only an indication of our ignorance.

A FUNCTIONAL EXPLANATION OF FEEDING IN XIPHINEMA INDEX

The odontostyle of Xiphinema index has a longitudinal slit, i.e. it is not a closed tube. As a model we can take a circular tube, constructed of rigid material and subjected to bending forces. Reference to textbooks on strength of materials shows that due to an unequal distribution of the shearing stresses in the cross-section of the tube, a bending moment will induce a torque. Translated from the model to reality, it means that the presence of a longitudinal slit in the odontostyle allows X. index to rotate its stylet while pushing it into a root. It is obvious to anyone who has tried to push a nail into soft wood that rotating the nail helps a lot. Hence I consider that the presence of a slit in the odontostyle is of adaptive value.

Looking at the general structure of the pharynx we can ask the question "Why has X. index no spherical bulb?" Indeed, a spherical bulb is generally described as a powerful sucking organ and its presence in X. index would be most useful. A careful comparison of a spherical and a cylindrical pharynx model has shown (unpublished results) that a spherical bulb develops less suction than an equivalent cylindrical structure. On the other hand, a spherical bulb generates a higher pressure, and this explains why a bulb is situated at the caudal end of the pharynx, where the contents of the lumen have to be pumped into the gut. Indeed, it is the very need for a high suction pressure that prevented X. index from developing a bulb. The need for positive pressure had to be generated by an other mechanism, in casu the special arrangement of the pharyngeal muscles, some of which can close the pharynx lumen (3). These muscles need a positive transmural pressure inside the wall of the pharynx and this pressure depends upon the relative volume of the pharynx lumen: the larger the volume of the lumen, the larger the pressure in the pharynx wall. In X. index the relative volume is very small and an additional pressure generating mechanism is needed. It can be found in the presence of longitudinal muscles attached to the outer wall of the pharynx.

NEMATODE MOVEMENT

A suitable model for a nematode is a slender elastic rod subjected to longitudinal compressive forces. If such a rod buckles under the axial thrust it takes on a form which is called "the elastica". J. Calcoen (1) has demonstrated that this model agrees quite well with the observations she made on moving nematodes. It is also clear that the elastic properties of a nematode depend upon its turgor pressure, which in its turn depends upon the volume of the animal. Hence it can be expected that there will be an interaction between the feeding of a nematode and its locomotion. In discussing this interaction somewhat further, it will also be necessary to go into more detail to show more clearly how one can use models while studying nematodes.

From a mechanical point of view a feeding nematode is a self inflating balloon. As a model we can take a cylindrical structure with helical fibres in its outer wall, forming an angle of $90^{o} - \alpha$ with the longitudinal axis of the cylinder; this is the wellknown model of Harris and Crofton (2). To make this calculation easier, the helical fibre can be thought of as the hypotenuse of a rectangular triangle wrapped around the cylinder (Fig.1).

Also for the ease of calculations it can be assumed that the cylinder is at its maximal volume and that it will diminish its volume (i.e. defecation of our model nematode), hereby defining

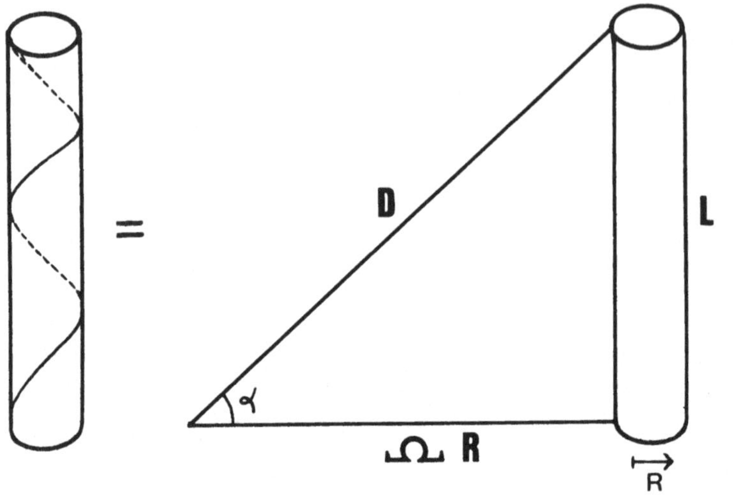

Fig. 1

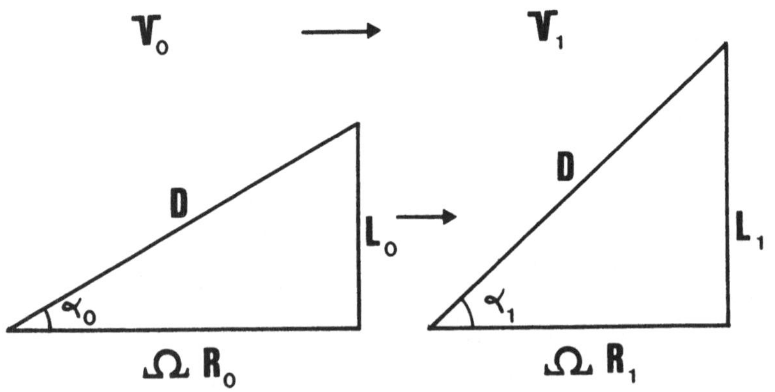

Fig. 2

$$V = V_0 - V_1 \tag{1}$$

The geometry of the system is shown in Fig. 2.

Further assuming that the internal pressure will be maximal at V_0, the fibre will be oriented so as to bear the stresses most efficiently and without assistance from the body muscles. As the longitudinal stress in a cylinder with internal pressure is equal to half the hoop stress this means that

$$\operatorname{tg} \alpha_0 = \tfrac{1}{2} \quad \text{or} \quad \alpha_0 = 26^{\circ}40'$$

We can now calculate α_1 in function of α_0 and of

$$\frac{\Delta V}{V_0} = 1 - \frac{V_1}{V_0}.$$

Indeed, in the Harris and Crofton (2) model we have

$$L_i = D \sin \alpha_i$$

$$R_i = \frac{D \cos \alpha_i}{\Omega}$$

and

$$V_i = \pi R_i^2 L_i$$

so that

$$\frac{V_i}{V_o} = 1 - \frac{\cos^2 \alpha_i \sin \alpha_i}{\cos^2 \alpha_o \sin \alpha_o} \tag{2}$$

Solving this equation for different values of $\frac{V_i}{V_o}$ gives:

$\frac{\Delta V_1}{V_o}$	α_1
0	$26^{o}40$
0.1	$22^{o}10$
0.2	$18^{o}20$
0.3	$15^{o}40$

We will now put muscles into the model nematode. It is clear that if we take longitudinal muscles they will shorten by the same amount as the length of the cylinder. If we put the muscles parallel with the fibres, the muscle length will not change.

This provides the model shown in Fig. 3 where 1_o = the length of the muscle and $90^o - \beta_o$ = angle that the muscles make with the longitudinal axis.

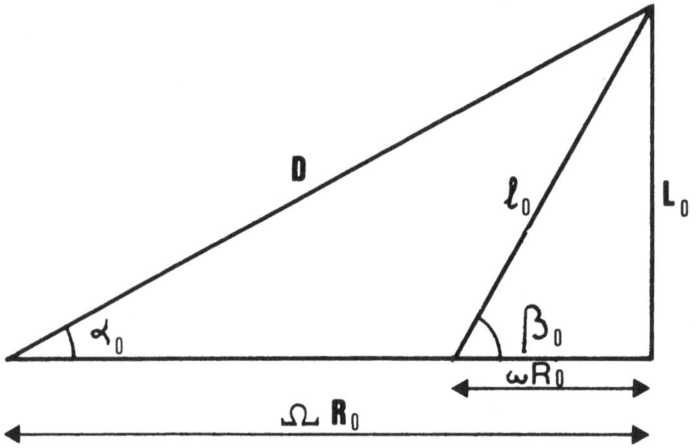

Fig. 3

Keeping in mind that both the fibres and the muscles are wrapped around the same cylinder, that only the fibre length remains constant, and that we know $\frac{\Delta V}{V_o}$, α_o and α_1 we can calculate $\frac{\Delta 1}{1_o}$ for the muscle:-

$$\frac{\Delta 1}{1_o} = 1 - \frac{\cos\alpha_1 \cos\beta_o}{\cos\alpha_o \cos\beta_1} \tag{3}$$

Solving gives the results shown in table 1:

Table 1 : $\dfrac{\Delta 1 / 1_o}{\Delta V / V}$

β_o	0	0.1	0.2	0.3
89°	0	0.176	0.242	0.435
85°	0	0.148	0.276	0.389
80°	0	0.147	0.281	0.379

If the tension a muscle can produce is taken to be a linear function of its contraction, and taking the maximal contraction to be 1/3 of the resting length, it can be stated

$$F = -3\frac{\Delta l}{l_o} + 1 \qquad (4)$$

This formula allows us to calculate the available muscular tension F_{max}, assuming that $\frac{\Delta l}{l_o} = 0$ for $\beta = \beta_o$, with the results as shown in Table 2.

Table 2 : F_{max}

β_o	$\frac{\Delta V}{V_o}$			
	0	0.1	0.2	0.3
89°	1	0.472	0.124	0
85°	1	0.556	0.172	0
80°	1	0.559	0.157	0

This available tension has to be split in 2 parts : a tension needed to equilibrate the tension in the cuticular fibres and the remainder which can be used for locomotion. The muscular tension needed to equilibrate a tension 'a' in the direction given can be readily calculated by $\alpha_o = 26°40'$ (Fig. 4).

Let us call this tension F_1; its value is shown in Table 3.

Table 3 : F_1

β_o	$\frac{\Delta V}{V_o}$			
	0	0.1	0.2	0.3
89°	0	0.085a	0.148a	0.199a
85°	0	0.088a	0.152a	0.204a
80°	0	0.093a	0.159a	0.212a

Finally we can calculate the ratio R = needed tension (F_1) : available tension (F_{max}), as shown in Table 4.

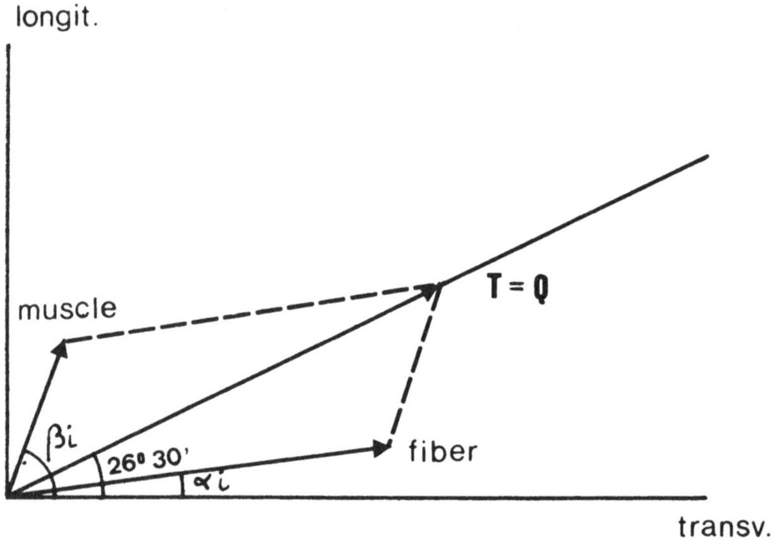

Fig. 4

Table 4 : R

β_0	$\frac{\Delta v}{V_0}$			
	0	0.1	0.2	0.3
89^0	0	0.180a	1.194a	∞ a
85^0	0	0.158a	0.884a	∞ a
80^0	0	0.166a	1.025a	∞ a

This result means that for each $\frac{\Delta v_1}{V_0} < 0.3$ there exists an angle β, $80^0 < \beta \leqslant 90^0$, which gives a maximal separation between the two functions.

I think this model offers a first approximate explanation for the fact that in many nematodes the somatic muscles are not exactly longitudinal, but form a small angle with the longitudinal direction.

Also, the same model probably explains the presence of helical constriction fibres around the pharynges of, for instance, the Belondiridae. A study of these pharynges will probably show that there exists a relation between the muscle direction and the relative lumen volume. In the case of X. index this relative volume is very small and the muscles should be longitudinal.

REFERENCES

1. Calcoen, J. (1972). Funktionele anatomie van de voortbeweging bij Nematoden. Licentiaatsthesis - Vrije Universiteit Brussel.

2. Harris, J.E. & Crofton, H.D. (1957). Structure and Function in the nematodes: internal pressure and cuticular structure in Ascaris. J. exp.Biol. 34: 116-130.

3. Roggen, D.R., Raski, D.J. & Jones, N.O. (1967). Further electron microscopic observations of Xiphinema index. Nematologica, 13: 1-16.

4. Rudwick, M.J.S. (1964). The inference of function from structure in fossils. Brit. J. Phil. Sci. 15: 27-40.

DISCUSSION

The function of the dorsal longitudinal split in longidorid stylets provoked considerable interest and discussion. Several participants reported observations of stylet bending during feeding and Roggen agreed that two observed features, viz. greater thickness on the ventral side of the stylet in Xiphinema and the concentration of bending close to the lip region, would increase the twisting moment resulting from stylet bending, thus aiding penetration. Coomans, while agreeing with the function of the slit, suggested that it had a phylogenetic origin, as it was also present in dorylaims with short and stout odontostyles. It was agreed that phylogeny and functional morphology are often complementary, but Roggen suggested that original ideas concerning function may be inhibited by too strongly phylogenetically-orientated thinking. Good ideas can arise from many diverse origins.

It was agreed that the twisting of tylenchid stylets (which lack a dorsal slit) could not arise from a simple bending force and that such twisting could have a different function e.g. Hemicycliophora arenaria has been observed to rotate its stylet when withdrawing from a plant cell and Wyss suggested that this may be done to free it from the plaque formed around the head during feeding.

Roggen agreed that the nematode body was not a classic elastica shape, as it was not a perfect cylinder. However, their dynamics had much in common and he hoped that more sophisticated models, now under construction, would improve our knowledge of the mechanisms of nematode movement.

CYTOGENETICS AND REPRODUCTION IN XIPHINEMA AND LONGIDORUS

A. DALMASSO

I.N.R.A., Station de Recherches sur les Nematodes,

Antibes, France.

INTRODUCTION AND METHODS

The general characteristics of the various reproductive processes of Rhabditida and Ascaridida were described by Bělař (2), Nigon (7) Nigon and Robert (8), Lin (68), and others, but there has been little information on phytophagous nematodes until the last fifteen years. Since 1960, tylenchids have also been studied, first by Triantaphyllou (9) and more recently by others. Data on dorylaimids are recent and few, the main reason for this being the difficulty of keeping them alive and multiplying them under laboratory conditions.

Gametogenesis of Longidoridae is dependent on environmental factors such as climate or food, and takes place only when new roots are available to the nematode. Under unfavourable conditions it ceases. Longidorids live and may reproduce for several years and this may compensate for the slowness of multiplication.

Observations on the activity of the ovaries throughout the year provide information on reproduction periods. Basic chromosome numbers of each species and genus are of considerable value in the study of phylogenetical relationships (3).

Methods

Nematodes are dissected on glass slides in normal saline solution, by cutting at the vulva for females and near the spicules for males. Reproductive tracts are then pushed out by the internal pressure and fixed by putting a drop of Carnoy or acetic sublimate on them with a micropipette. Excess Carnoy fixative is removed by dipping the slides in water, or when sublimate is used by dipping in sodium hypochlorite followed by water. After hydrolysis by normal hydrochloric acid for 12 min at 58°C the nuclei are stained by submersion in fuchsin. Excess fuchsin is washed out with a sulphurous solution. The cytoplasm is stained by immersion in "vert sulfo" for 2-3 sec. Finally, the slides are dipped in water again and then after alcohol dehydration mounting is done in euparal or Canada balsam.

SPERMATOGENESIS

Xiphinema and *Longidorus* males have two tubular apposed testes which converge in a common vas deferens. In both genera the two testes are well developed.

Spermatogenesis in *Xiphinema*

Spermatogenesis of longidorids is easily followed as most stages are well represented amongst the numerous cells that fill the two long testes. However, some rarer stages are often lacking in single specimens and therefore several mounts should be investigated. Great differences occur between the spermatogenesis patterns of individuals from the same population, these variations in gonad activity indicating differences in nutrition or microclimatic conditions experienced by individuals.

In *Xiphinema diversicaudatum* spermatogonial divisions occur in the apical portion of the testes, but they are rare. Therefore this part stains less strongly than the adjacent parts of the gonads. The oogonial nuclei are about 2 to 3 μ in diameter and cytoplasm is scarce.

Nuclei in the meiotic prophase are found in the part of the testis next to the apical portion. Nuclei in the first 50 to 100 μ stain deeply with fuchsin. Further on the spermatocytes in prophase become larger mainly because the nuclei become larger; these then take up less and less stain. Also the cytoplasm increases in volume (see also oogenesis). Spermatocytes become polyhedric while nucleoli disappear. Chromosomes contract and gradually become distinct. This differs from oogenesis where, before diakinesis, nuclei of primary oocytes fail to take up stain for quite a long period; this is related to the presence of a large nucleolus. At this time the youngest spermatocytes I show 12 coloured bodies that are transformed into 10 bivalent chromosomes during diakinesis. The size of the cells is 5-6 μ and nuclear diameters of vesicles are 3-3.5 μ. Chromo-somes are small, 0.3 μ. Each anaphase I plate shows 10 chromosomes. Secondary spermatocytes are smaller and mitotic divisions are difficult to distinguish. Numerous spermatozoa are stored in the distal part of testes and some are located in the vas deferens.

The preceding description also applies to X. mediterraneum, although males are very rare in this species, but only five chromosomes have been observed.

Spermatogenesis in Longidorus

Spermatogenesis of Longidorus is similar to that of Xiphinema. It has been studied in three common European species: L. elongatus, L. vineacola and L. macrosoma.

The germinal zone consists of about thirty spermatogonia. In a few cases, fourteen chromsomes may be counted. As in the genus Xiphinema, the meiotic prophase includes several zones. The first one stains heavily as it consists of many small spermatocytes having little cytoplasm. The next zone appears less coloured because of the increase in the size of the cells. Chromosomic material is pushed close to the periphery of the nucleus by a large nucleolus. Probably this stage is of short duration as it is represented by few spermatocytes and is absent in many testes.

Prediakynesis is characterised by a considerable increase of the cytoplasm. In L. vineacola spermatocytes are 5-6 μ long. Concentration of the chromosomes mostly begins with 8 or 9 bodies, but there can be more in the youngest cells. This sequence may be interpreted as diplotene. Finally, chromosomes condense into 7 bivalents.

Diakynesis is brief and only some testes show good figures. In the metaphase 7 rounded double chromosomes are readily discerned; 1.3-1.8 μ diameter for L. vineacola and smaller for L. elongatus and L. macrosoma. Spermatocytes then advance to the anaphase in which two chromosomes separate into two plates each one with 7 units.

The chromosome pattern tends to be circular. At first, the two plates have a central chromosome, whose migration to the periphery is retarded. At this stage, the diameters of the chromosomes measure approximately 0.4 μ and those of the plates 2-2.5 μ in the three species.

During diakynesis and until the last maturation division spermatocytes have an hexagonal shape. During the prophase and until the telophase of L. vineacola, they measure 5-6.5 μ while they are only about 4 μ during the second division. Hence observations again become difficult.

The homotypical prophase and metaphase are brief. Perhaps anaphases and telophases are easier to observe. The figures of these last stages are very similar to those of meiosis and 7 chromosomes may be counted. In L. vineacola their size is only about 0.2 μ whereas the plates are 1.5-1.7 μ in diameter.

Spermatogenesis is more interesting here than in Xiphinema because of the transformations that occur in the nuclear vesicle. Spermatids lost their hexagonal shape, the cytoplasm shrinks and nuclei become fusiform. Then the internal vesicle structure appears reticulate by Schiff colouration; more advanced spermatids enlarge with some swellings as if the nucleus were turgescent. Later the nucleus is spherical again and the chromatin tends to accumulate in a network just under nuclear membrane. Finally spermatids return to an elongate structure. The cytoplasm has been largely evacuated when spermatogenesis is completed. This differs from Parascaris equorum, Trilobus longus, etc., where spermatozoa possess a large cytoplasmic mass. As in Xiphinema, spermatozoa occupy a large part of the testes and some occupy the vas deferens.

OOGENESIS

Female reproductive tracts appear more variable in Xiphinema than in Longidorus. In the former genus there are differences between different groups of species. Thus, for example, the ovary of X. mediterraneum is characterized by a thick external sheath and by considerable differences in physiological aspects during active and quiescent periods. Some of these anatomical differences, such as the Z organ, seem to be correlated with the reproduction process (see Luc in these Proceedings).

Xiphinema

In Xiphinema, oogenesis is of short duration because eggs are deposited immediately after the first maturative division is complet -ed. Thus few stages are represented in the genital tract by oogonia and primary oocytes (Fig. 1). A short germinal zone occupies the

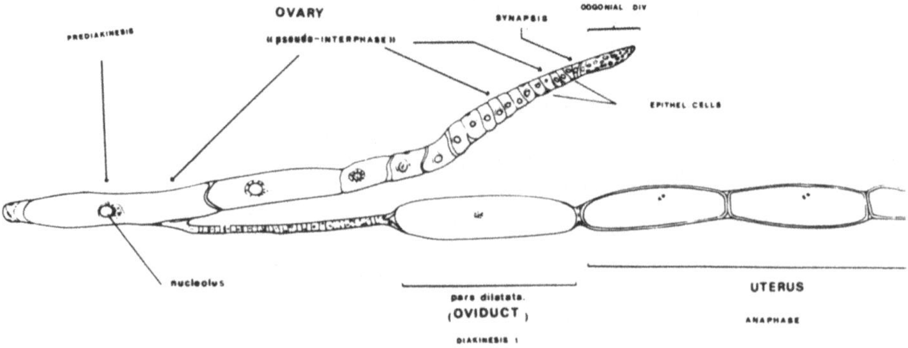

Fig. 1. Oogenesis in Xiphinema index

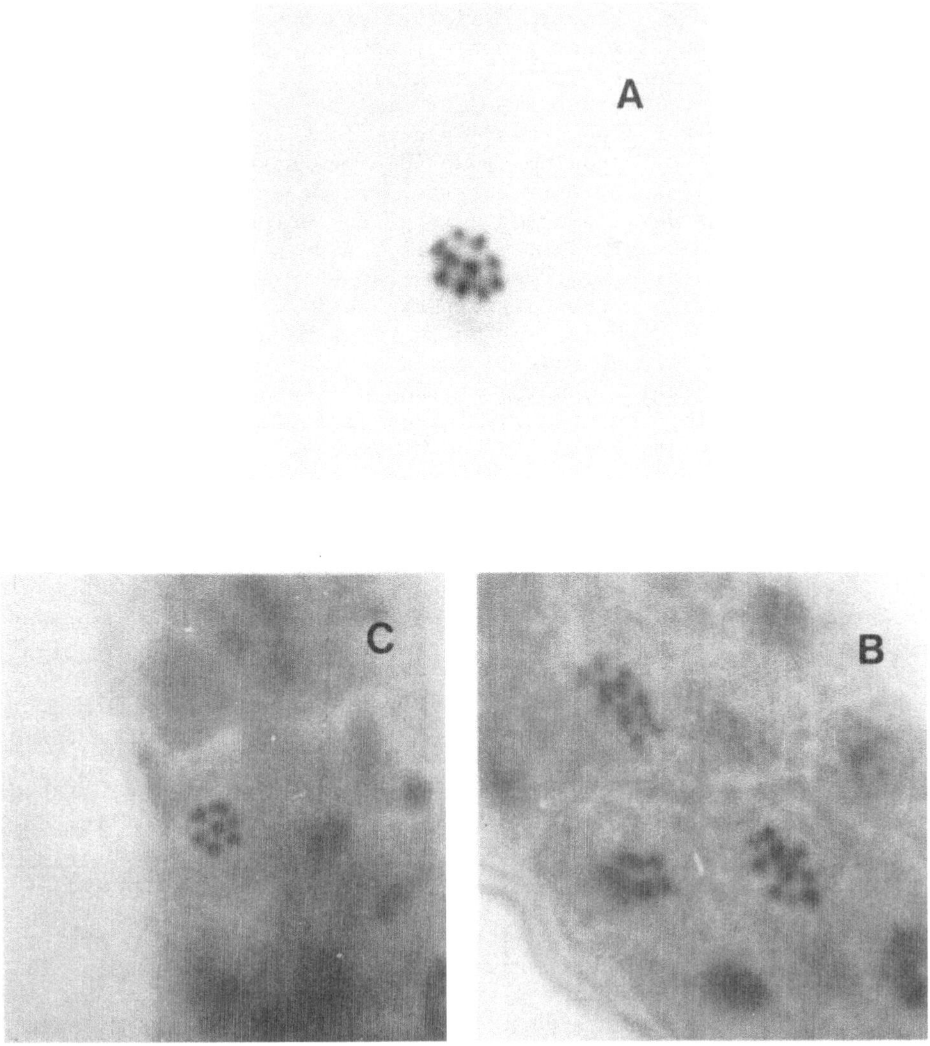

PLATE I: A – Metaphase of the meiosis during oogenesis of X. index (about X 5000)
B – Spermatogenesis of X. diversicaudatum (about X 5500)
C – Spermatogenesis of L. elongatus (X 3800)

apex of the ovary. It occupies about a tenth of the ovary and contains 10-30 oogonia. Mitotic divisions are less frequent than in spermatogenesis and few gametes are produced. On one occasion 20 chromosomes were counted in this stage in X. index. When there is no reproduction the ovaries are empty in the X. americanum species group ("small spiral Xiphinemas") except for the germinal **zone. The** X. index - X. diversicaudatum group ("large hooked Xiphinemas") ovaries contain 10-12 primary oocytes. These cells have a large nucleus that does not stain and a large nucleolus. Although the meiotic process has only just begun it has been stopped by poor environmental conditions. The nucleus appears as in the interphase, but in reality it is a pseudointerphase that follows a short synapsis and precedes diakynesis. So in large hooked Xiphinema the "germinative push" is already prepared but this is not so in the small spiral Xiphinema. During periods of reproduction differences persist between the two groups. The ovaries of X. index and X. diversicaudatum contain about 20 oocytes in an advanced pseudointerphasic stage, whereas in X. mediterraneum there are not more than 4 or 5. This may be correlated with the low fecundity of the latter species.

During the post synaptic pseudo-interphase, the capacity of the nucleus to take on stain decreases and only a pale pink patch remains at the periphery of the nuclear membrane. The first signs of ovarian activity are observed in the blind sac of the ovaries and in the "pars dilatata" of the oviduct where there are some oocytes in diakynesis. These are dark in living material in contrast to the clear oldest ones corresponding to the anaphase of the first maturative division, that are located in the uterus. **Diakynesis** and anaphase show very distinct plates. During the first one, 5 bivalents have been counted in X. mediterraneum and 10 in X.index and X. diversicaudatum. X.index chromosomes do not exceed 0.5 µ; those of X. diversicaudatum seem slightly smaller. The oocyte sheath is still very thin. Oocytes in the "pars dilatata" of the oviduct may be in metaphase or early meiosis anaphase. During the anaphase the plate is oriented at 45° to the oocyte axis. When the anaphase begins it becomes more difficult to wash out the fuchsin by a sulphurous solution; therefore the cytoplasm remains coloured pink or red when the oocyte has reached the uterus.

The permeability of the oocyte sheath may have changed and in fact the sheath seems to be thicker. Similar observations can be made in other phytophagous nematodes, but not always at the moment of oogenesis. Finally the resulting 10 or 20 chromosomes, depending on the species, are distributed over two homologous plates. Egg laying occurs at the end of the meiosis except when the female is dying.

In X. diversicaudatum, males are frequent and females are fertilised abundantly. The genital tract of females is relatively

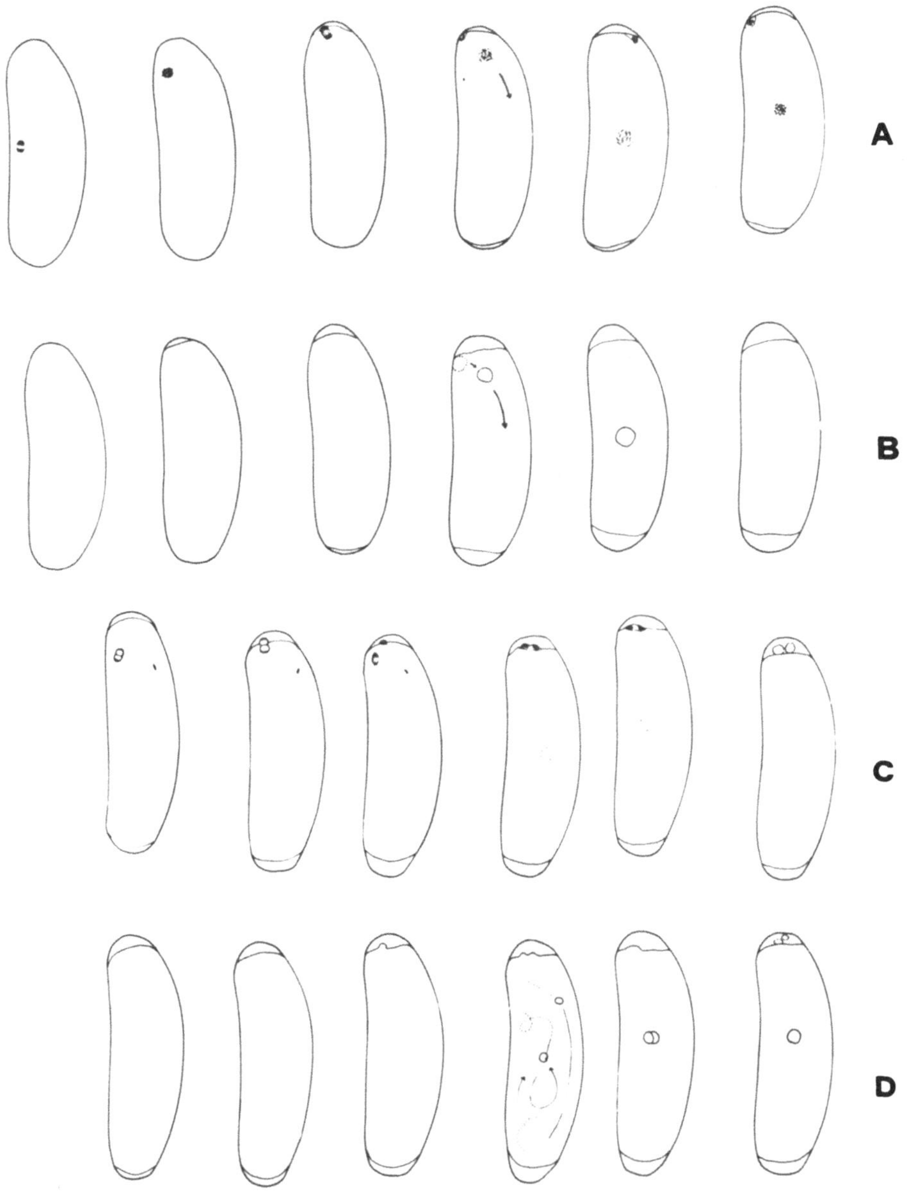

Fig. 2. Maturation process in X. index: A, as seen by Feulgen colouration; B, as seen in living material. Maturation process in X. diversicaudatum: C, as seen by Feulgen colouration; D, as seen in living material.

elaborate, especially the uterus which is separated from the oviduct by a strong sphincter. The uterus begins with a large pouch, the proximal part of which consists of thick lining cells with large vacuoles, whereas the distal part (forming the bottom of the pouch) is a dense epithelium. The two types of tissue are closely connect-ed. Spermatozoa are probably maintained on the nutritious epithelium (see also ref. 1). This pocket is strongly dilated at the passing of the oocyte which is then surrounded by a sleeve of spermatozoa; these are not pushed from their place by the oocyte. Preferential location is not common in dorylaimids where both the oviduct and the uterus are often full of spermatozoa (cf. ref 4). X. diversicaudatum possesses a pseudo Z differentiation consisting of 12 to 15 characteristic cells with a dense granular cytoplasm and a central chromophobe vesicle. This differentiation, in a manner homologous with the columella of tylenchids, seems more frequent in species where fecundation occurs (see also X. ingens). There are some exceptions; for example X. neovuittenezi has no pseudo Z differentiation but correlated with this there is no preferential location of spermatozoa in the uterus pouch. On the other hand, there is no fecundation in X. turcicum which has a pseudo Z differentiation.

One spermatozoan enters each oocyte while it passes through the oviduct pouch, but remains quiescent close to the entrance point until the egg is laid. The final maturation takes place in the soil. In spite of some parental link between X. index, X. pyrenaicum and X. diversicaudatum, spermatozoa have never been found in the female genital tract of the first two species when occurring together with the last named, which is amphimictic. Hence sexual isolation appears to be complete.

Longidorus

In Longidorus there are many analogies with Xiphinema oogenesis: germinal apical zone with few mitotic divisions, prophase characterized by a long pseudo-interphase period both outside and during the reproductive period. Large oocytes are only present when germinative pushes occur. The size of these oocytes renders observations difficult. Perhaps 7 chromosomes occur in L. macrosoma at anaphase I just before egg laying.

Oocytes are fertilized by one spermatozoan during the passage through the oviduct, as in X. diversicaudatum. There is no specialized uterine tissue for sperm storage and no Z differentiation.

Oocyte maturation in soil (Fig. 2)

From the preceding observations it is concluded that oocyte laying occurs at the end of the meiosis anaphase for all the species of longidorids studied.

RECAPITULATIVE TABLE

Species	Localities	chromosomes number	
X.mediterraneum	Cagnes-sur-Mer	n = 5	meiotic parthenogenesis
X.diversicaudatum	Antibes	n = 10	amphimictic
X.index	Several localities	n = 10	meiotic parthenogenesis
L.macrosoma	Cagnes-sur-Mer	n = 7	meiotic process
L.elongatus	Saint-Emilion	n = 7	meiotic process (sometimes parthenogenesis)
L.vineacola	Saint-Emilion	n = 7	meiotic process

In Xiphinema diversicaudatum, eggs are produced in spring and summer. Maturing oocytes must be observed in vivo and then fixed at a determined stage or at a chosen time. Immediately after oviposition, anaphase I plates and spermatozoa are not visible in the living oocyte. The two pronouclei become visible only 18 h later. The second maturative division takes place in that interval as shown by Feulgen colouration. This homotypic division has a short metaphase. The achromatic spindle is 2-3 μ in diameter and more elongate than the meiotic one. A batch of 10 chromosomes represents the first polar body. When maturation is completed, the spermatozoan starts to form itself into a vesicle and becomes visible on the living oocyte; 18 h have passed since oviposition. The 2 pronuclei begin to move along a cycloid path in the apparently very fluid cytoplasm. Until 35 h after egg laying they remain in interphasis. Only the two polar bodies, localized at the periphery, take up stain. Then cyclosis stops, the two pronuclei move slowly, drawing very close together. Amphimixis occurs 1-2 h after movement stops. During this time and before the first cleavage the two chromosomic stocks are still separate. Telophase I as the caryokinetic phase is very long (8, 10).

In X. index, anaphase plates of meiosis are also invisible in living laid eggs (cf. X. diversicaudatum). After 10-12 h, the nuclear vesicle reappears as a small clear ball at the animal pole. The diploid number has already been re-established then, as can be observed in treated oocytes. In vivo the nuclear vesicle remains near the air chamber for a few hours. Then it moves to the centre of the oocyte which it reaches in about 24 h, but no polar body is visible yet. One, sometimes two, appear in the air chamber during embryogenesis. In contrast with amphimictic species there is only one nuclear vesicle instead of two. Cytological mounts have been made on this chronologic basis to complete previous observations. Chromatic condensation phases obtained by levels of staining correspond with the times of disappearance of the nucleus in vivo. This demonstrates that the anaphasis of meiosis persists for several hours when reaching the animal pole of the oocyte. The spindle does not break and finally the two anaphasis plates move together again, giving a new diploid nucleus. Endomeiotic parthenogenesis is known in Psychidae (Lepidoptera) such as Luffia ferchaultella or Selonebia lichenella, but has never been reported in nematodes. Twelve to 14 hours later a new anaphasic spindle is formed. It stays near the oocyte membrane in oblique position, just under the air chamber of the pro-egg. Counting the chromosomes in the two small plates is very difficult. Observations at this stage confirm previous conclusions: during this second maturative division a polar body has never been seen, although it would have been conspicuous if meiosis had been a normal one. The duration of this process still corresponds to the time the nucleus is invisible in the living oocyte. After 15 h, the re-establishment of the interphasic stage can be observed as the diploid pronucleus moves to the centre of the oocyte; a polar body is now visible.

The process of maturation ends when the nucleus reaches the centre. At the metaphase of the first cleavage 20 chromosomes are counted. There is no confusion with metaphase II because of the eccentric position of the plate in the latter case.

L. macrosoma shows few differences with X. diversicaudatum maturation, in spite of the voluminous rich cytoplasm and the smallness of the nucleus (2 μ diameter) making observations very difficult.

In vivo two pronuclei are formed and combined in a diploid nucleus about 60-100 h after the presumed moment of laying. Some cases of pseudogamy may occur, which would explain inequality in the numbers of males and females especially in some meridonial populations of X. diversicaudatum where few males are produced.

The most important point is that the long time of maturation in L. macrosoma is associated with a low nucleoplasmatic ratio, as in X. diversicaudatum, whereas maturation and development are faster in X. index which has a higher nucleoplasmatic ratio. Possibly a high nucleoplasmatic ratio also induces parthenogenesis. The totality of the embryogenic cycle as well as post embryogenic development seems influenced in this manner in Longidoridae. But other examples are easily found in many other families of nematodes (e.g. in

Meloidogyne).

CONCLUSIONS

Cytological studies of Longidoridae provide fundamental information on the biology and taxonomy of these nematodes.

There are evident differences between Xiphinema and Longidorus genomes and this supports the division into two subfamilies as defined in 1969 and 1970 (5). The chromosomic stock is relatively constant in both genera: x = 7 in Longidorus and x = 5 or (2 x 5) in Xiphinema.

In Tylenchida aneuploidy is frequent. Two types of parthenogenesis are common. In Longidoridae mitotic parthenogenesis is unknown and that seems a general rule in Dorylaimida. From a genetic point of view, X. index appears to be one of the most evolved species. On the other hand, the large group of hooked Xiphinemas (X. diversicaudatum group) is super evolved in comparison to the small spiral Xiphinemas (americanum group). Unfortunately little is known about the numerous and diversified tropical species.

Another aspect is the correlation between the respective size of the egg and of its nucleus and the species development. X.index possesses a relatively large nucleus, which is logical because it has many chromosomes, but X. diversicaudatum has an equal number and yet the two chromosomic stocks are not quite equivalent. As noted before, chromosomes and correlated nuclei in X. index are larger than those of X. diversicaudatum. If any polysomy or polyploidy is not discernible by counting, then D.N.A. surplus may exist at a genetic level and is reflected in the biology of the species.

REFERENCES

1. Anderson, R.V. & Darling, H.M. (1964). Embryology and reproduction of Ditylenchus destructor with emphasis on gonad development. Proc. Helm. Soc. Wash. 31, 240-256.

2. Bělař, K. (1923). Uber dem Chromosomenzyklus von parthengenetischen Erdnematoden. Biol. Zentrbl., 43, 513-518.

3. Berge, J.B. et al (1973). Nouvelles données phylogéniques sur les Tylenchides. C.r. Acad. Sc. Paris, 276, serie D, 3307-3309.

4. Coomans, A. (1964). Structure of the female gonads in members of the Dorylaimina. Nematologica, 10, 601-622.

5. Dalmasso, A. (1970). Contribution à l'étude des Longidoridae de France (Nematoda - Dorylaimida). Fac. Sci., Nice, These, 189 pp.

6. Lin, T.P. (1954). The chromosomal cycle in Parascaris equorum (Ascaris megalocephala) oogenesis and diminution. Chromosoma, 6, 175-198.

7. Nigon, V. (1949). Les modalités de la reproduction et le déterminisme du sexe chez quelques nématodes libres. Ann. Sci. Natur. Zool. 11, 1-132.

8. Nigon, V. & Robert, M. (1952). Contribution à l'étude de la gamétogenèse chez Parascaris equorum Goeze. I - Formation des tétrades durant l'ovogenèse de Caenorhabditis elegans. Chromosoma, 7, 129-169.

9. Triantaphyllou, A.C. (1971). Genetics and cytology. In B.M. Zuckerman, W.F. Mai and R.A. Rohde (eds.) Plant Parasitic Nematodes vol.2. Acad. Press, New York.

10. Triantaphyllou, A.C. & Hirschmann, H. (1962). Oogenesis and mode of reproduction in soybean cyst nematodes, H. glycines. Nematologica 7, 235-241.

DISCUSSION

Asked if intraspecific variation of chromosome numbers might point to the existence of biotypes, Dalmasso replied that within several Meloidogyne spp. many chromosome accessories or garnitures exist, but these cannot be correlated with biotypes, although such a correlation does not a priori seem impossible. No variation in chromosome numbers is known in Longidorus spp.

Mrs. Grimaldi de Zio reported that she had found a population of Xiphinema index with six chromosomes that were larger than usual; but Lippens suggested that because the chromosomes are exceedingly small, two lying close together might be mistaken for a single larger one. Dalmasso, however, stated that the chromosomes in X. index can clearly be seen so that there should be little doubt of their identity; in Longidorus it is difficult to see 7 in the egg, but the shape is clear.

Cohn asked if there are any differences in the rate of embryological development between species, and Dalmasso replied that in X. index the development period is shorter than in X. brevicolle; in L. macrosoma and other Longidorus species it is very long (J.J. M.Flegg).

Lippens proposed the introduction of the infraspecies concept for morphologically and biologically different populations within the complexes L. elongatus and X. americanum, which he suggested should be called supraspecies. Luc said he supported this view as it clearly identified X. americanum as a group, but he said there appeared to be many problems in introducing this into regular taxonomy. Hackney asked how the criteria for designating infraspecies could be standardised; how, for example, could such infraspecies be distinguished from forms that owe their peculiar characters to the influence of some particular host plant? Lippens pointed out that species have a time dimension, and suggested that it is possible that parthenogenetic forms are already genetically separate, though we may not recognise this because of our own limitations of investigation and in establishing criteria for their separation from bisexual forms. Seinhorst felt that there was confusion between the cataloguing of names and the things that are named. In his opinion the present question is "are we able to separate the forms on morphological characters?". When Hackney observed that within taxa there are characters more appropriate or less appropriate, Seinhorst replied that this was not so, and such observations resulted from wrong teaching!

Asked whether there is a mechanism which prevents the sperm from being pushed out of the uterus, Dalmasso replied that in Xiphinema the sperm seems to be stored in a restricted, specialised part of the uterus; this is not so in Longidorus where the sperm is more diffuse.

In a discussion on the possibility of crosses between species, Dalmasso reported that he had been able to cross Aphelenchoides species with two and with three chromosomes. The hybrids were morphologically different and their characters were not very stable, but they were fertile although reproduction was much reduced. Dalmasso suggested that hybridisation might occur in Trichodorus sensu strictu, in which genus males are numerous.

Short Report

EMBRYOLOGY AND HATCHING OF TRICHODORUS SIMILIS AND LONGIDORUS ELONGATUS

Urs Wyss (Institut für Pflanzenkrankheiten und Pflanzenschutz der Technischen Universität Hannover, BRD.)

Studies, including time-lapse filming, on the embryology and hatching of Trichodorus similis(1) and Longidorus elongatus(2) at 25 ± 1°C produced the following results:

Trichodorus similis

Freshly laid eggs are 70-82 µ long by 37-42 µ wide (n = 50) and still in the single-celled stage. About 12-15 h after deposition, the egg protoplasm starts to separate from the egg shell and later exhibits pronounced movement. An inner egg layer does not become visible at this, nor at a later stage of the embryonic development. Shortly before the two pronuclei fuse, the protoplasmic movement comes to a rest and the egg protoplasm then occupies approximately 70% of the total egg volume. About one hour after nuclear fusion, the first cleavage, transverse to the longitudinal axis of the egg, gives rise to two blastomeres of equal size in which protoplasmic movement is again resumed. The plane of division of the first somatic cell is nearly always parallel to the longitudinal axis of the egg, whereas the first parental germinal cell commonly divides in a plane oblique to it. After the four-celled stage, characterized by a stable rhomboidal conformation of the blastomeres, the four divisions of the third cleavage are longitudinal and their sequence is not strictly determined. Time-lapse filming clearly shows the rhythmical alternation between rest and division periods. It also demonstrates the continuation of protoplasmic movement in the newly formed cells and reveals that after each cleavage the blastomeres are drawn together into a compact arrangement and then occupy a minimum space, as shown by the egg's protoplasm prior to the first mitotic division.

Gastrulation occurs after the fifth cleavage, apparently by invagination of the endoblasts. About 4.5 days later, the vermiform larva elongates very rapidly. At first the developing larva moves vigorously and almost continually, but with increase in size and consequent restriction, movement becomes more and more sporadic.

One day prior to hatching, the stomodaeum is fully developed and functional. At this stage, occasional pulsations occur in the basal bulb of the oesophagus that force fluids into the intestine. Oesophageal bulb pulsations rarely exceed 40 and are confined to the relatively short periods of quiescence, subsequent to longer periods of larval movement. About 12 h before hatching, the egg shell, hitherto rigid, gradually becomes thinner and flexible. It is finally ruptured at a point upon which the greatest pressure is just exerted by the moving larva. The stylet is never used to puncture the flexible egg 'shell'. Under aseptic conditions the time from egg-laying to hatching of the first stage juvenile covers an average of 12 days at 25 ± 1°C.

Longidorus elongatus

Newly deposited eggs are 159–182 µ long and 55–65 µ wide (n = 50) and always undivided. Embryology and hatching differ in several respects from that of T. similis. A male pronucleus is not present (parthenogenetic reproduction); an inner egg layer always becomes visible, when the egg protoplasm starts to contract from the egg shell; the first somatic and parental germinal cells are usually of unequal size; the two divisions of the second cleavage are perpendicular to the longitudinal axis of the egg and give rise to a tandem arrangement of the four blastomered. Endo- and ectoderm can be distinguished from the 128-celled stage onwards.

Again oesophageal bulb pulsations, that force fluids into the intestine, start about one day prior to hatching during periods of larval inactivity. These pulsations are continuous, though irregular in rhythm, and can last up to 20 min. As long as they persist, the tip of the stylet of the first stage larva remains protruded 2–3 u beyond the oral aperture. A few hours prior to hatching, the now thinner and flexible egg 'shell' usually separates from the inner egg layer. At this stage, larval movement is still restricted. The head of the larva moves forward and backwards and finally ruptures, without stylet action, firstly the inner egg layer and then the egg 'shell'. Under sterile conditions the time of egg-laying to hatching of the first stage larva is 9 days at 25 ± 1°C.

REFERENCES

1. Wyss, U. (1973). Trichodorus similis (Nematoda). Embryonalentwicklung. Film E 1910 der Enc. Cin. Gottingen, 15 pp.

2. Wyss, U. (1974). Longidorus elongatus (Nematoda). Embryonalentwicklung. Film E 2046 der Enc. Cin. Gottingen (In press).

CELL STRUCTURE AND CELL FUNCTION - AN INTRODUCTION TO ULTRA-STRUCTURE

PETER L. LIPPENS

Instituut voor Dierkunde,

Rijksuniversiteit Gent, Belgium.

ORGANISATION OF THE CELL

The unit of a metazoan organism is the cell. The cell membrane is composed of a lipid bilayer, structural proteins and enzymes, and separates two thermodynamically different systems - the environment and the interior of the cell. Desmosomes (Fig. 1, A, C) are special cell-to-cell contacts. In nematodes there are three main types - the zonula adhaerens which is mainly for support or strengthening, the zonula occludens and the septate desmosome, both of which seal off the intercellular spaces.

The nucleus (Fig. 1, A, B) is the centre of organisation of cellular activity. It contains nucleoproteids; the condensed peripheral chromatin (pc) is thought to represent inactive materials. The nucleolus (n) serves in storing RNA. The nucleus (N) is surrounded by the nuclear envelope (ne), which is interrupted at the nuclear pores (np), through which cyto- and nucleo-plasm are in contact. The diaphragm (dia) probably acts as a filter. The outer membrane of the envelope is confluent with the rough endoplasmic reticulum (rER) which is covered by ribosomes. In general, ribosomes are the site of protein synthesis. The protein is transported through the cisternae of the rER (Fig. 1, E) to the Golgi region, where it is accumulated and sequestered. Membrane-bound vacuoles split off from the Golgi apparatus and are transported through the cell (Fig.1, A). The usual type of Golgi body is the dictyosome (Fig. 1, F), but its structure may also be vesiculate or more elaborate. The Golgi region probably also serves in several other, yet unknown, processes.

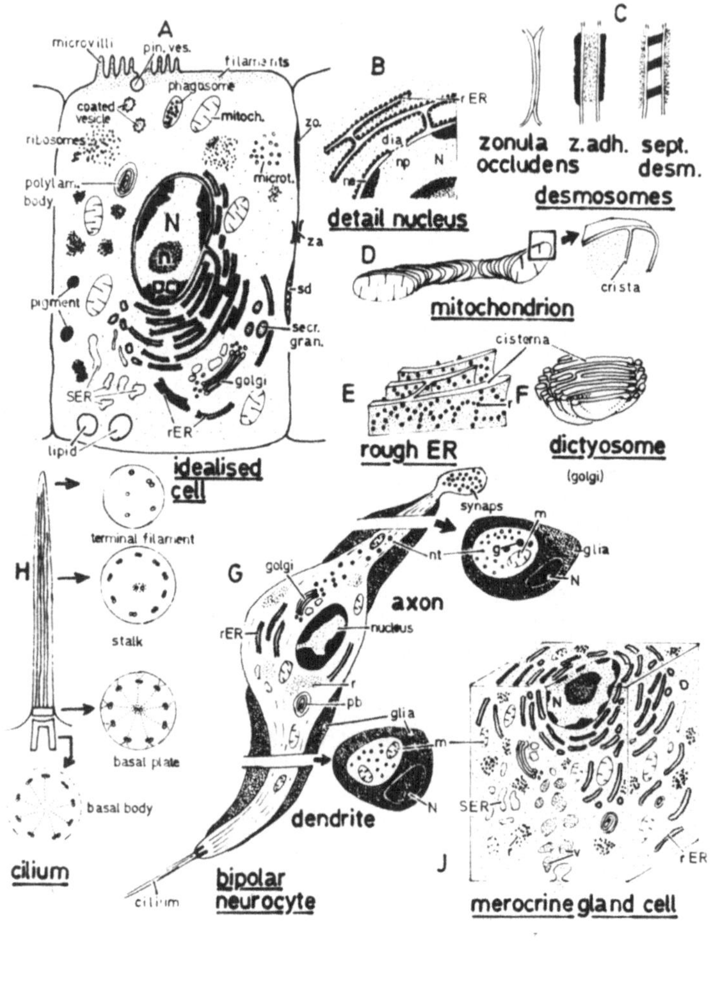

Fig.1. Cell structure of free-living nematodes. dia: diaphragm; g: granules; m: mitochondrion; microt.: microtubules; N: nucleus; n: nucleolus; ne: nuclear envelope; np: nuclear pore; nt: neurotubules; pb: polylamellar body; pc: peripheral chromatin; r: ribosomes; rER: rough endoplasmic reticulum; sd, sept. desm.: septate desmosome; secr. gran.: secretory granule; SER: smooth endoplasmic reticulum; v: vacuoles; za, z. adh: zonula adhaerens; zo: zonula occludens.

FUNCTION

Protein synthesised by free ribosomes (r), is used in metabolism in the cell itself rather than exported.

Smooth endoplasmic reticulum (SER), not covered by ribosomes, may have several functions. In nematodes it probably is involved in fat metabolism, transport of non-proteinaceous material and, in muscles, in transmission of stimuli. There is, however, no direct evidence to support this statement. Also microtubules (microt.) may serve in transport and transmission; often, they function as a cytoskeleton.

Mitochondria (Fig. 1, D) are the main source of cellular energy. The enzymes of the terminal phosphorilation are thought to be located upon the cristae, whereas the enzymes of the Krebs cycle are located in the matrix. Three-dimensional reconstruction of mitochondria in several tissues of free-living nematodes revealed that they may be large, often branched structures (Fig. 1, D).

Glycogen and fat are the main storage products. Lipids are especially abundant in the intestinal epithelium of rhabitids and tylenchids, but in dorylaims they are scarce. In dorylaims particularly glycogen is stored, mainly in the muscles and epidermis.

Little is known about food uptake in nematodes. The intestinal epithelium is usually covered by microvilli in which often cytoplasmic filaments are observed (Fig. 1, A). As will be reported in a later paper, pinocytosis associated with coated vesicles occurs in some marine nematodes. The material brought into the cell by phago- and pino-cytosis is digested in phagosomes (= multivesicular bodies) which contain acid phosphatase. Another type of lysosome, the cytolysosome, is involved in destruction of (probably defect) cell organelles; this type of structure is not yet reported from nematode tissues. In my opinion, some of the polylamellar bodies which are so often observed in nematodes, may represent cytolysosomes.

In some specialised cell types pigment granules are found, as for example in eyespots of marine nematodes.

ORGANELLES

Different cell types are characterised by predominance of certain cell organelles rather than by special features. In gland cells for instance, the ER is predominant; also the Golgi apparatus is usually well-developed and the nucleus enlarged, often containing a distinct nucleolus (Fig. 1, J).

Neurocytes (Fig. 1, G) are characterised physiologically rather than morphologically. A typical neurone consists of a cell body which contains the nucleus, one or more afferent offshoots (dendrites), and an efferent branch (axon). An unipolar neurone has an axon only, a bipolar cell has a dendrite and an axon, multipolar cells have an axon and several dendrites. In many nematodes the structure of axon and dendrite is quite similar, as both offshoots contain mainly microtubules, called neurotubules (nt), and some mitochondria (m), which are often very small. Many neurocytes in free-living nematodes are accompanied by slender, elongated cells which are regarded as glial elements (glia).

The contact between axon and effector is called synapse; here the axon is enlarged, containing numerous synaptic vesicles which are thought to contain the humoral transmitter.

The dendrites of many sensory neurones have ciliary processes. A cilium (Fig. 1, H) consists of a stalk that contains 9 peripheral (often doublet) and 2 central longitudinal microtubules. Near the tip of the cilium, the pattern of the tubules becomes irregular as not all tubules are of equal length. Strongly modified tip regions are known as terminal filaments. At the base of the cilium (basal plate) the peripheral tubules and the central tubules are connected by protoplasmic filaments. The cilium is anchored upon the basal body which in fact, is a specialised centriole. The basal body is composed of 9 peripheral (often triplet) tubules and the centrals are absent. In nematodes, the configuration of the filament complex is often irregular, reflecting the loss of the capacity of motility.

Muscles are characterised by their cytoplasmic filaments. Up till now, all nematode muscles appear to be obliquely striated. Details of these structures are given in the following paper in these Proceedings on the ultrastructure of Dorylaimidae.

ULTRASTRUCTURE OF DORYLAIMIDAE

PETER L. LIPPENS

Instituut voor Dierkunde

Rijksuniversiteit Gent, Belgium.

INTRODUCTION

It is likely that Longidoridae evolved from free-living, probably dorylaimoid, ancestors. From this point of view, the study of Dorylaimidae is like looking in the past, giving a better insight in the adaptive changes that occurred during the evolution of these phytoparasites.

Comparative anatomy indicates that the structure of the cheilostomatal and stomodeal regions of Longidoridae may be derived by allometry from the corresponding regions in Dorylaimidae. Also, the histological features of Longidoridae and Dorylaimidae are strikingly similar.

Several Dorylaimidae, noticeably *Aporcelaimellus* and *Eudorylaimus* spp., are relatively easily maintained *in vitro*. Some species have a relatively short life cycle and are therefore more favourable subjects for fundamental research than longidorids. In the genus *Aporcelaimellus* there are many characters which are considered to be primitive in the group of Dorylaimoidea; in particular, the anterior part of the stomodaeum is of a plesiomorph type. This genus is used as a model for Dorylaimidae in the context of these Proceedings.

HISTOLOGY OF THE BODY WALL

The body wall is composed of the cuticle, the epidermis (or hypodermis) and the somatic musculature.

Somatic musculature (Fig.1)

The somatic musculature is shallow coelomyarian, composed of a single layer of cells longitudinally arrayed. The muscles are separated from the epidermis by an area of connective tissue, the basal lamina (bl), which invaginates between adjacent muscle cells. The cell membrane of the muscles is simple, trilaminate.

The contractile part is composed of fine and thick cytoplasmic filaments, approximately 8 nm and 22 nm diameter respectively. The electron-dense Z-bands (Z) are more or less perpendicularly oriented with regard to the cell membrane. The area between two neighbouring Z-bands constitutes a sarcomere. Each sarcomere is divided in I- and A-bands, corresponding to the regions of fine and thick filaments respectively. The A-band consists of two zones: an area where fine and thick filaments overlap, and an area which is entirely composed of thick filaments. The latter region is known as the H-band. Bridges can be seen between the fine and thick filaments in the A-band; these bridges are fundamental to the "sliding filament theory" (6) of muscular contraction. In Dorylaimoidea the fine filaments in the I-bands tend to aggregate and appear as clusters (5, 8). Morphologically similar dense areas also occur in the corners of the contractile part.

Invaginations of the plasma membrane (T) run obliquely through the contractile region, forming a system of tubules that may be compared to the T-tubules of vertebrate and insect muscles. In the latter case the function of the T-tubules is thought to be related to the transmission of the depolarisation of the cell membrane inside the muscle cell. In Aporcelaimellus the T-tubules tend to run opposite the Z-bands of the adjacent muscle cell. The functional significance of this association is not yet understood.

Especially in carnivorous Dorylaimidae, such as Aporcelaimellus spp., the sarcoplasmic reticulum (SER) is relatively well-developed; the cisternae run mainly alongside the cell membranes and the Z-bands, but in addition, irregularly spaced cisternae cross the cell in random direction.

Each cell contains only one nucleus. The sarcoplasm (Sa) contains many mitochondria (m) that are often very large and which contain numerous closely-packed cristae. Free ribosomes are abundant; also some cisternae of rough endoplasmic reticulum are found. The main storage product is glycogen (gly), and fat is scarce (8).

The ultrastructural features of the muscles of Longidorus elongatus (5) differ from those of Aporcelaimellus by the poor development of sarcoplasmic reticulum and T-tubules. The basal lamina in Longidorus also covers the muscle cells at the pseudocoelomic side. The number and size of mitochondria are significantly less and the

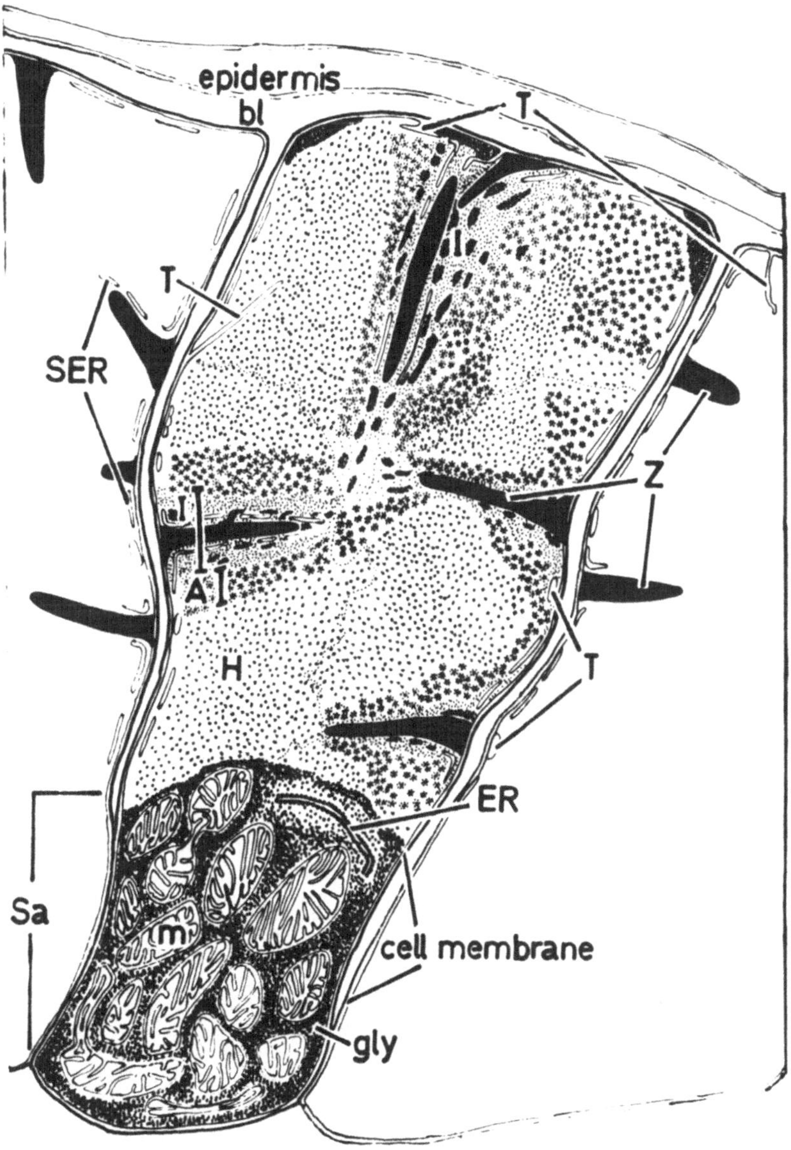

Fig.1. Cross-section through a somatic muscle cell of Aporcelaimellus. A: A-band, bl: basal lamina, ER: rough endoplasmic reticulum, gly: glycogen, H: H-band, I: I-band, m: mitochondrion, Sa: sarcoplasm, SER: smooth endoplasmic reticulum, T: T-tubule, Z: Z-band.

storage of glycogen is less extreme. In other respects the musculature architecture is identical.

Obviously, the differences are related to the mode of life. Some Aporcelaimellus spp. are predators, mainly of small nematodes. In cultures they appear somewhat less sluggish than has been reported in the literature. L. elongatus on the other hand, is a phytoparasite, for which rapid reactions to stimuli involving muscular contraction and coordination are less crucial for survival, hence the less developed system of T-tubules and sarcoplasmic reticulum. The increased number of mitochondria, the density of their cristae, and the large amount of glycogen in Aporcelaimellus is indicative of the energy consumption.

Body cuticle

The body cuticle (Fig. 2) is composed of seven zones, complemented with an infracuticle that reaches from the labial region to the level corresponding to the base of the odontophore in its retracted position (Fig. 9: ic). The structure of adults and intermoults is identical, but in the first juvenile stage (Fig. 2: J_1) the infracuticle is absent (4). The cuticle is composed of fibrill -ar material, covered by a three-layered membrane (zone 1). Except in the cortical layer (zone 2), the fibrillar material is condensed into fibres of which structure and orientation is different in the successive zones. In the zone 6, for instance, the fibres are organised in successive lamellae.

An expanded system of lacunae was found in the cuticle (4), consisting of canals which run obliquely from the epidermis to the base of the cortical layer (zone 2). Some of these canals are empty, others contain coarse granular material (1a), striated fibres, or epidermal projections. The canals are interconnected by the interfibrillar spaces. Zone 5 is almost entirely lacunar. It was suggested (4) that the lacunar system is important in the flexibility of the cuticle as it permits displacements of the fibres with reference to each other.

The structure of the body cuticle of Xiphinema (10, 12) and of Longidorus macrosoma (1) is strikingly similar to that of Aporcelaimellus. Number and sequence of the different zones are identical, and only differences in relative thickness occur. L. elongatus, however, may show a somewhat aberrant pattern in the zonation of the cuticle (1), as one of the layers seems to be absent. Recently, dorylaimoid patterns of cuticular architecture have also been found in some other genera of Dorylaimidae and Nygolaimidae.

It was pointed out (4) that the structure of the body cuticle probably could be used in evaluating the relationships between the higher categories of Dorylaimina. Especially the infracuticle may

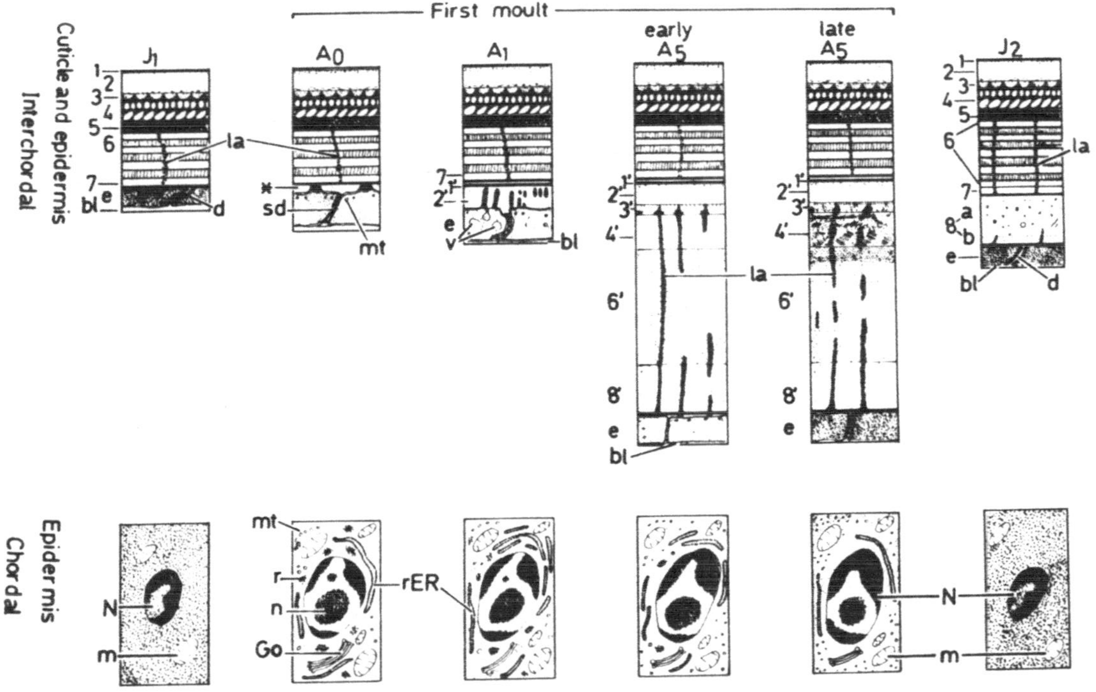

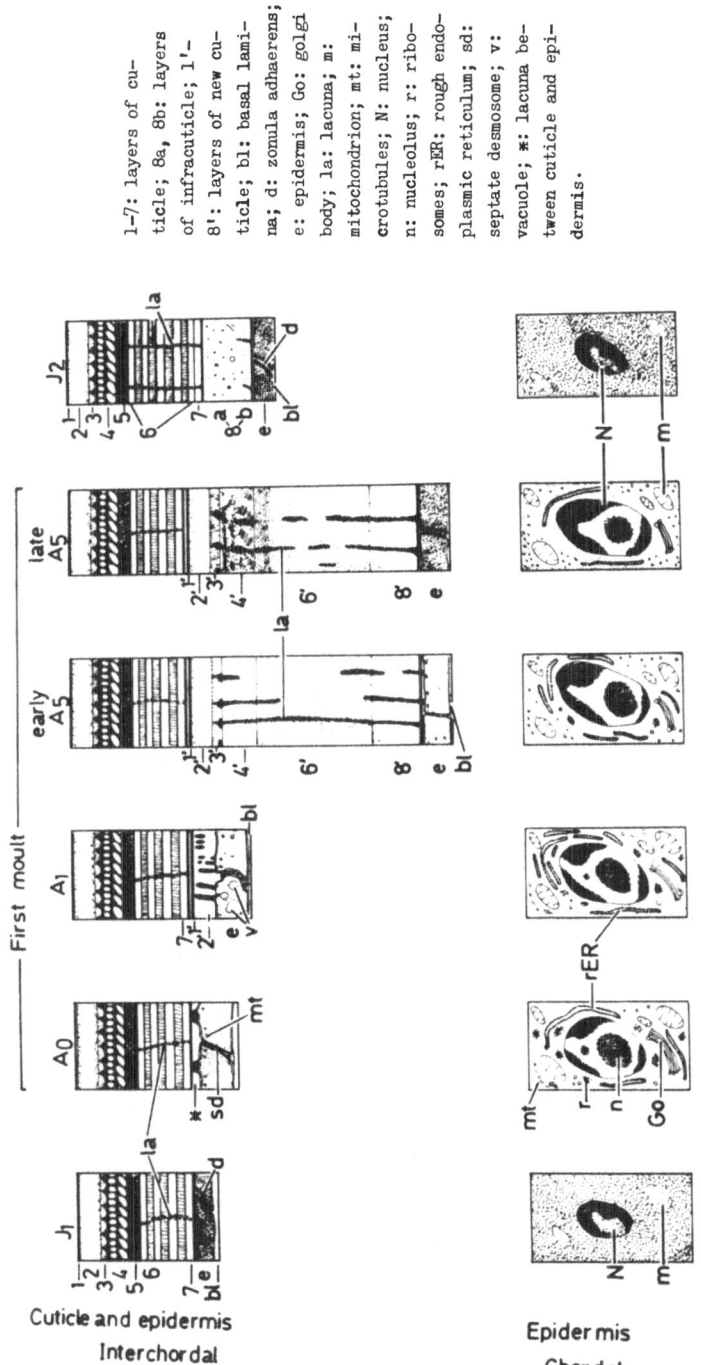

1-7: layers of cuticle; 8a, 8b: layers of infracuticle; 1'-8': layers of new cuticle; bl: basal lamina; d: zonula adhaerens; e: epidermis; Go: golgi body; la: lacuna; m: mitochondrion; mt: microtubules; N: nucleus; n: nucleolus; r: ribosomes; rER: rough endoplasmic reticulum; sd: septate desmosome; v: vacuole; *: lacuna between cuticle and epidermis.

Fig. 2. Cross-sections through cuticle and epidermis of the successive stages of the first moult in *Aporcelaimellus obtusicaudatus*.

be important in this respect. Such a structure has not been found outside Dorylaimoidea and Nygolaimoidea. As the infracuticle is absent in the first stage juveniles of (at least) Aporcelaimellus and Discolaimus, it could well be a secondary formation, even an autapomorphy.

The body wall structure of trichodorids and also of leptonchids is noticeably different from that in Dorylaimoidea. The somatic musculature in these groups is platymyarian instead of coelomyarian. The architecture of the contractile part in Trichodorus muscles is much simpler (2, 5). Also the zonation of the cuticle of Paratrichodorus allius (9) shows a considerably different pattern when compared to Dorylaimoidea.

First moult (Fig. 2)

The epidermis of Aporcelaimellus is fully cellular. During moulting the epithelium is a typical gland. It has been found (4) that moulting of the body cuticle and stomodeal cuticle are synchronised, by which the division in A- and B-stages (3) was extended to the body cuticle. A stage, chronologically prior to the A_1-stage was found, which is, a proviso, referred to as A_0-stage.

The A_0-stage is the initial phase of the moult, corresponding to the decreasing activity of the nematode. The epidermal cells become hypertrophic; their cytology indicates active protein synthesis. The cuticle becomes detached from the epidermis, probably by the formation of a moulting fluid. There are some indications that the moulting fluid is produced by the epithelial cells. A peculiar change was observed concerning the interchordal desmosomes. During the intermoult stages these are represented by zonulae adhaerentes (d), whereas during the moult only septate desmosomes (sd) are observed.

During the A_1-stage cuticle formation has already begun. The osmiophilic membrane (zone 1') and the cortical layer (zone 2') have been laid down. Secretory vacuoles (v) originating from the epithelial cells were seen to empty their contents in the newly formed cuticle.

In the A_5-stage the new cuticle is apparently fully formed but not yet differentiated into zones. Offshoots from the epidermis run through lacunae in the cuticle (1a) to the base of the transverse grooves that constrict the cortical layer (zone 2'). The epidermal offshoots gradually retract during the late A_5-stage, leaving canals in the cuticle which are the origin of the lacunar system. Meanwhile differentiation initiates. The filamentous components condense, forming protofibres that eventually will give rise to the fibrillar architecture. As a consequence of the condensation, spaces are formed between the fibres, by which the final structure of the lacunar system is completed. The retraction of the epidermal projections apparently comes to a stop within the infracuticle, since they are still found in this layer in intermoults and adults.

ANATOMY AND HISTOLOGY OF CHEILOSTOME AND STOMODAEUM

The ultrastructure of the anterior body region of Dorylaimidae was given for Aporcelaimellus (8). Recent investigations of other genera of the family indicates that the model is satisfactory.

The composition of the labial cuticle is similar to that of the body cuticle, but it lacks an infracuticle (Fig. 9). Toward the oral aperture, the cortical layer (zone 2) diminishes and zone 5, a part of the lacunar system, becomes strongly developed (Fig. 3). The mouth leads into the cheilostome (chs) which connects with the oesophastome; this junction acts as a guiding ring, but a differentiated sclerotised ring does not exist. The cheilorhabdia project beyond the guiding ring as a cuticular cylinder (cc) (Figs 4, 9). Dorsally and ventrally, the cylinder bears a crest (dk and vk)

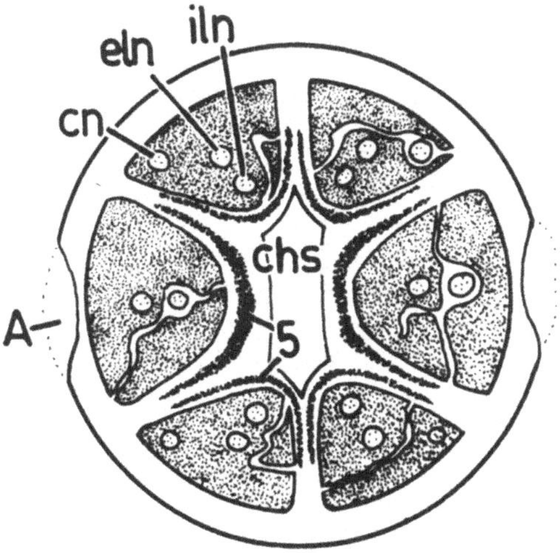

5: zone 5 of cuticle,

A: amphid,

chs: cheilostome,

cn: cephalic nerve,

eln: external labial nerve,

iln: internal labial nerve.

Fig. 3. Cross-section through the labial region of Aporcelaimellus obtusicaudatus.

which detaches from it posteriorly (Figs 4, 9). A thin layer of epidermis (e), continuous with that of the labial region, follows the outline of the cylinder (Figs 4, 9) and reaches the guiding ring at the inner side of the cylinder (Fig. 9). From the base of the cylinder this epidermal epithelium proceeds posteriad toward the oesophagus (Figs. 5-9).

The oesopharhabdia of the anterior part of the oesophastome consist of 3 zones. The inner zone (igs) is a homogeneous cuticular layer, continuous with the odontophore (Op) (Figs 4-5, 9). Together with the middle zone (gs) it represents the guiding sheath which is surrounded by the third zone, viz. the hydrostatic substance (h) (Figs. 4-6, 9). The hydrostatic substance is enclosed in a fluid-filled space and consists of numerous granular and fibrillar elements. The guiding sheath has essentially the same composition, but the components are very densely packed in this layer (8). The guiding sheath reaches posteriad to the level of implantation of the odontostyle (O) (Figs 5, 9), so that further back, the odontophore is lined by the hydrostatic substance (Figs 5B, 6, 9), until

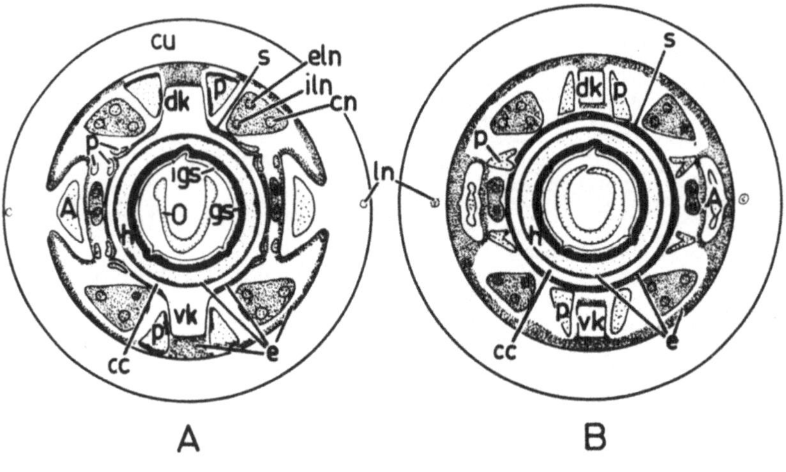

Fig. 4. A. obtusicaudatus. Cross-sections through anterior (A) and posterior (B) part of stylet-aperture. A: amphid, cc: cuticular cylinder, cn: cephalic nerve, cu: cuticle, dk: dorsal crest, e: epidermis, eln: ext. labial nerve, gs: guiding sheath, h: hydrostatic substance, igs: inner lining of guiding sheath, iln: int. labial nerve, ln: ext. lateral nerve, O: odontostyle, p: protractor, s: stomodeal membrane, vk: ventral crest.

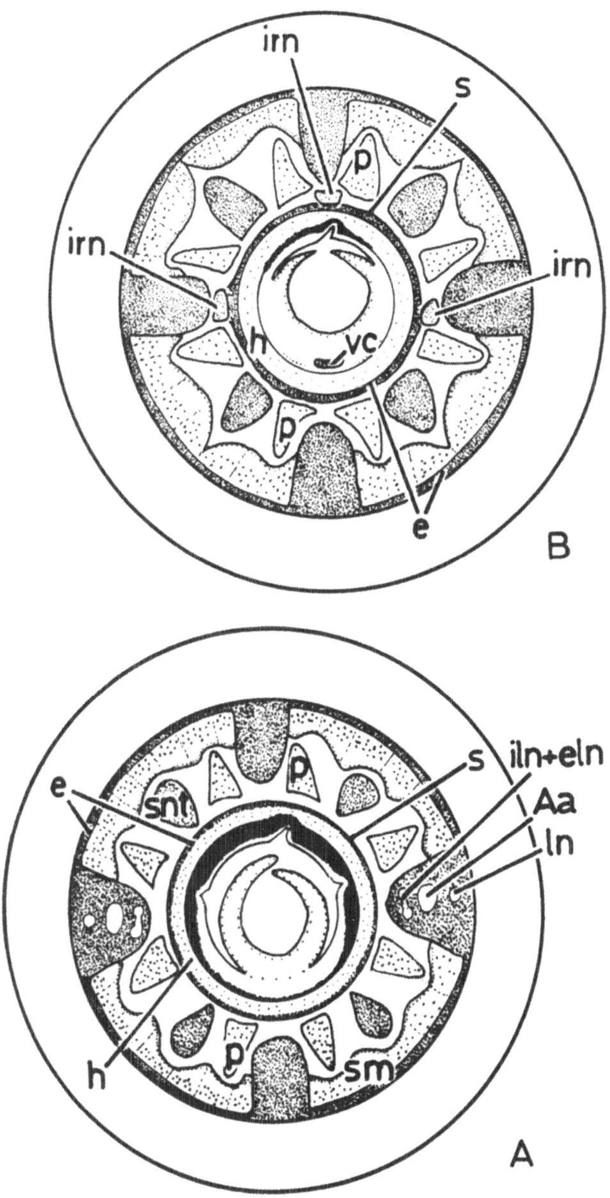

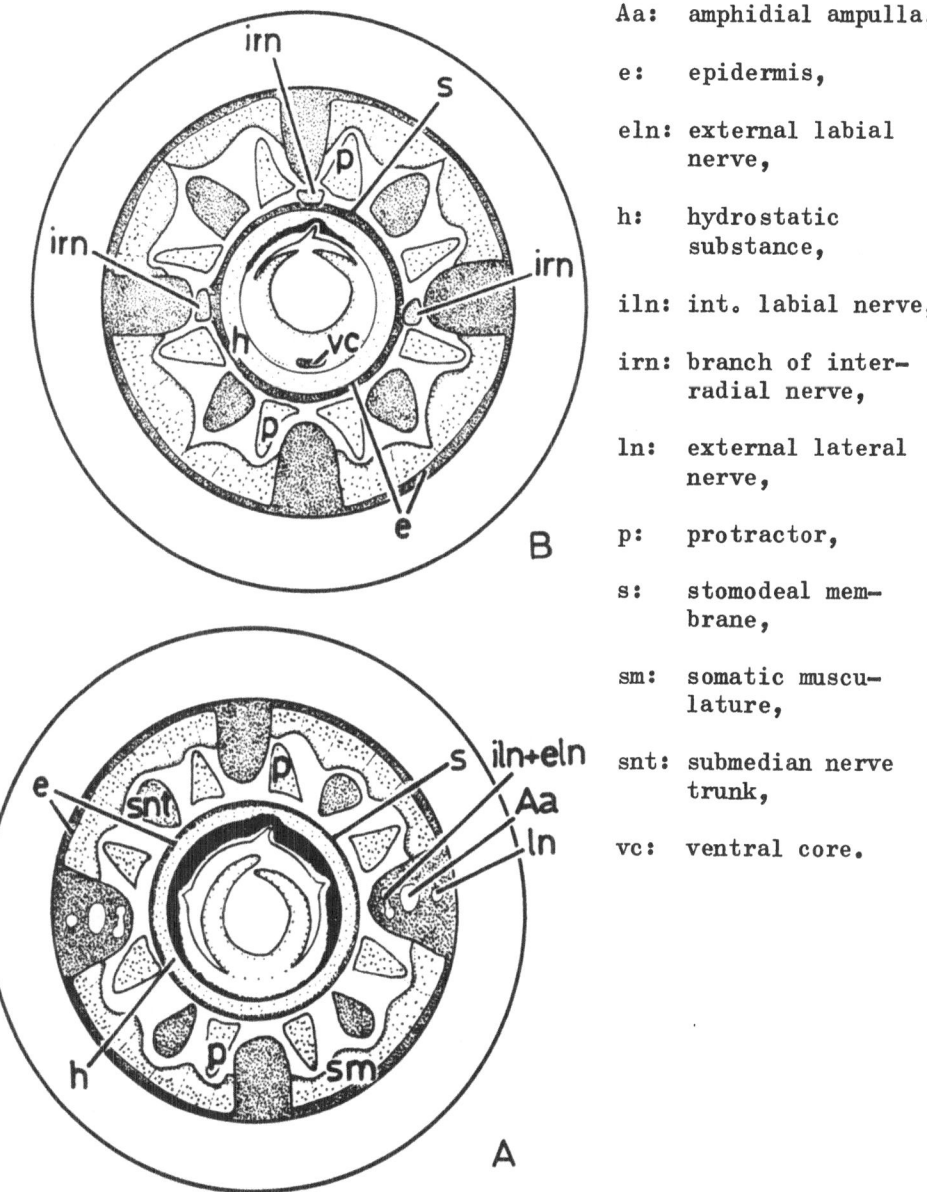

Aa: amphidial ampulla,

e: epidermis,

eln: external labial nerve,

h: hydrostatic substance,

iln: int. labial nerve,

irn: branch of inter-radial nerve,

ln: external lateral nerve,

p: protractor,

s: stomodeal membrane,

sm: somatic musculature,

snt: submedian nerve trunk,

vc: ventral core.

Fig. 5. A. obtusicaudatus. Cross-sections through area of insertion of odontostyle upon odontophore. A- anteriad, B- posteriad.

it meets the oesophagus.

The odontostyle is implanted upon a ventral sector of the odontophore (Figs 5, 9). This orientation, which is unusual in Dorylaimoidea, is due to a rotation of the entire stomodaeum in the anterior part (3). The feature is not constant, as specimens occasionally were found in which the odontostyle is implanted in a subventral sector.

The odontostyle is a tube-like structure with a dorsal longitudinal slit (Figs 4-5, 9), of which the enclosing rims overlap somewhat (Figs 4B, 5A, 9). In the anterior half of the odontostyle the slit is widened, forming the spear aperture (Figs 4A, 9). The tip is solid and composed of some extremely hard material.

Each of the interradial sectors of the odontophore contains a tissue core (Figs 6B, 9) of which the ventral (vc) is the largest. The tissues in the cores consist of a branch of the interradial oesophageal nerve which is surrounded by epidermis.

The second part of the oesophastome is the apex of the oesophagus, including the basal portion of the odontophore (Figs 7-9). The tissue cores of the odontophore open here (Figs 7, 9); only the central part of the odontophore separates the oesophageal tissues from the lumen (Figs 8,9). The epidermis bordering the hydrostatic substance penetrates the oesophagus alongside the odontophore (Figs 8,9). The stomodeal membrane (s) which externally covers the oesophagus (Figs 7-9) runs anteriad till the guiding ring (Figs 4-6, 9).

The oesophageal tissues consist of six nerves, which radially alternate with six sets of muscles (Figs 8, 9, 10A-B). The muscles of the anterior part run almost parallel to the body axis when the stylet is retracted (Fig. 9). During protrusion, these muscles become perpendicularly arrayed. The muscles contain two types of filaments; the thin filaments are situated near the lumen and near the oesophageal wall, with the thick filaments in between. Thus, one striated sarcomere is indicated (8).

The nerves are to be divided in an interradial (irn) and a radial (rn) group of three. The interradial nerves penetrate the tissue cores. Branches of these nerves pass into the stomodeal membrane (Fig. 7) and after rotation (Fig. 6), connect with the dorsal and lateral nerve trunks (Figs 5B, 11).

The terminology of Inglis (7) was adopted recently (8) since the division into cheilostome and stomodaeum is not arbitrary, but corresponds to a secondary and primary invagination respectively. The rhabditoid terminology leads to confusion by the ill-defined concept "stoma". It has complicated interpretation of the odontophore, as this structure on the one hand is structurally similar

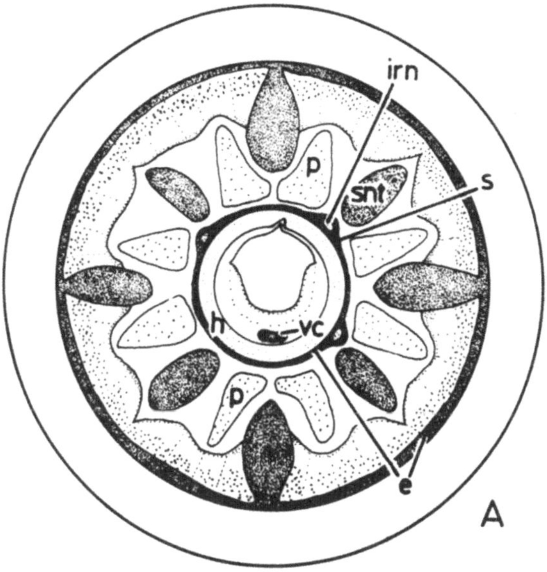

e: epidermis,

h: hydrostatic substance,

irn: branch of interradial nerve,

Op: odontophore,

p: protractor,

s: stomodeal membrane,

sc: subdorsal core,

snt: submedian nerve trunk,

vc: ventral core.

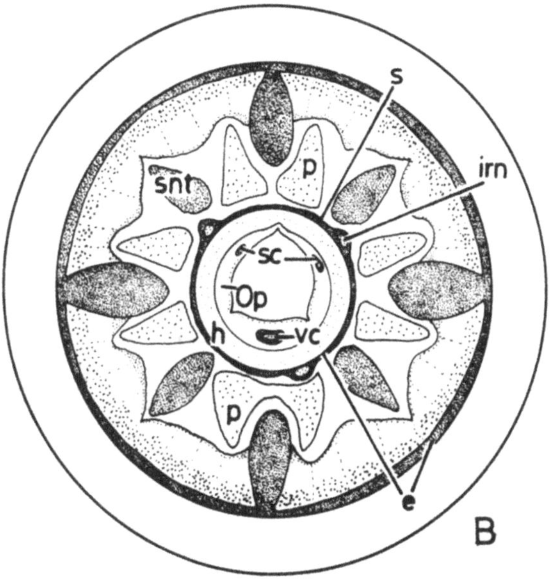

Fig. 6. A. obtusicaudatus. Cross-sections through anterior part of the odontophore. A- level of rootlets of the odontostyle; B- level near the oesophagus.

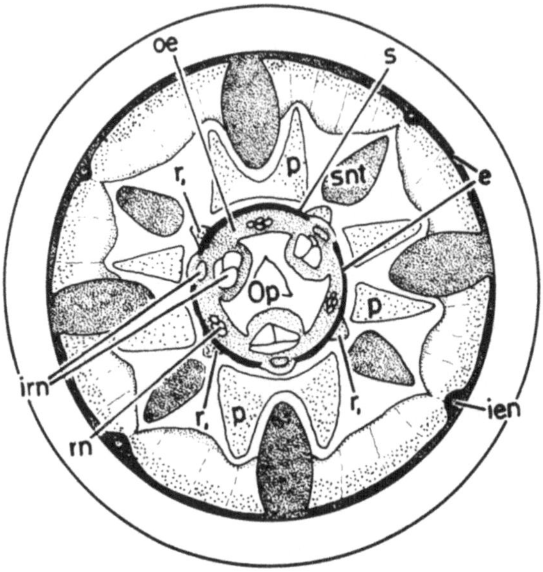

e: epidermis,

ien: intra-epithelial nerve,

irn: interradial nerve,

oe: oesophagus,

Op: odontophore,

p: protractor,

r_1: retractor,

rn: radial nerve,

s: stomodeal membrane,

snt: submedian nerve trunk.

Fig. 7. A. obtusicaudatus. Cross-section through apex of oesophagus.

to the oesophageal lining, but on the other hand is shed together with the "stomatal" parts. Moreover, the inner lining of the guiding sheath (a stomatal part!) is a prolongation of the odontophore. Using Inglis' interpretation, the guiding ring is the border-line between cheilostome and oesophastome. The guiding apparatus is, from this point of view, to be considered as a secondary evolved formation.

The stylet is protruded by the action of eight protractor branches (p). These branches are attached to a kind of framework. The dorsal and ventral pairs are inserted, subdorsally and subventrally respectively, on the crests of the cuticular cylinder (dk and vk), and also on the body wall (Figs 4A, 9). Each sublateral protractor has three insertions (Fig. 4A): one upon the crest of the amphidial cuticle, and one on either side of the corresponding arm of the separation of the lips; these insertions are quite long and extend alongside the cuticular cylinder. Posterior to the amphidial crest, the heads fuse, forming one sublateral protractor (Fig. 4B). Near the oesophagus (Figs 6B, 7) the dorsal and ventral

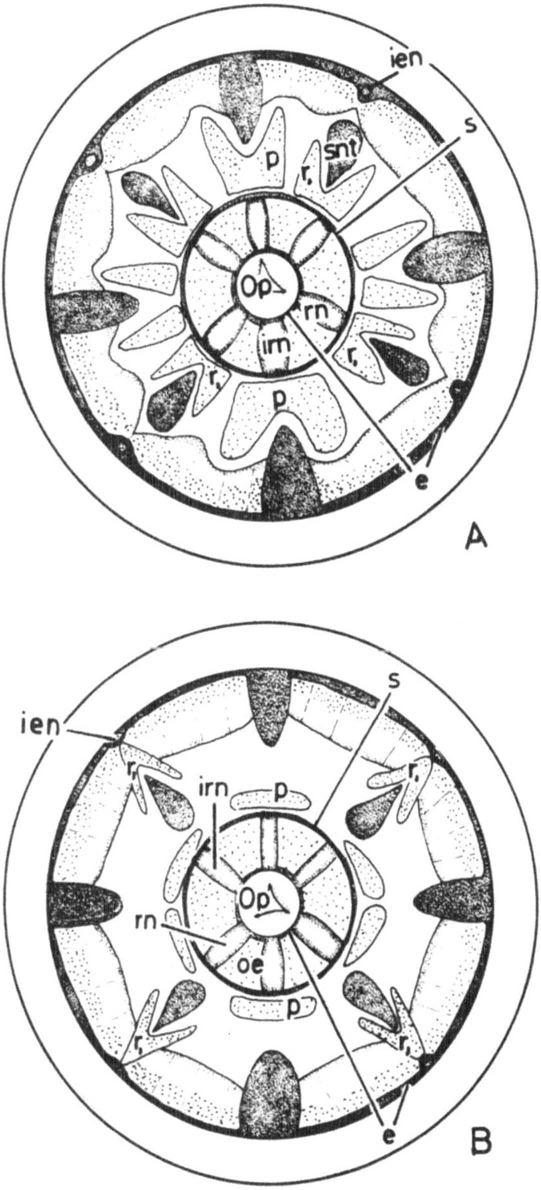

e: epidermis,

ien: intra-epithelial nerve,

irn: interradial nerve,

oe: oesophagus,

Op: odontophore,

p: protractor,

r_1: retractor,

rn: radial nerve,

s: stomodeal membrane,

snt: submedian nerve trunk.

Fig. 8. A. obtusicaudatus. Cross-sections through anterior part of the oesophagus. A- level approx. 5 µm posterior to that of Fig. 7, B- level approx. 8 µm further posteriad.

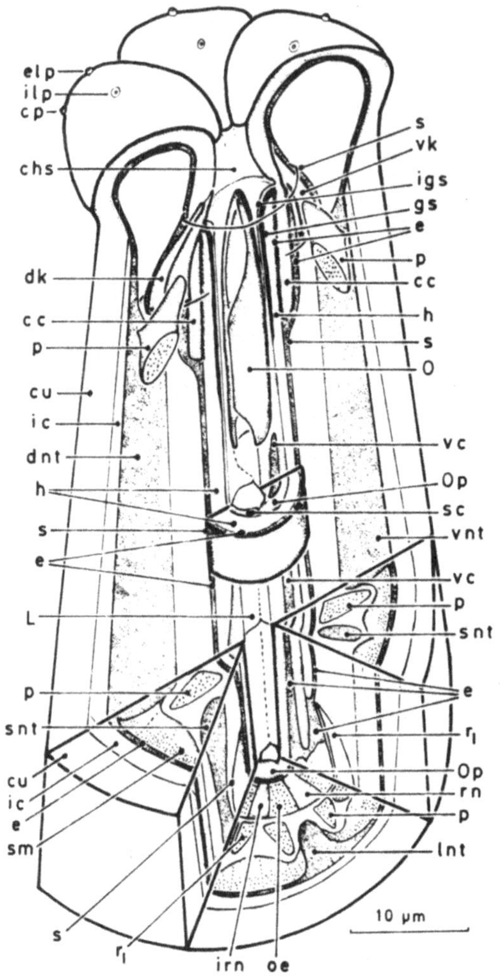

cc: cuticular cylinder, chs: cheilostome cp: cephalic papilla, cu: cuticle, dk: dorsal crest, dnt: dorsal nerve trunk, e: epidermis, elp: external labial papilla gs: guiding sheath, h: hydrostatic substance, ic: infracuticle, igs: inner lining of guiding sheath, ilp: internal labial papilla, irn: interradial nerve, L: lumen, lnt: lateral nerve trunk, O: odontostyle, oe: oesophagus, Op: odontophore, p: protractor, r_1: anterior retractor, rn: radial nerve, s: stomodeal membrane, sc: subdorsal core, sm: somatic musculature, snt: submedian nerve trunk, vc: ventral core, vk: ventral crest, vnt: ventral nerve trunk.

Fig. 9. Reconstruction of the anterior body region of _Aporcelaimellus obtusicaudatus_.

pairs of the protractors fuse, so that six muscles run posteriad alongside the oesophagus (Fig. 8).

The stylet is retracted by four muscles which are inserted upon the stomodeal membrane (s) near the base of the hydrostatic substance (Figs 7, 9). These muscles (r_1) run diagonally to the body wall in submedian position. Where they cross the submedian nerve trunks (snt), the latter pass through the retractors (Figs 8, 9). The ultrastructure of the retractors indicates that they are derived from the somatic musculature (8).

A second set of retractors is inserted upon the oesophagus near the nerve ring (Fig. 10). Anteriorly, the three muscles are located opposite the radial nerves (rn) (Fig. 10A), but their positions shift spirally, so that near the posterior ganglia they are located opposite the interradial nerves (irn) (Fig. 10B). The muscle which is here in dorsal position, divides into 2 branches (Fig. 10C). The four muscle strips run in pairs through the pseudocoel, reaching the body wall at both sides of the lateral chords.

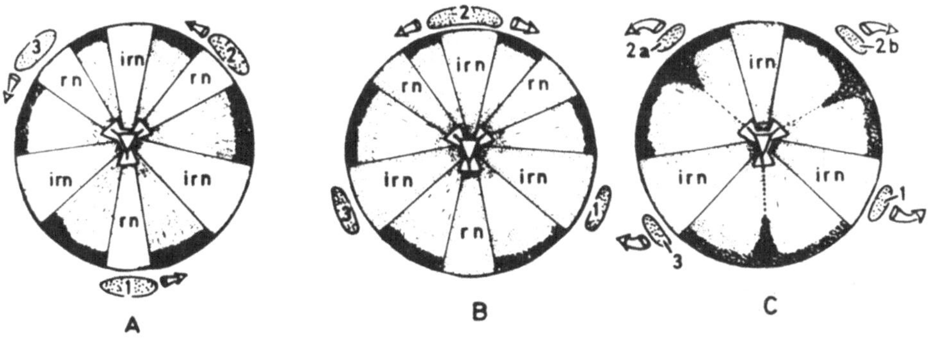

Fig. 10. A. obtusicaudatus. Cross-sections through oesophagus at level of anterior ganglia (A), nerve ring (B), and posterior ganglia (C). 1-3: retractors; irn: interradial nerve; rn: radial nerve.

RECONSTRUCTION OF THE CEPHALIC NERVOUS SYSTEM

The cephalic nervous system consists mainly of eight nerve trunks. Additionally, four nerve strands (ien) run in submedian position through the epidermis (Figs 7, 8). The dorsal and ventral trunks are mixed nerves, composed of sensory and motor nerves; the sensory endings lead to somatic receptors. The lateral trunks (Fig. 11) are sensory; the branches lead to somatic receptors (sr) of which one is clearly reduced (v. sr). Near the base of the hydro-

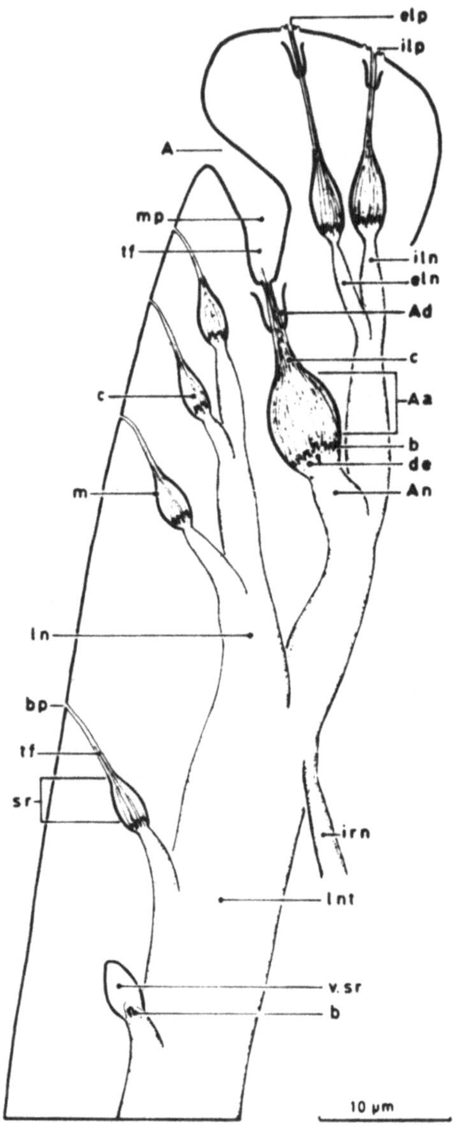

A: amphid,
Aa: amphidial ampulla,
Ad: amphidial duct,
An: amphidial nerve,
b: basal body,
bp: body pore,
c: sensory cilium,
de: dendritic process,
eln: ext. labial nerve,
elp: ext. labial papilla,
iln: int. labial nerve,
ilp: int. labial papilla,
irn: interradial nerve,
ln: ext. lateral nerve,
lnt: lateral nerve trunk,
m: membrane,
mp: mucoid-plug,
sr: somatic receptor,
tf: terminal filament,
v.sr: vestigial somatic receptor.

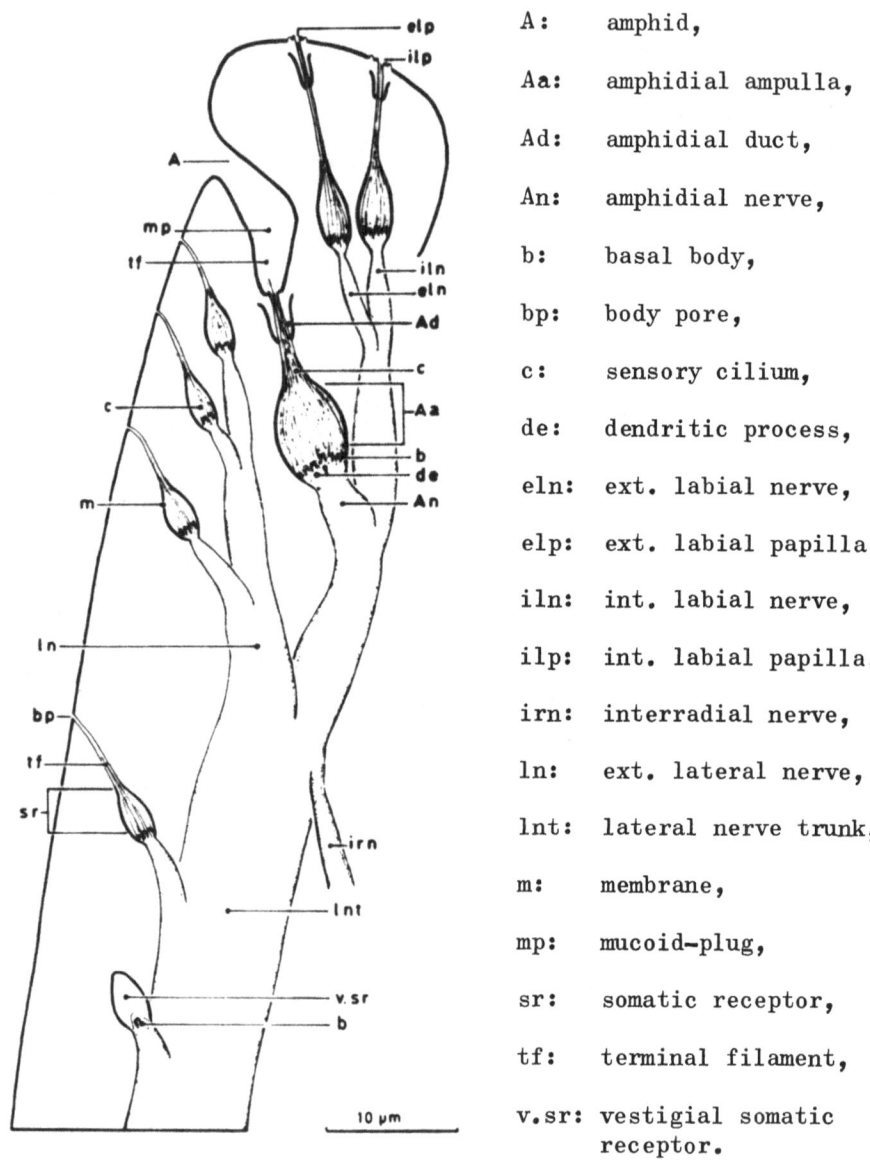

Fig. 11. A. obtusicaudatus. Reconstruction of the lateral nerve trunk and its branches.

static substance the trunk divides in two branches. The outer branch, the external lateral nerve (ln), leads to three or four somatic receptors (sr), ending in the most anterior of these. The inner branch divides once more (Figs 4A, B, 5A, 11), into the amphidial nerve (An) and the labial nerve, the latter dividing into an internal (iln) and an external branch (eln) opposite the guiding ring (Figs 4, 11). The submedian trunks (snt) run from the frontal ganglia to the labial region, where each of these divides in a cephalic nerve (cn) and an internal (iln) and external labial nerve (eln) (Figs 3, 4). These trunks are not connected with somatic receptors (8).

The composition of all receptors is essentially alike. They consist of three (somatic receptors and cephalic nerves) four (labial nerves) or 15-22 (amphidial nerves) dendrites, surrounded by one or more glial elements. The extremities of the dendrites are enclosed in a membrane-bound ampulla (Fig. 11). The basal bodies (b) are located in the apices of the dendritic processes (de) with the sensory cilia (c) on top. The cilia have a well-developed basal plate. The fibrillar configuration of the cilia is extremely variable and usually highly irregular. Their distal ends are modified in terminal filaments (tf) that sometimes are branched (8). The ampulla of the receptor leads into a membrane-bound duct that passes into the cuticle, and which is surrounded by a sheath of epidermis. The duct opens to the exterior through a pore. The pore of a somatic receptor (bp) is simple; that of a papilla is surrounded by one or two concentric cuticular ridges. The cuticularised amphidial duct (Ad) is distally widened (pouch). The duct forms 2 blindfoldings in the amphidial crests (Fig. 4B), but the development of these blindfoldings is variable. The "pore" of the amphid is a more or less oval, transverse slit at the base of the labial region (Fig. 11). A similar type of receptor-structure was observed in Xiphinema (10).

The amphid, papillae and body pores often contain a mucoid-plug (mp), of which the origin is still unknown. Neither in the surrounding epidermis, nor in the glial elements was any trace of glandular activity detected.

REFERENCES

1. Aboul-Eid, H.Z. (1969). Electron microscope studies on the body wall and feeding apparatus of Longidorus macrosoma. Nematologica 15, 451-463.

2. Bird, G.W. (1970). Somatic musculature of Trichodorus porosus and Criconemoides similis. J. Nematol. 2, 404-409.

3. Coomans, A. & van der Heiden, A. (1971). Structure and formation of the feeding apparatus in Aporcelaimus and Aporcelaimellus (Nematoda: Dorylaimida). Z. Morph. Tiere 70, 103-118.

4. Grootaert, P. & Lippens, P.L. (1974). Some ultrastructural changes in cuticle and hypodermis of Aporcelaimellus during the first moult (Nematoda: Dorylaimoidea). Z. Morph. Tiere (in press).

5. Hirumi, H., Raski, D.J. & Jones, N.O. (1971). Primitive muscle cells of nematodes: Morphological aspects of platymyarian and shallow coelomyarian muscles in two plant parasitic nematodes, Trichodorus christiei and Longidorus elongatus. J. Ultrastr. Res. 34, 517-543.

6. Huxley, H.E. & Hanson, J. (1960). The molecular basis of contraction in cross-striated muscles. In G.H. Bourne (ed.) "The structure and function of muscle", Vol. 1, 183-227.

7. Inglis, W.G. (1966). The origin and function of the cheilostomal complex in the nematode Falcaustra stewarti. Proc. Linn. Soc. Lond. 177, 55-62.

8. Lippens, P.L., Coomans, A., De Grisse, A.T. & Lagasse, A. (1974). Ultrastructure of the anterior body region of Aporcelaimellus obtusicaudatus and A. obscurus. Nematologica 20, (in press).

9. Raski, D.J., Jones, N.O. & Roggen, D.R. (1969). On the morphology and ultrastructure of the oesophageal region of Trichodorus allius Jensen. Proc. helminth. Soc. Wash. 36, 106-118.

10. Roggen, D.R., Raski, D.J. & Jones, N.O. (1967). Further electron microscopic observations of Xiphinema index. Nematologica 13, 1-16.

11. Taylor, C.E., Thomas, P.R., Robertson, W.M. & Roberts, I.M. (1970). An electron microscope study of the oesophageal region of Longidorus elongatus. Nematologica 16, 6-12.

12. Wright, K.A. (1965). The histology of the oesophageal region of Xiphinema index Th. & All., 1950, as seen with the electron microscope. Canad. J. Zool. 43, 689-700.

DISCUSSION

The question of terminology was raised with reference to the use of the terms hypodermis, mesodermis and ectodermis, and their analogy in describing the origin of various tissues and structures in nematodes. Lippens replied that one cannot directly apply termin-ology devised for describing vertebrate tissue directly to inver-tebrates; and in invertebrates it seems reasonable to refer to hypodermis when a cuticle is present, and ectodermis when there is no cuticle. In arthropods the term hypodermis is the secreting layer which gives rise to various layers of the cuticle; with reference to nematodes it is recognised that hypo means "under", and that there is no dermis. Lippens felt that a better descriptive term might be "tegument", which refers to the outer covering layer as in Ashelminthes where there is no cuticle, but meanwhile there seemed to be no particular problem in using the term "hypodermis".

Questioned about the structure and function of the infracuticle, Lippens suggested that it might be a connection of cuticle and muscle to the hypodermis. He had observed striated fibres, origin-ating in the hypodermis, pushing into the infracuticle.

Loof commented that G. Thorne had given particular emphasis to the double character of the amphids in his classification of the dorylaims and asked whether the peculiar or distinctive anatomy of the amphids *Aporcelaimus* was characteristic of the dorylaims. Lippens explained that the body cuticle surrounds the amphids for some distance within the body and that a ridge formed by the body cuticle as seen in the amphidial canal was not typical of *Longidorus* or *Xiphinema*. The blind foldings of the amphidial canal are not constant.

Luc asked if the anterior (cervical) body pores could be con-sidered of taxonomic value. Lippens replied that it appeared that only the first pore was reasonably constant in relation to the amphidial aperture, and therefore they were of limited taxonomic value. It was asked if there was anything in *Aporcelaimellus* analogous with the excretory system of other dorylaims and Lippens replied that a vestigial pore is present near the nerve ring which appears different from other pores and which does not connect with any gland. Lippens pointed out that in Dorylaimida the anterior pores are usually connected to nervous tissue and not to glands, and speculated that perhaps terrestrial nematodes,in contrast to marine nematodes, have lost the glandular function of the anterior pores. He said it is possible that the caudal pores may have a glandular function but this has not been investigated. Robertson commented that in *Longidorus* and *Xiphinema* the so-called excretory pore appears as a much larger body pore containing up to 7 cilia, as compared with 3 for other body pores; the function of this larger pore is unknown.

Asked about the nature of neurosecretions Lippens replied that there are at least three types of neurosecretory cells in *Aporcelaimellus* and these produce proteins of at least four different kinds. It is not known whether the mucus associated with the amphids is related to neurosecretions.

THE STRUCTURE AND MUSCULATURE OF THE FEEDING APPARATUS IN <u>LONGIDORUS</u> AND <u>XIPHINEMA</u>

W.M. ROBERTSON and C.E. TAYLOR

Scottish Horticultural Research Institute

Invergowrie, Dundee, Scotland

INTRODUCTION

Modifications from the basic dorylaimoid pattern in the structure of the feeding apparatus and oesophageal digestive tract in *Xiphinema* and *Longidorus* nematodes are correlated with a development of the plant parasitic habit. The two genera are characterised by the long hypodermic-like stylet which enables them to feed on the deeper tissues of plant roots and yet remain ectoparasites. Several species of both genera are vectors of plant viruses and the possible association of virus particles with the feeding apparatus prompted several electron microscope studies on the ultrastructure of the oesophageal region (1, 3, 4, 5, 8, 10, 11, 13). Electron microscopy studies have been extended to include several non vector species to see if any differences exist between them and the vector species that could be associated with the ability to transmit viruses.

MATERIALS AND METHODS

Nematode species were obtained locally and the more remote populations from the soil samples as part of the NATO survey (9). Active specimens with optically dense gut contents (*i.e.* assumed to be well fed) were fixed for 2 hours in cold 3% glutaraldehyde buffered with 0.1M sodium cacodylate ($4^{o}C$, pH 7.3) and then rinsed overnight in sodium cacodylate buffer. Specimens were postfixed in 1% osmium tetroxide ($4^{o}C$) buffered to pH 7.3 with 0.1M veronal buffer. The nematodes were rinsed in veronal buffer, embedded in agar and then placed in 20% ethanol. Dehydration was completed in a graded ethanol series at $4^{o}C$ over 2 hours and transferred through

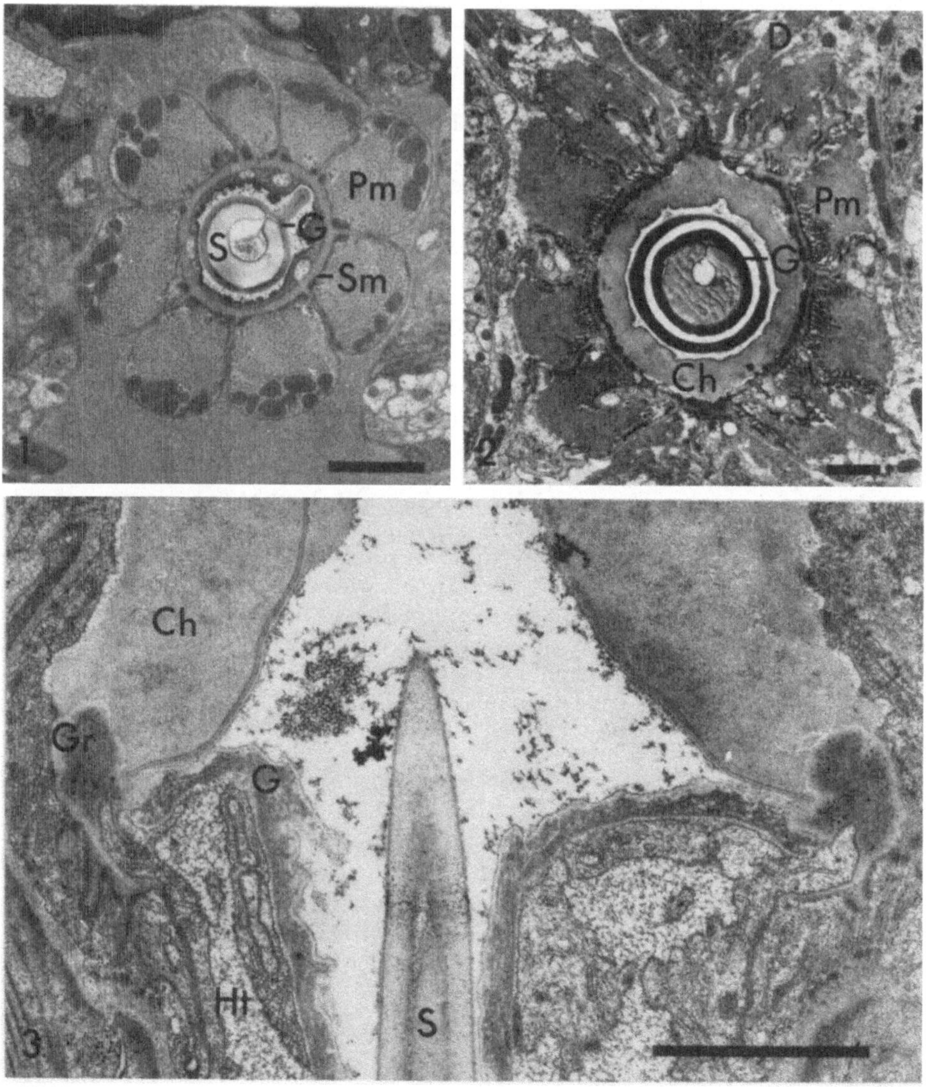

PLATE I.

Fig. 1. T.S. of odontostyle (S) of L. attenuatus. Guide sheath (G), support membrane (Sm), protractor muscles (Pm).

Fig. 2. T.S. of odontostyle (S) of X. diversicaudatum. Cheilostome (Ch), dilatores buccae (D).

Fig. 3. L.S. of guide ring area in L. goodeyi. Cheilostome (Ch), odontostyle (S), guide ring (Gr), hydrostatic tissue (Ht)

Bars represent 1 μm.

two changes of 1, 2 epoxy propane. Infiltration was by a 1:2 mixture of 1,2 epoxy propane : 'Araldite' for 15 hours followed by 'Araldite' alone for two days. Polymerisation was carried out at 60°C for three days. Specimens were sectioned with glass or diamond knives. Sections were stained in alcoholic uranyl acetate and lead citrate using the bulk staining technique of Robertson and Roberts (7) and examined at 50KV in a Hitachi HS-8.

THE STYLET AND GUIDING SHEATH

The mural tooth of the primitive dorylaimoid nematode has evolved in Xiphinema and Longidorus to form a long hypodermic-like odontostyle. It is protracted through the mouth by the eight protractor muscles which have a branched anterior attachment to the cheilostome and somatic muscle areas. Posteriorly the protractor muscles attach to the base of the odontophore (stylet extension).

The odontostyle is morphologically similar in Longidorus and Xiphinema but structurally different, and probably more robust in Xiphinema (Plate I, Figs. 1,2). Essentially it is a needle-like tube, usually about 90 µ long in the adult nematode although this varies between species, and with a lumen of approximately 450 nm diameter; a narrow slit runs throughout its length to the point where it becomes embedded in the cuticle of the odontophore.

In Longidorus spp. a cuticular guiding sheath surrounds the odontostyle for its entire length (Plate I, Fig. 1). At its anterior end the guiding sheath becomes attached to the guide ring by a short, flexible connection. The guide ring is an annulus of electron -dense granular material fused into the base of the cheilostome (Plate I, Fig. 3). The cheilostome of Longidorus is much shorter than that in Xiphinema, but in both genera with the spear fully retracted, the guide ring provides a fixed point, and the distance from the head end to the guide ring provides a specific measurement which can be usefully applied in taxonomy. In Xiphinema, however, the guide ring moves forward slightly when the stylet is fully protracted (12 and see below).

In Longidorus species, between the cylindrical support membrane for the protractor muscles and the guiding sheath, is an amorphous, 'hydrostatic substance' and reservoirs of the 'hydrostatic substance' associated with secretory-like tissue are found in the hypodermal cords at the level of the guide ring (10). It is thought that the substance maintains the shape of the support membrane acting against the turgor pressure of the body. Because of the length of the cheilostome, Xiphinema does not appear to require the 'hydrostatic substance' to the same extent, if at all, and so far no reservoirs have been found.

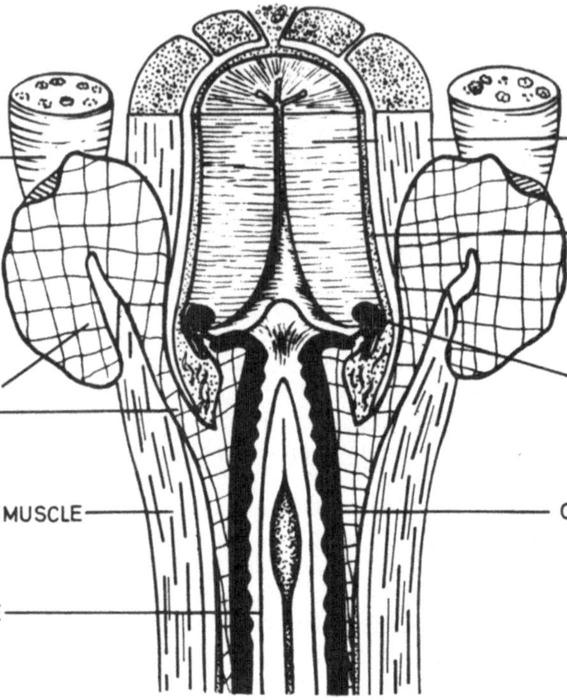

Fig. 1. Diagram of the relationship of odontostyle, guiding sheath and cheilostome in Longidorus elongatus (X 4000)

In Xiphinema the cheilostome is thick-walled and ends posteriorly with the guide ring; it is much longer than in Longidorus and extends posteriorly, almost to the level of the odontophore. The guiding sheath is doubled within the cheilostome and in cross sections is seen as a single structure only below the level of the guide ring. (At its most anterior end the fold in the guiding sheath is seen as the secondary guide ring, and depending on the state of protraction or retraction of the stylet its position within the cheilostome is somewhat variable.

In both Xiphinema and Longidorus the guiding sheath becomes gradually thinner posteriorly and where it eventually fuses with the odontophore it is only about 40 nm thick compared with 150 nm at the anterior end. In studying X. diversicaudatum it was found that the guiding sheath was auto-fluorescent and by this means it was possible to trace it for about one-third of the length of the odontophore. That of L. elongatus did not fluoresce, indicating that it differs in chemical structure from that of X. diversicaudatum this has a earing on the ability of L. elongatus to transmit rasp-

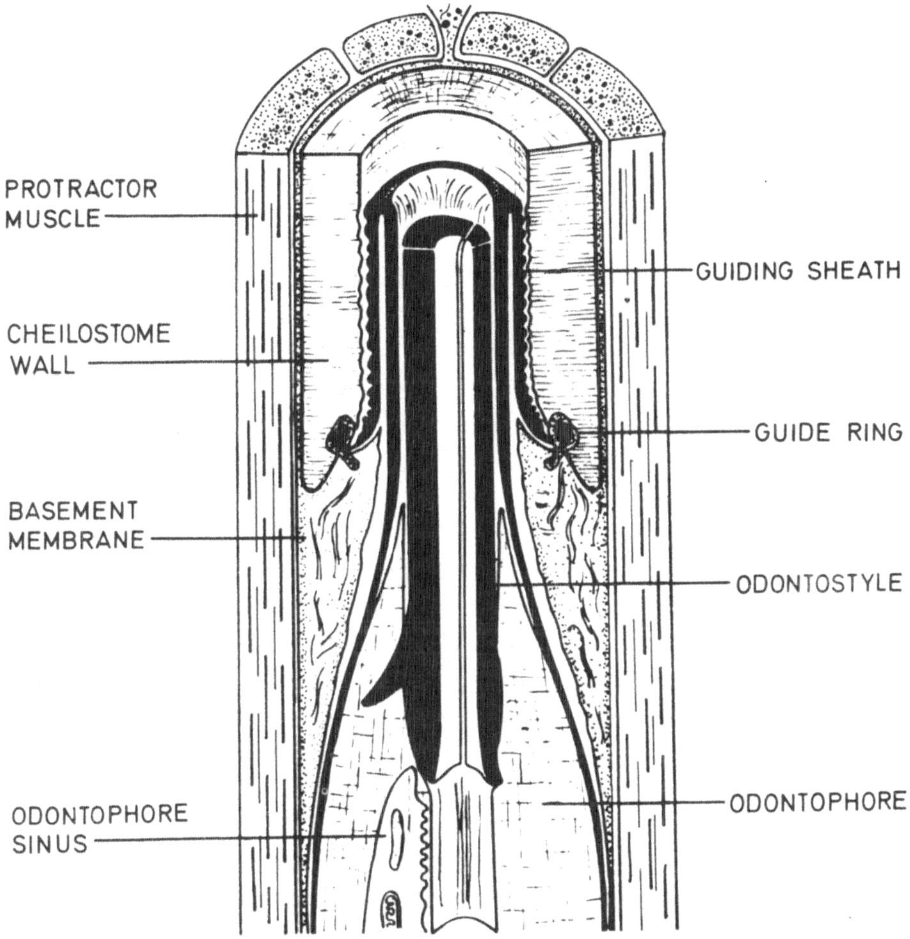

Fig. 2. Diagram of the anterior part of the odontostyle and guiding apparatus in *Xiphinema diversicaudatum* (X 3500)

berry ringspot and tomato black ring viruses by their becoming specifically associated with the guiding sheath in this species.

In *Longidorus* the odontostyle, when protracted, moves forward mostly within the guiding sheath, which folds into an anterior loop from the guide ring (cf. *Xiphinema*) but only for a short distance into the posterior part of the cheilostome lumen. As protraction continues, the thin posterior part of the guiding sheath forms

LONGIDORUS

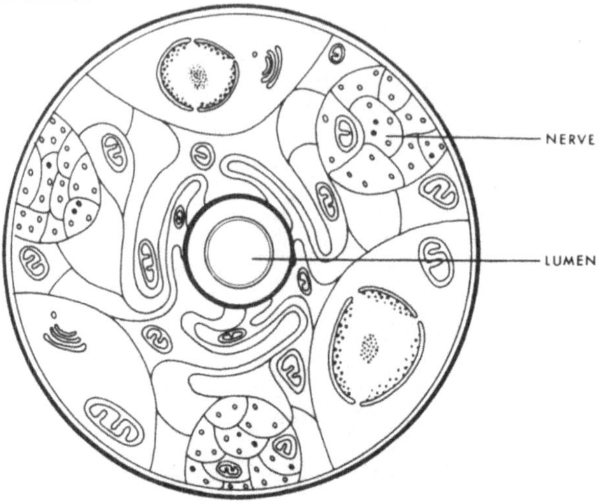

3

XIPHINEMA

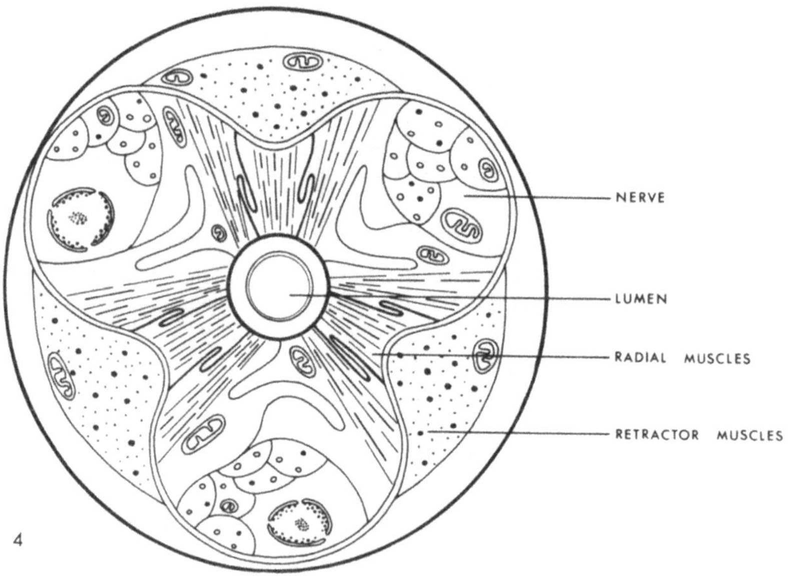

4

Fig. 3. Diagram of T.S. of typical arrangement of slender oesophagus in those Longidorus species without radial muscles e.g. L. elongatus (see also text)

Fig. 4. Diagram of T.S. of oesophagus in Xiphinema and those Longidorus species with radial muscles e.g. L. macrosoma.

concertina-like folds around the neck of the odontophore. In Xiphinema most of the length of the odontostyle lies within the lumen of the distal part of the cheilostome and only a small part is enclosed within the guiding sheath. As the odontostyle is protracted the odontophore moves forward through the guide ring; this is accompanied by some forward movement of the guiding sheath, and hence the so-called secondary guide ring. As the odontophore moves forward into the cheilostome it tends to compress longitudinally the cheilostome wall; on retraction of the odontostyle the wall is restored to its normal length by the contraction of the dilatores buccae muscles, which thus function more as suspensory ligaments rather than to widen the cheilostome (2, 10). Dilatores buccae muscles are not present in Longidorus species.

THE ODONTOPHORE (Plate II)

The odontophore is formed by the cuticularisation of the most anterior part of the oesophagus; three arms radiate from the cuticle surrounding the oesophageal lumen and progressively close over anteriorly to form the restricted periphery of the odontophore. In Xiphinema species these arms are particularly conspicuous and form the posterior flanges which are easily seen in the light microscopy of whole nematodes (Plate II, Fig. 4). Running parallel to the lumen of the odontophore are three sinuses containing tissue continuous with that of the oesophagus. The ventral sinus is the largest and extends anteriorly to a point of weakness in the cuticle of the odontophore; during the moulting process the replacement stylet moves forward from the oesophagus into the anterior stomodaeum through the sinus. The odontophore is, of course, morphologically a part of the anterior oesophagus but it functions as a supporting and ejecting mechanism for the odontostyle, and hence the name odontophore was derived from the Greek odous - a tooth, pherein - to bear (11).

ANTERIOR OESOPHAGUS (Text-figs.3,4)

The anterior oesophagus in Longidorus and Xiphinema is a flexible tube with a lumen of approximately 750 nm bounded by a 270 nm thick cuticle. In transverse sections unmyelinated nerve axons are identifiable with the large nuclei, mitochondria and other cell organelles of the glial cells accompanying them.

Radial muscles are present in the anterior oesophagus for only about 5 μ distance from the base of the odontophore in L. elongatus, L. attenuatus, L. goodeyi, L. leptocephalus and L. vineacola. In L. macrosoma, L. profundorum, L. africanus, L. siddiqi and L. caespiticola, and in Xiphinema, radial muscles are present throughout the length of the oesophagus in varying degrees. (Text-fig. 4).

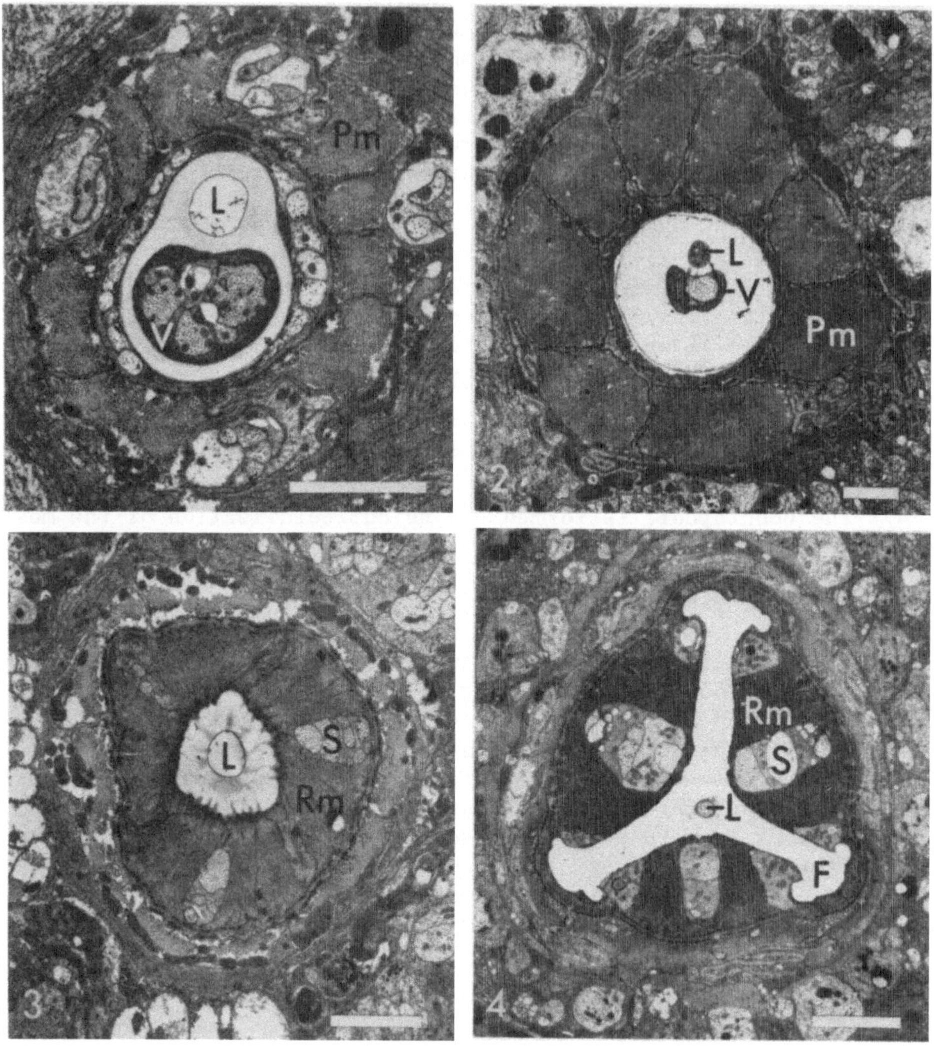

PLATE II.

Fig. 1. T.S. of anterior odontophore in L. macrosoma. Lumen of food canal (L), ventral sinus (V) and protractor muscles (Pm).

Fig. 2. T.S. of anterior odontophore in X. diversicaudatum.

Fig. 3. T.S. of posterior odontophore in L. macrosoma. Lumen (L), radial muscles (Rm), sinus process (S).

Fig. 4. T.S. or posterior odontophore in X. diversicaudatum. Note large flanges (F). Other letters as in Fig. 3.

Bars represent 2 µm.

OESOPHAGEAL BULB (Text fig.5)

For most of its length the lumen of the bulb is triradiate with the cuticle thickened to form three pairs of triangular platelets. Radial muscles extend from the platelets to the peripheral wall. Contraction of these muscles causes the lumen to open to a triangular cross section. This is accompanied by the complementary relaxation of the three sets of peripheral muscles which are arranged equidistantly around the circumference of the bulb throughout its length. This complementary action of the radial and peripheral muscles maintains the bulb at constant volume and turgor pressure. Contraction of the peripheral muscles tends to decrease the volume of the bulb and increase its turgor pressure and with the radial muscles relaxed this results in the closure of the lumen through the natural elasticity of the cuticle (11). The oesophageal bulb acts both as a pump to inject saliva into the root and for the intake of plant sap which is forced into the intestine through the one-way oesophago-intestinal valve.

Secretory glands in the oesophageal bulb are well-developed. The dorsal gland orifice opens into the lumen just anterior to the platelets, the latero-ventral glands open within the platelet area.

ODONTOSTYLE PROTRACTOR AND RETRACTOR MUSCLES

Eight odontostyle protractor muscles originate as four pairs in the fore-cheilostome region, each pair deriving from a quadrant of muscle. They extend posteriorly to the base of the odontophore and attach to the oesophageal wall where their pull is transferred to the cuticular base by short radial muscles. (Plate II, Figs. 6, 7)

In Xiphinema three sets of odontostyle retractor muscles are identified, extending from the base of the odontophore and splitting to form four muscles which attach to the body wall cuticle in the region of the oesophageal bulb (Text-fig.9).

In Longidorus various combinations of odontostyle retractor muscles are present (Text-fig. 6, 7, 8). In L. elongatus, for example, which is a species without oesophageal radial muscles, except for the most anterior part, odontostyle retractor muscles as such are absent. Instead, a band of longitudinal muscle encircles the oesophagus about halfway along its length; posteriorly the band becomes U-shaped in cross-section and eventually divides into four segments which diverge and attach to the body wall in the region of the oesophageal bulb, similar to the odontostyle retractor muscles of Xiphinema (Text-fig. 6). The function of the muscles in this Longidorus species appears to be the withdrawal of the anterior oesophagus, forming the loop, and thus probably also assists in retracting the odontostyle.

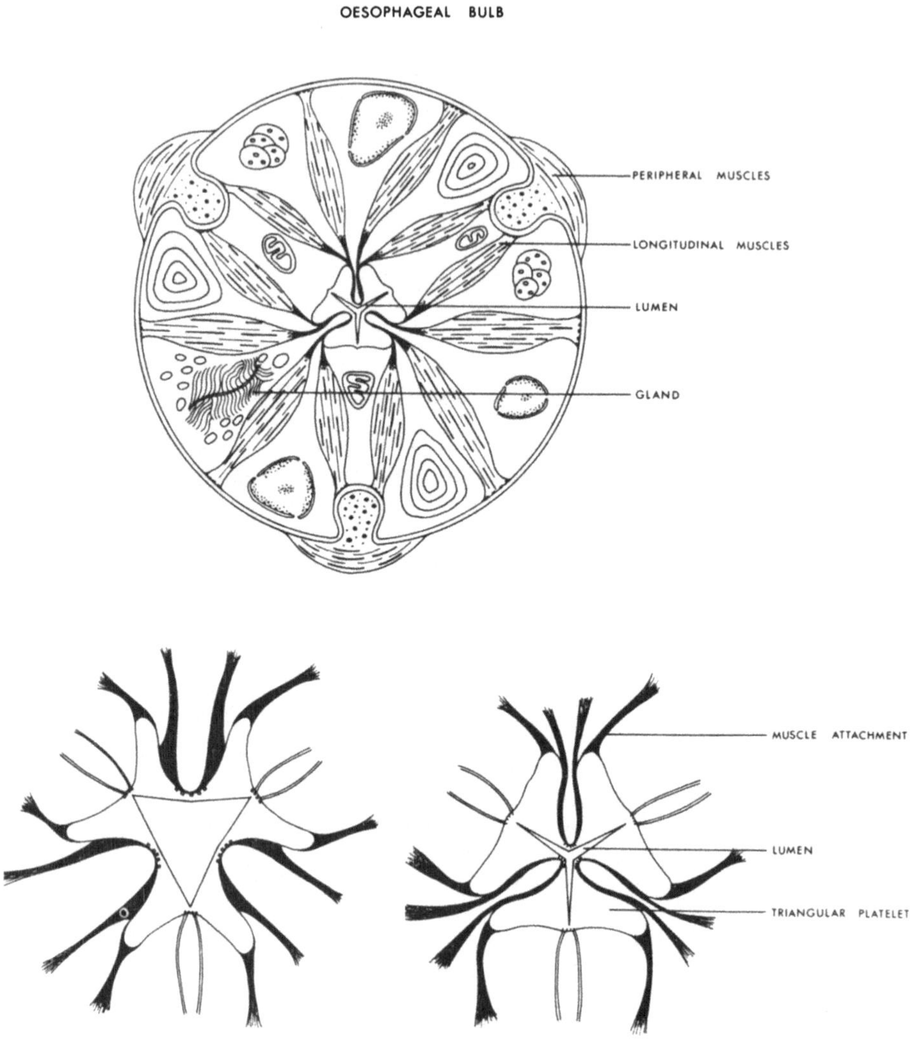

Fig. 5. Diagram of T.S. of oesophageal bulb in Longidorus and Xiphinema.
The lower diagram illustrates the open and closed position of the lumen (see text).

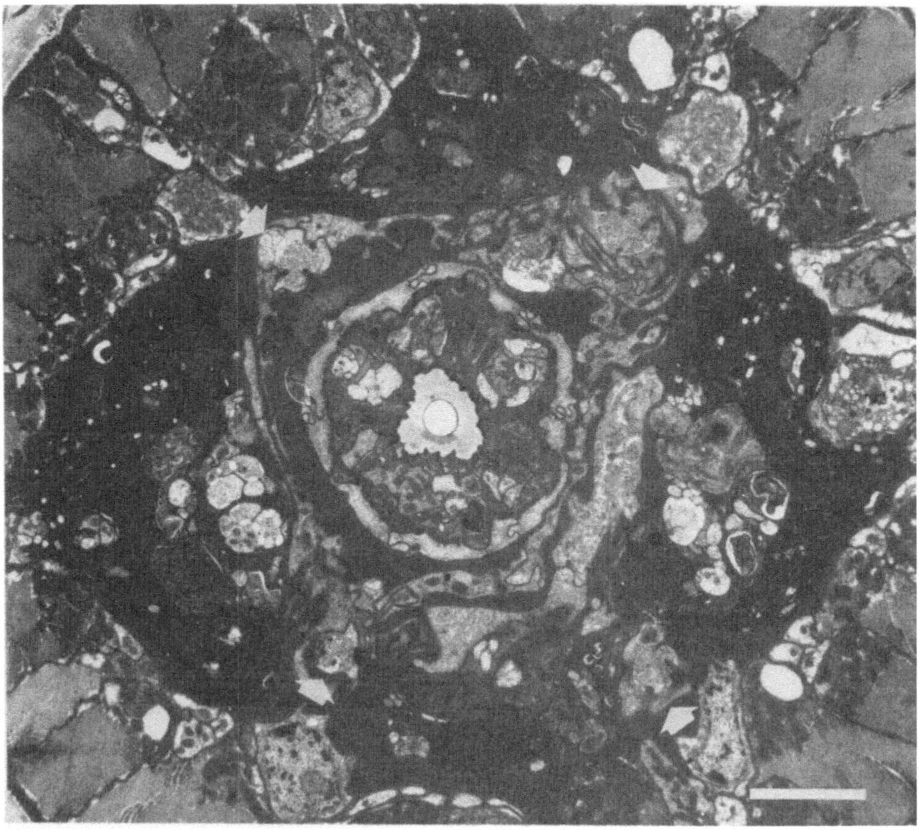

PLATE III.

T.S. of L. attenuatus at the base of the odontophore. Note the four odontophore retractor muscles (arrowed).

Bar represents 2 µm.

In a second group of Longidorus without oesophageal radial muscles, including L. attenuatus, L. goodeyi, L. leptocephalus, L. vineacola, two sets of retractor muscles are present (Text-fig.7). One set surrounds the oesophagus about halfway along its length and is similar to the retractor muscles as seen in L. elongatus. In addition, however, a second set of four retractor muscles extend from the base of the odontophore; these diverge, passing outside the nerve ring, to become attached to the body wall only some 30 μ posteriorly and well anterior to the band of muscle encircling the oesophagus (Plate III, fig.8).

In a third group of Longidorus, including L. macrosoma, L. profundorum, L. caespiticola, L. africanus, L. brevicaudatus, radial muscles are present throughout the length of the anterior oesophagus (as in Xiphinema) and stylet retractor muscles attach to the anterior oesophagus about 20 μ posterior to the base of the odontophore, extending posteriorly inside the nerve ring to the body wall in the region of the oesophageal bulb (Text-fig.8); this is similar to the situation in Xiphinema except for the difference in the anterior attachment to the oesophagus rather than to the base of the odontophore. In this species the contraction of the oesophageal radial muscles is complementary to the contraction of the odontostyle retractors. The radial muscles draw in the peripheral wall of the oesophagus (the lumen of the oesophagus remains open and circular in cross section at all times!) and this allows for the expansion of the contracting retractors and their easier movement through the nerve ring.

ASSOCIATED SENSORY AND NERVE STRUCTURES

In the labial area of Longidorus and Xiphinema are six inner and six outer labial papillae, four cephalic papillae arranged subdorsally and subventrally and two amphids placed laterally. Six nerve trunks innervate these structures, one pair in each of the dorsal and ventral cords and one in each lateral cord. The papillae contain cilia and Longidorus species are typified by having eight microtubule doublettes in the ring whereas Xiphinema have nine. The amphidial openings in Longidorus and Xiphinema are also a distinguish -ing feature of both genera, Longidorus having very small pores and Xiphinema large transverse slits.

The six nerve trunks connect posteriorly with the nerve ring and also carry the nerve connection for the somatic receptors within the body pores. Usually the body pores contain two or three cilia but the number varies between one and seven. As with amphids, the body pores also have secretory tissue associated with them.

Recently it has been found in L. leptocephalus that the central part of the feeding apparatus is linked to the hypodermal cord by

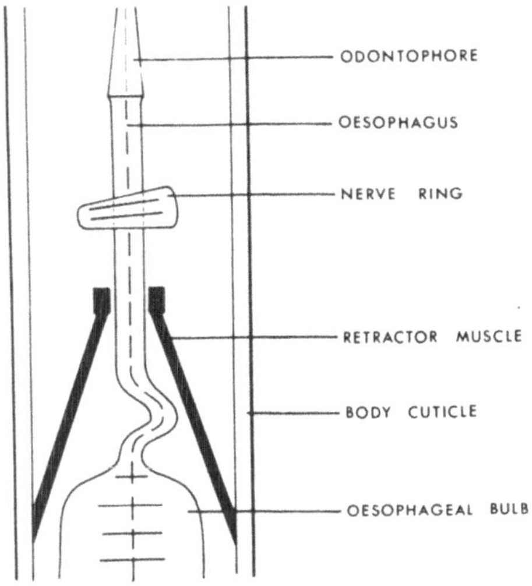

6 L. elongatus

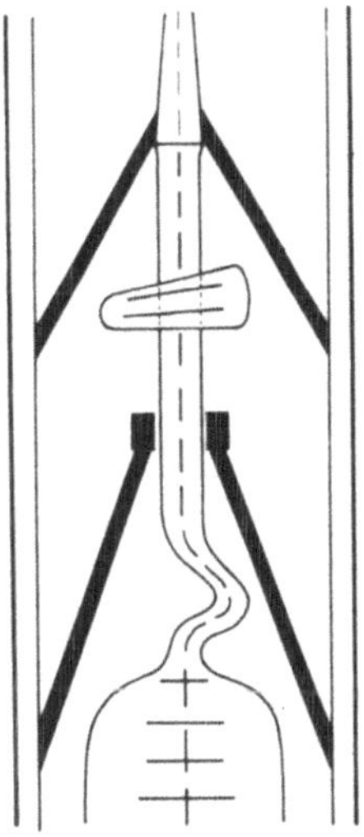

7 L. attenuatus

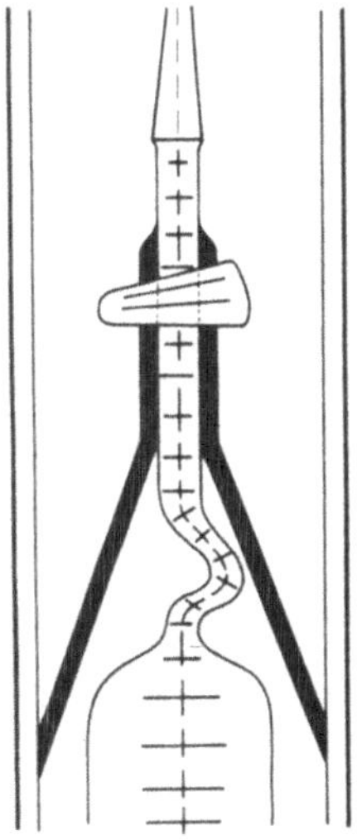

8 L. macrosoma

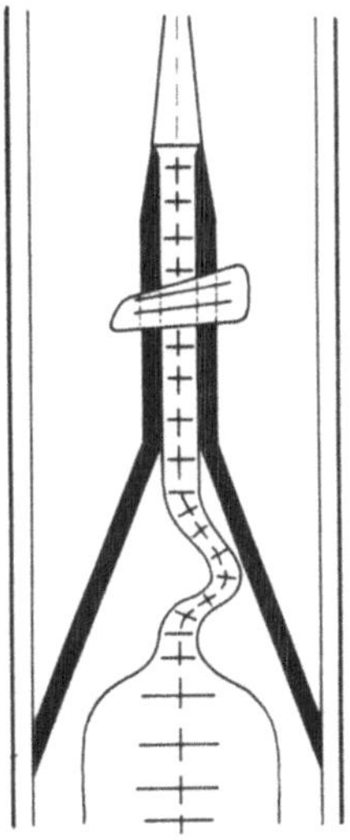

9 Xiphinema spp.

Figs. 6, 7, 8, 9. Diagrams of the retractor muscle arrangements found in Longidorus and Xiphinema species.

five nerve fibres, which have part of their length free in the pseudocoel to allow for odontostyle protraction and retraction. These fibres possibly provide the connection between the nerve ring and two sets of nerve processes in the feeding apparatus which come together in the region of the odontophore. The first set, comprising four small nerve processes, pass forward along the length of the anterior stomodaeum in close association with the support membrane for the odontostyle protractor muscles, to innervate the cheilostome. The second set of nerve processes in the three sinuses of the odontophore are closely associated with pore-like areas which penetrate almost to the food canal. The location and structure of these pore-like innervated areas suggest they have a gustatory function. Such a facility is probably essential for the nematodes accurately to locate their feeding sites deep in plant roots.

The three nerve processes from the sinuses after linking with the other nerves, pass posteriorly, via three ganglia within the central oesophagus, presumably to innervate the radial muscles and salivary glands in the oesophageal bulb (6).

REFERENCES

1. Aboul-Eid, H.Z. (1969). Electron microscope studies on the body wall and feeding apparatus of *Longidorus macrosoma* *Nematologica* **15**, 451-463.

2. Coomans, A. (1964). Stoma structure in members of the Dorylaimina. *Nematologica* (1963) **9**, 587-601.

3. Lopez-Abella, D., Jimenez Millar, F. & Garcia Hidalgo, F.(1966). E. structura submicroscopica del esofago muscular de *Xiphinema* (Nematoda). *Boln R. Soc. esp. Hist. nat.* **64**, 177-185.

4. Lopez-Abella, D., Jimenez Millar, F. & Garcia Hidalgo, F. (1967). Electron microscope studies of some cephalic structures of *Xiphinema americanum*. *Nematologica* **13**, 283-286.

5. Lopez-Abella, D., Jimenez Millar, F. & Marin, M. (1964). Algunas estructuras submicroscopicas de *Xiphinema* y *Peledora*. *Boln R. Soc. esp. Hist. nat.* **62**, 379-384.

6. Robertson, W.M. (1974). Ultrastructure of virus vector nematodes. *Ann. Rep. Scot. Hort. Res. Inst.* 1973. 72-73

7. Robertson, W.M. & Roberts, I.M. (1972). A simple device for the bulk staining and storage of ultrathin sections on grids. *J. Microsc.* **95**, 425-428.

8. Roggen, D.R., Raski, D.J. & Jones, N.O. (1967). Further electron microscope observations of Xiphinema index. Nematologica 13, 1-16.

9. Taylor, C.E. & Brown, D.J.F. (1973). NATO-finance survey (Longidorus and Xiphinema). Ann. Rep. Scot. Hort. Res. Inst. 1972. p.74-75.

10. Taylor, C.E. & Robertson, W.M. (1971). Ultrastructure of the guide ring and guiding sheath in Xiphinema and Longidorus. Nematologica 17 303-307.

11. Taylor, C.E., Thomas, P.R., Robertson, W.M. & Roberts, I.M. (1970). An electron microscope study of the oesophageal of Longidorus elongatus (de Man). Nematologica 16, 6-12.

12. Williams, J.R. . The position of the spear guiding ring in Xiphinema species. Nematologica 12, 467-469.

13. Wright, K.A. (1965). The histology of the oesophageal region of Xiphinema index Thorne and Allen, 1950 as seen with the electron microscope. Can. J. Zool. 43, 689-700.

DISCUSSION

Lippens doubted whether Robertson's opinion of the function of the pores in the sinus walls was correct, and commented that in Aporcelaimus the equivalent tissue to that described in Xiphinema was not nervous, but appeared as lacunae filled with hypodermal tissue. Robertson replied that in Xiphinema and Longidorus the tissue is certainly nervous but the pore like areas differ between the two genera. Coomans suggested that the organs as described may be gustatory, and Robertson commented that this could be so although there was a membrane between the lumen of the oesphastome and the nervous tissue; he referred to the gustatory organ in the food pump of aphids which is associated with deep feeding.

Cotten said that he had observed the elongation of the oesophageal bulb referred to by Robertson but had been unable to detect any change in diameter. Robertson pointed out that the change in diameter of the bulb would at most only be sufficient to accommodate the volume of the lumen and therefore would be very little.

Asked for further information on the movement of the replacement stylet, Robertson explained that in Longidorus during the moult the replacement stylet is moved to its final position by several groups of a type of filament through the ventral sinus of the odontophore which is softened during moulting. The old stylet is trapped by the lips of the old skin and is pulled out as the newly moulted nematode moves backward prior to emergence from the old skin through the split in the region of the oesophagus. Coomans commented that the filaments associated with the replacement stylet represented a new observation, but he had always understood that there was a connection between the wall of the odontophore and the stylet because the new stylet forms in the odontophore sinus and then moves back. Pitcher suggested that the replacement stylet could be moved forward by peristalsis, and both Robertson and Coomans agreed that this could be so.

THE ULTRASTRUCTURE OF THE FEEDING APPARATUS IN TRICHODORIDS

W.M. ROBERTSON and C.E. TAYLOR

Scottish Horticultural Research Institute

Invergowrie, Dundee, Scotland.

INTRODUCTION

Trichodorids (Trichodorus and Paratrichodorus spp.) share with Longidorus and Xiphinema the ability to transmit plant viruses but there have been relatively few electron microscopy studies of its feeding apparatus (1, 2, 3, 4). Like Xiphinema or Longidorus, trichodorids feed by probing plant roots with their stylet. However, the mode of feeding is different between the two nematode groups (5) and this is reflected in observed differences in the ultrastructure of the feeding apparatus.

Methods for fixing and embedding specimens were the same as those used for Longidorus and Xiphinema (see preceding paper by Robertson and Taylor in these Proceedings).

CHEILOSTOME

The cheilostome is formed by an infolding of the body cuticle. For part of its length it is shaped presumably to fit the periphery of the spear during feeding (Plate I; also cf. ref 5). Anteriorly, as seen in longitudinal section, it appears to be deeply invaginated, which suggests that the lips may be everted to make a seal between the nematode and plant root (3). However, this may be an artefact produced by the strong reaction of the cuticle to fixatives used in electron microscopy. The cheilostome has triradiate symmetry and posteriorly is triangular in shape with one point of the triangle indicating the ventral side (Plate I). Posteriorly the cheilostome joins the anterior pharynx, the cuticle of the cheilostome reducing in thickness to form the pharyngeal lumen.

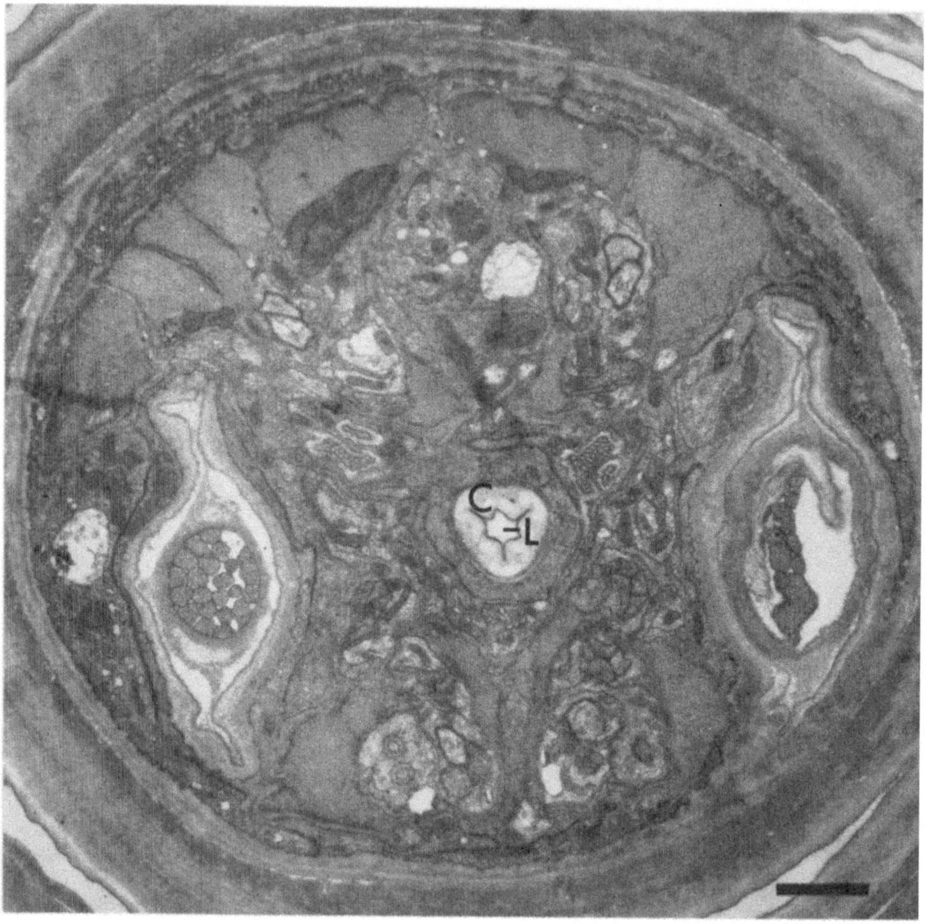

PLATE I.

T.S. through the cheilostome (C) of T. pachydermus showing the triradiate lumen (L). The four muscles compress the cheilostome longitudinally during the feeding cycle.

Bar represents 1 μm.

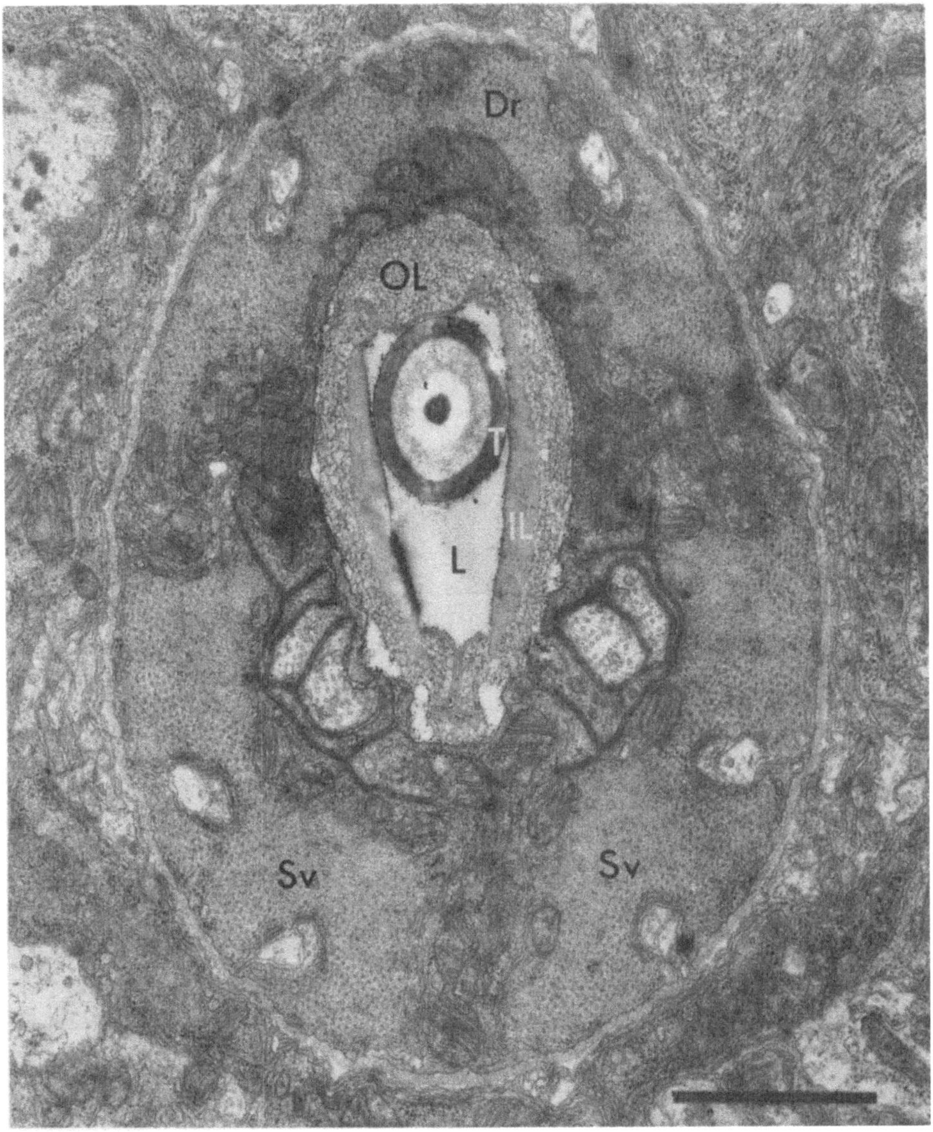

PLATE II.

T.S. through the anterior pharynx showing the tooth (T) inside which lies the posterior onchiostyle. The inner (IL) and outer layers (OL) of the cuticular lining to the lumen (L) can be seen and also the subventral (SV) and dorsal (Dr) protractor muscles attached to the basement membrane.

Bar represents 1 μm.

PHARYNGEAL REGION

The onchiostyle in trichodorids is essentially similar in its basic construction to the odontostyle and odontophore in Longidorus and Xiphinema. The tooth of trichodorids has no opening to the exterior and is completely solid anteriorly. Posteriorly it has a lumen which accommodates the posterior onchiostyle (considered homologous with the odontophore) (Plate II). The posterior onchiostyle is a cuticular structure with a dorsal groove (Plate III) which is evident about one-third of its length from the base and gradually closes over anteriorly to form a single sinus. Posterior to the groove the onchiostyle tapers and protractor muscles are attached dorsally and laterally (Plate IV). The cuticle of the pharynx is fused to the ventral side of the posterior onchiostyle for most of its length. Posterior to the region where the protractor muscles attach to the onchiostyle the cuticle thickens and in cross section is triangular, similar to the anterior oesophageal lumen.

The cuticular pharyngeal wall in trichodorids is complex in shape and structure. It is considered to be equivalent to the guiding sheath of Longidorus and Xiphinema but in these two genera the sheath is mostly detached from the underlying tissue. In trichodorids the guide sheath works in conjunction with the spear to actively pump in plant sap (see Wyss in these Proceedings) whereas in Longidorus and Xiphinema it acts as a seal around the odontostyle. In trichodorids two layers forming the cuticle are clearly identifiable anteriorly. The outer layer, distal from the lumen, is an open matrix of variable thickness (Plate II). This surrounds an inner layer which is not homogeneous in structure throughout its circumference; dorsally and ventrally it is attenuated and invaginated and laterally it is thickened and straight. In well fixed specimens the lateral sides of the cuticle appear to have membranous attachments. Posteriorly, the pharyngeal cuticle fuses with the onchiostyle beginning on the dorsal side and eventually excluding the posterior onchiostyle from the pharyngeal lumen. The posterior onchiostyle is fused to the pharyngeal cuticle on its dorsal side exterior to the lumen for most of its length.

In the anterior pharynx a group of three protractor muscles are situated, one dorsal and two subventral (Plate II). Posteriorly, the contractile parts of the muscles extend around the enlarged circumference of the basement membrane and are separated from the tooth by the sarcoplasm. In the region of the posterior onchiostyle the protractor muscles gradually fuse with the onchiostyle base (Plate IV, Fig. 1). The position of the protractor muscles relative to the peripheral basement membrane is the reverse of that seen in Longidorus and Xiphinema where the muscles are attached to it distally. The protractor muscles act directly on the posterior onchiostyle whereas in Longidorus and Xiphinema they are attached to the

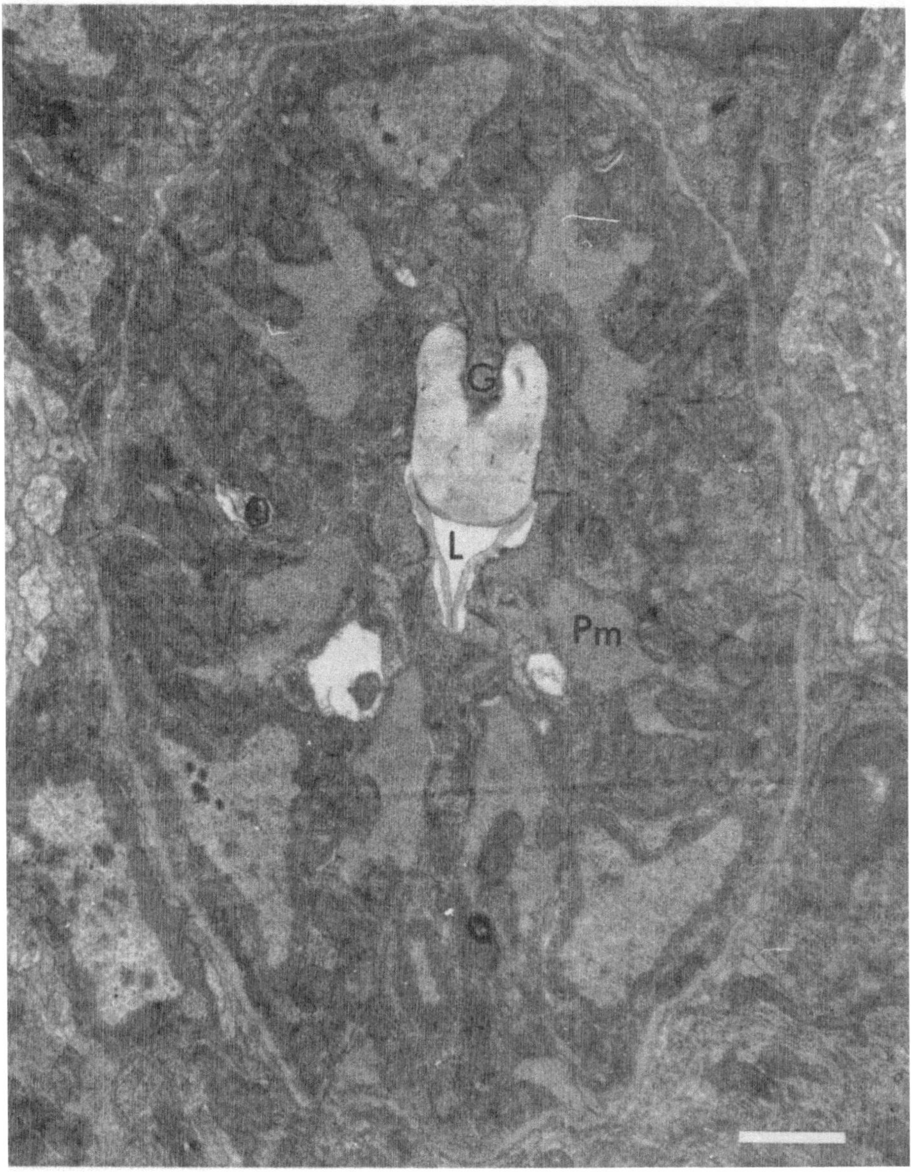

PLATE III.

Section through the posterior onchiostyle showing the dorsal groove (G), the protractor muscles (Pm) and the pharyngeal lumen (L).

Bar represents 1 μm

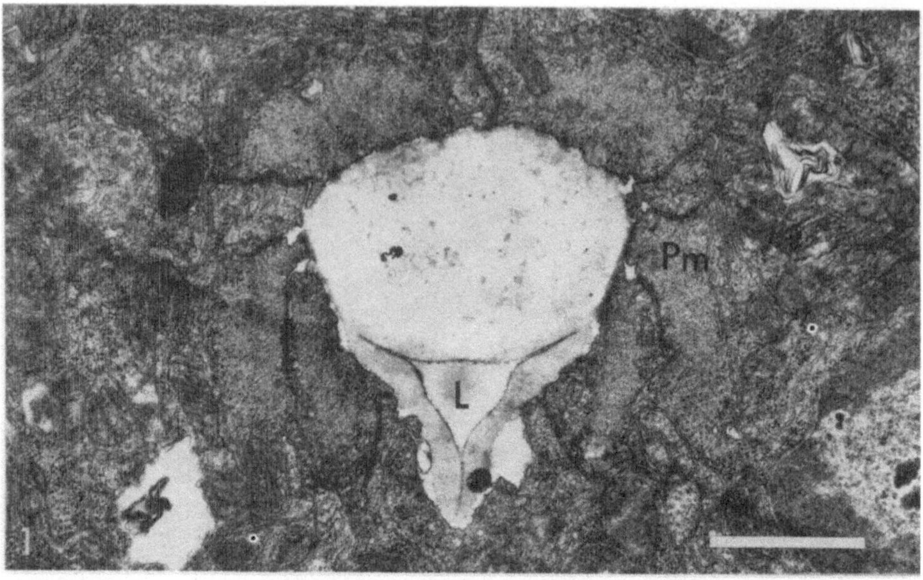

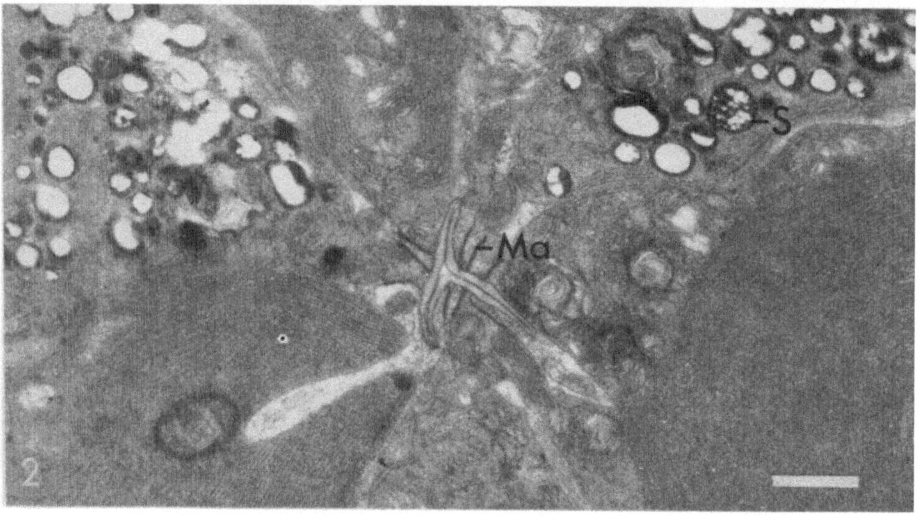

PLATE IV.

Fig. 1. Section through the posterior onchiostyle base showing the protractor muscles attached dorsally and laterally to its base. Pharyngeal lumen (L).

Fig. 2. T.S. through the posterior oesophagus showing the muscle attachment (MA) to the cuticle forming the triradiate lumen. Note the secretory globules (S) produced by the glandular tissue.

Bars represent 1 µm.

basement membrane and a second set of radial muscles transfer the protraction force to the base of the odontophore. No retractor muscles have been found as yet in trichodorids. Probably there is an increase of hydrostatic pressure within the closed system of the pharynx during protraction which at least assists the retraction of the onchiostyle when the protractor muscles relax.

OESOPHAGEAL REGION

The triangular lumen of the oesophagus is opened by the action of radial muscles throughout its length whereas in *Longidorus* and *Xiphinema* the anterior oesophageal lumen is circular and remains open (Plate IV, Fig. 2). The oesophageal lumen enlarges to form the 'cardia' (5), in the posterior lobe of the oesophagus which overlaps with the intestine. The difference between the oesophagus and the oesophageal bulb is not as well defined as in *Longidorus* or *Xiphinema*. In trichodorids the anterior oesophagus is slender to enable it to pass through the nerve ring, but posteriorly it gradually enlarges to form an oesophageal bulb. In *Longidorus* and *Xiphinema* the anterior oesophagus is also slender and passes through the nerve ring but is longer to allow almost the entire length of the odontostyle to be protracted. In some trichodorid species the anterior oesophagus forms a slight loop to allow onchiostyle protraction.

REFERENCES

1. Bird, G.W. (1971). Digestive system of *Trichodorus porosus*. *J. Nematol.* **3**, 50-57.

2. Hirumi, H., Chen, T.A., Lee, K.J., & Maramorosch, K. (1968). Ultrastructure of the feeding apparatus of *Trichodorus christiei*. *J. Ultrastructure Res.* **24**, 434-453.

3. Raski, D.J., Jones, N.O. & Roggen, D.R. (1969). On the morphology and ultrastructure of the oesophageal region of *Trichodorus allius* Jensen. *Proc. helminth. Soc. Wash.* **36**, 106-118.

4. Taylor, C.E. & Robertson, W.M. (1970). Location of Tobacco Rattle Virus in the nematode vector *Trichodorus pachydermus* Seinhorst. *J. Gen. Virol.* **6**, 179-182.

5. Wyss, U. (1971). Der Mechanismus der Nahrungsaufnahme bei *Trichodorus similis*. *Nematologica* **17**, 508-518.

DISCUSSION

Wyss queried the function of the ventral rectractor muscles in Trichodorus. Robertson said he visualised the pharynx as a closed sac; protraction of the muscles lying on the periphery would tend to increase the internal pressure whilst their relaxation would lead to retraction of the stylet. Wyss offered the view that retrac -tion could be caused by pressure around the muscles in the pseudo-coelome - as the muscles increase in size they would exert pressure in the pseudocoelome, and the stylet would retract when the muscles relaxed. Coomans commented that the body cavity in Trichodorus is relatively large and therefore any pressure could probably be dispersed, but Roggen felt there could be truth in both views.

In a discussion on fixation of Trichodorus, Robertson said that the osmotic pressures of the fixatives normally used cannot be adjusted sufficiently well to avoid undesirable effects. He suggested that investigation would be worthwhile and that there may be critical values in fixation for Trichodorus as seen with avian tissues.

FEEDING OF TRICHODORUS, LONGIDORUS AND XIPHINEMA

URS WYSS

Institut für Pflanzenkrankheiten und Pflanzenschutz der

Technischen Universität Hannover, BRD

INTRODUCTION

The discovery of plant root ectoparasitism by nematodes is as recent as 1951, when Christie and Perry (2) experimentally proved that a *Trichodorus* sp. (later named *T. christiei*) produces stubby-root symptoms on the roots of beet and sweet corn. Ten years later, the genus *Trichodorus* again attracted much attention when, in the same year, three species (26) were shown to be vectors of tobacco rattle virus. Direct damage to plant roots, caused by an ectoparasite of the family Longidoridae (*Longidorus*, *Paralongidorus* and *Xiphinema* spp.), in this case *Longidorus* sp., was first recorded in 1954 (14). Three years later, Schindler (24) gave a detailed account of the pathogenicity of *X. diversicaudatum*. In 1958 it was first dis-covered that a nematode of the genus *Xiphinema* can transmit a virus disease in plants (12). This discovery considerably encouraged the search for nematode vectors of soil-borne plant viruses. To date several nematode species, all belonging to the dorylaimoid genera *Trichodorus*, *Longidorus* and *Xiphinema*, are established vectors of plant viruses.

Detailed observations of the feeding of nematode vectors of plant viruses will answer numerous questions by providing information on subjects such as: orientation responses to plant roots, preferred feeding regions, exploration and selection of stylet penetration sites, feeding phases, minimum feeding periods for virus acquisition and inoculation, morphology and function of the feeding apparatus as well as responses of root cells and tissues to the feeding.

This paper outlines the present knowledge of the feeding and its effects in the genera known to be vectors of plant viruses. It will become evident that information is still fragmentary and requires considerable amplification.

GENUS TRICHODORUS

Feeding Behaviour

When the feeding behaviour of Trichodorus viruliferus on white extending apple roots was observed in an underground laboratory, it was noticed that up to 100 or more nematodes followed in the region of the elongation zone, 1-3 mm behind the root tip, while the root continued to grow. When nematode numbers increased rapidly, a marked decline in root growth usually occurred; some of the nematodes moved away and the remaining ones finally transferred their attack to the apical region. When root growth had ceased, the nematodes dispersed into the soil (17).

Nematode aggregation around the tips of growing roots appears to be common for Trichodorus spp. Chemicals, secreted by microorganisms surrounding the roots, are unlikely to be the attractive agents, as root tip aggregation was also observed for Trichodorus spp. on seedling roots in sterile agar culture (32). On two of the plants examined, Nicotiana tabacum 'Samsun' and Fragaria vesca var. semperflorens, the nematodes fed at first irregularly along the root-hair region of rapidly growing tap roots. They then gradually moved to the zone of elongation, where nematode numbers and feeding intensity were most pronounced. A massive attack in this region led to a drastic reduction of the root's growth rate. Most of the nematodes then left the root in search of other feeding sites, usually emerging lateral roots, which, when attacked by several nematodes, stopped growing within one or two days. It appeared that the nematodes were able to locate the emerging root initials through orientation responses and therefore did not accidentally find the new feeding site.

A correlation between feeding intensity on growing root tips and egg production probably exists. In sterile agar culture, T. similis reproduced on seedling roots of N. tabacum and F. vesca, the tips of which were without exception attacked intensively. T. similis did not, however, reproduce on seedling roots of Brassica rapa var. silvestris, where the nematodes fed along the root hair region and did not show the phenomenon of root tip aggregation (32). Pitcher and McNamara (19) concluded that root tip colonies constitute the main breeding site of T. viruliferus on mature apple trees. From observations on the feeding behaviour of T. christiei on tomato roots in agar culture, Högger (13) concluded that young tissues, such as actively growing root tips, provide the most suitable feeding sites. Having observed the massing of T. similis around the root tips of plants that induced reproduction, Wyss (32) suggested that the undifferentiated cells in the elongation zone of root tips probably constitute the most suitable nutrient medium for egg production. However, when single females of T. similis were kept on seedling roots of N. tabacum in sterile agar culture, it became evident that feeding was usually confined to epidermal cells and root hairs 0.5–2.0 nm behind the zone of elongation. In some cases females were seen to follow the growth of tap roots for up to 3 weeks within this region, producing an average of six eggs per week (Wyss, unpublished). All larval stages fed along the roots. Newly hatched larvae usually required several perforation attempts, before the puncture of a cell wall was accomplished.

Feeding Duration

The feeding period spent by T. similis females on an epidermal cell or on a root hair of seedling roots of B. rapa, F. vesca and N. tabacum in agar culture rarely exceeded 6 min. Quite often only a few seconds elapsed between the nematode's withdrawal from one cell and the resumption of feeding activity on a new or occasionally also on the same cell in close vicinity of the former feeding site. After several successive feeds, periods of inactivity lasting up to 20 min were common. In a few cases, young females were seen to ingest the contents of up to 15 cells (epidermal cells and root hairs) within 1 h feeding spells (32). The feeding periods of T. pachydermus, T. primitivus and T. sparsus were similar to those of T. similis and they correspond with the data given for T. christiei (1, 23, 37). It thus appears that the feeding duration of Trichodorus spp. on individual cells is usually only a few minutes. T. proximus is probably an exception, as feeding periods on seedling roots of St. Augustine grass (Stenotaphrum secundatum) varied considerably from a few minutes to up to 170 min (20).

Feeding Phases

A complete feeding cycle by T. similis is composed of five phases: exploration, perforation of the cell wall, salivation, ingestion and withdrawal from the feeding site (33). These phases could be recognized quite easily at low power (X 125) by observing the stylet thrusts, which were continuous throughout feeding. At the beginning of cell wall perforation the rate of stylet thrusts varied between 4–8 thrusts/s. This rate soon decreased and dropped to about one thrust per second towards the end of salivation. During ingestion the rate of stylet thrusts increased again to about 2/s. A similar rhythm of stylet thrusts was also observed for T. christiei (37). Based on 30 observations of completed feeding cycles, the mean total feeding time for females of T. similis feed- on epidermal cells and root hairs of B. rapa var. silvestris seedling roots was 3 min 07 s. The perforation of the cell wall comprised 24%, salivation 59% and ingestion 17% of this time (33). Identical observations on seedling roots of N. tabacum showed that the perforation of the cell wall was 11%, salivation 69% and ingestion 20% of the mean total feeding time of 3 min 25 s (n=30; Wyss, unpublished).

Observations of the feeding of T. similis females on epidermal

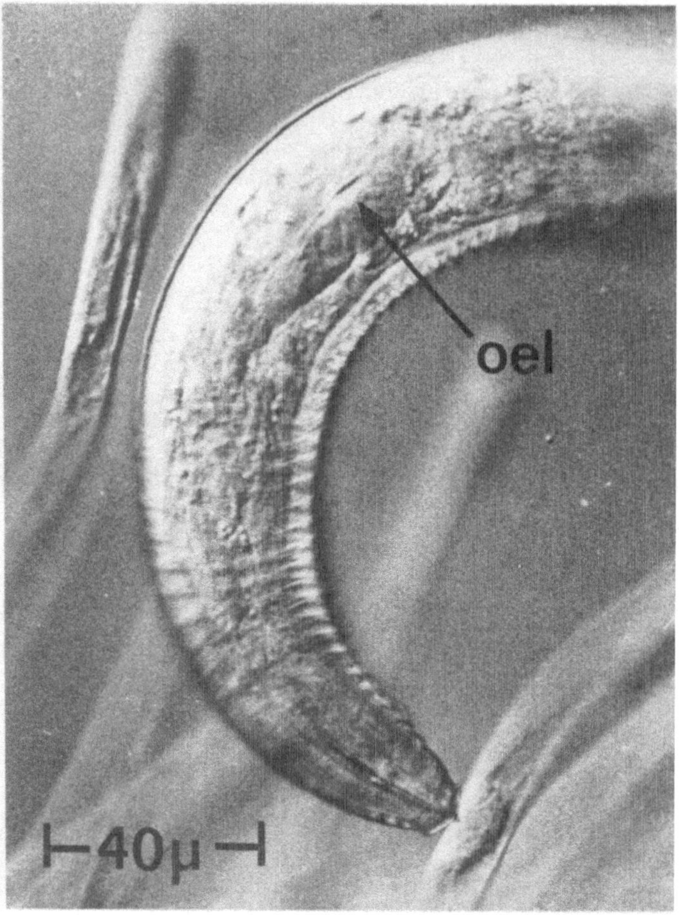

PLATE I: T. similis (female) feeding on root hair of Brassica rapa cv. silvestris. Oesophageal lumen (oel) in basal bulb dilated during feeding phases (here salivation).

PLATE II: T. similis (female) feeding on root hair of Nicotiana tabacum (A-C). A: 35s after cell wall perforation; nucleus (n) still finely granulated. B: initiation of ingestion, 110s after cell wall perforation; stylet deeply inserted into accumulated cytoplasm (cp), lining of pharyngeal lumen (pl) folded, nucleolus (nc) moves in swollen nucleus. C: withdrawal from feeding site, 135s after cell wall perforation; note feeding tube (ft) formed along wall of stoma (st) remains (attached to plant cell). D: Section of two parasitized epidermal cells of N. tabacum, showing feeding tubes (ft), coagulated cytoplasm (cp), and the disorganization of the nucleus (n).

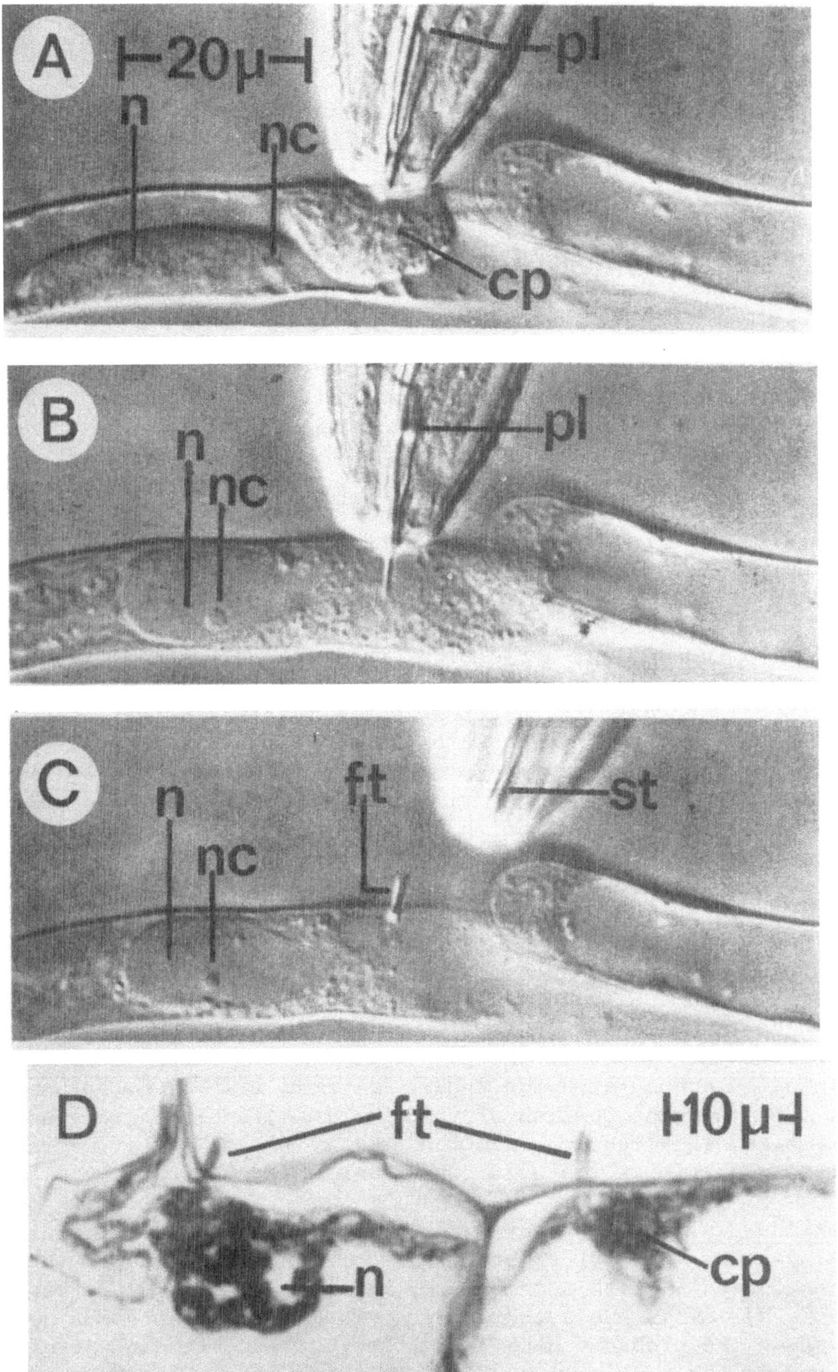
A
20μ
pl
n
nc
cp
B
pl
n
nc
C
n
ft
st
nc
D
ft
10μ
cp
n

cells and root hairs of seedling roots of B. rapa var. silvestris at high magnification (X 2000), using special observation chambers, Nomarski interference microscopy and cine film analysis (33, 34), revealed the following features:

Exploration. Characteristic for exploration is the side-to-side rubbing with the lips over a small area of the cell wall against which the lips are pressed. Usually no stylet thrusts occur during exploration.

Perforation of the cell wall. Having found a suitable spot for stylet penetration, the lips are pressed firmly, and most commonly at right angles, against the cell wall. Immediately before the first stylet thrust occurs, the stoma is drawn forward by special protractor muscles and brought into close contact with the cell wall. Perforation is then initiated by rapid stylet thrusts of about 6/s. After a few seconds the thrust rate gradually decreases. The stylet tip is always directed solely at a point only and does not perform a rasping action as mentioned for T. christiei (22, 23). Quite often perforation is abandoned before the stylet tip has penetrated the cell wall. In such cases a different site on the same cell or on new cells is then explored.

Salivation. After perforating the cell wall, the rate of stylet thrusts falls slowly to about 1/s. Throughout salivation the stylet tip is repeatedly inserted approximately 2 μ deep into the cell and the cytoplasm accumulates into a large mass around the feeding site. In contrast to a previous statement (33), salivary secretions are definitely not visible within the punctured cell. This was confirmed, using Smith interference microscopy. It is now evident that the sudden accumulation of cytoplasm at the perforation site had been misinterpreted as the release of a granular salivary globule. Probably Chen and Mai (1) had been misled in a similar way, when they reported a viscous substance being released by T. christiei into individual, isolated corn root cells.

Ingestion. Ingestion is always initiated by several slow and deep stylet thrusts, with the tip of the stylet being inserted about 8 μ into the accumulated and apparently more viscous cytoplasmic mass. Then a large portion of the cytoplasm, including organelles of any size, is withdrawn from the cell with each stylet retraction. The rate of stylet thrusts increases again to about 2/s.

Withdrawal. After nearly all of the accumulated cytoplasm has been ingested, the stylet thrusting stops and the stoma is then detached, usually with some effort, from the feeding tube (Plate II, figs. C, D). This tube, formed along the wall of the stoma during salivation and probably also during perforation, remains firmly anchored in the cell wall for an indefinite time. The tube, which functions as a suction tube (the stylet possesses no lumen), is most likely formed by hardened saliva rather than by any constituents of the parasitized cell. The tube is also formed within the stoma, when occasionally dead cells with coagulated cytoplasm are perforated and it extends into such cells around the stylet, when, under these conditions, the stylet is repeatedly inserted deeply into the cell for at least 15s. Zuckerman (37) noticed that the hole made in the cell wall by T. christiei appeared coated with a refractive substance which was probably secreted by the nematode.

Analysis of the Feeding Mechanism

The feeding mechanism of T. similis females was analysed by using frame by frame projection of cine film material (33). Stylet protraction is achieved by contraction of the onchiostyle protractor muscles, whereby the posterior part of the onchiostyle slightly shifts ventrally and the oblique slit in the middle region of the onchiostyle expands. The lining of the pharyngeal lumen, fused dorsally and also laterally to the onchiostyle about 18 μ behind the oral aperture where it forms the so called 'guide ring', is drawn forward as a fold (Plate II, fig.B) on each stylet protraction. The tip of the stylet remains inside the cell for a fraction of a second before it is retracted (Text-fig.1). Increased turgor pressure in the pharyngeal wall probably causes the retraction of the stylet upon relaxation of the protractor muscles. The retraction of the stylet is faster than its protraction (Text-fig. 1).

It became evident that contraction and relaxation of the onchiostyle protractor muscles and the oesophageal lumen muscles in the basal bulb of the oesophagus are simultaneous during all feeding phases. When the stylet is protracted, the oesophageal lumen is dilated about 10 μ in front of the oesophageo-intestinal valve (Text-fig. 1, Plate I). Upon stylet retraction, the lumen quickly narrows again (Text-fig. 1). The oesophageal lumen in front of the oesophageo-intestinal valve is dilated only during ingestion, when the contents of the food cell are forced into the intestine. The following feeding mechanism, based on a continual alternation of high and low pressure between the pharyngeal lumen and the oesophageal lumen in the basal bulb, appears to be functional:

Salivation. Upon protraction of the stylet, the pressure in the pharyngeal lumen increases, as the lining of the pharyngeal lumen is also drawn forward. At the same time, salivary secretions are probably pressed into the oesophageal lumen when the lumen muscles in the basal bulb contract and dilate the oesophageal lumen. Upon stylet retraction and simultaneous relaxation of the oesophageal lumen muscles, the increased pressure in the narrowing oesophageal lumen would force the secretions into the pharyngeal lumen, which is now under low pressure. The secretions are then injected into the cell when the onchiostyle protractor muscles next contract.

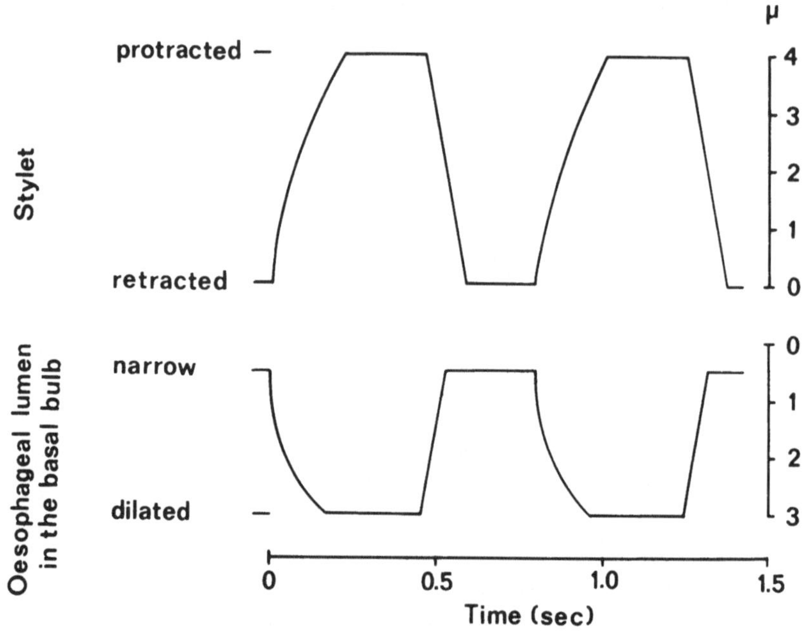

Fig. 1. Contraction and relaxation during salivation of the onchiostyle protractor muscles and the oesophageal lumen muscles in the basal bulb. Measurements were taken on the lining of the pharyngeal lumen, where it forms the so called 'guide ring'. The values on the y-axis thus do not represent depth of stylet penetration.

Ingestion. Upon retraction of the stylet, cell contents are sucked into the pharyngeal lumen. From there the contents pass into the oesophageal lumen when the stylet is protracted and the oesophageal lumen muscles contract. During subsequent stylet retraction the cell contents are forced through the oesophageo-intestinal valve into the intestine when the oesophageal muscles in the basal bulb relax. This ingestion mechanism is clearly demonstrated in the Encyclopaedia Cinematographica film E 2045 (35).

Visible Damage of Parasitized Roots

Visible damage of parasitized roots, caused by the gregarious feeding of *Trichodorus* spp., is revealed in the decline of root growth and finally in its cessation, when the nematodes transfer their attack to the apical meristem. New lateral root tips are then attacked and they become stubby and turn brown or black.

Russell and Perry (23) suggest that stubby-root production may result from the injection of oesophageal secretions by the nematodes into the meristem. This would explain how low populations of T. christiei produce severe stubby-root symptoms. However, as is the case in F. vesca var. semperflorens, root growth is sometimes resumed after Trichodorus spp. (predominantly T. similis), that had access to the meristem, have left the root (32). This indicates that cell division in the apical region is not always irreversibly stopped when the nematodes feed there.

Damaged areas show a distinct superficial discolouration (17, 20). Feeding under sterile conditions also produce superficial browning and the collapse of epidermal cells and inhibits root hair production on seedling roots of B. rapa var. silvestris, F. vesca var. semperflorens and N. tabacum (32). On heavily parasitized root tips epidermal cracks may be formed that extend into cortical cells, exposing these cells to nematode feeding, as observed in the host-parasite relation between apple roots and T. viruliferus (17).

Cellular Response to Feeding

In order to understand the processes of virus transmission, it is not only necessary to obtain an insight into the feeding mechanism of the virus vector, but also into the response of the cell to feeding. Preliminary observations confirmed that the cellular response to feeding in epidermal cells is identical to that of root hairs, in which the reactions can be seen more clearly. Therefore the cell's response to feeding by the nematode species T. similis was filmed in root hairs of N. tabacum 'Samsun', a good host of tobacco rattle virus, using Smith interference microscopy and time lapse techniques (35).

Reaction of the cell's protoplast during feeding. After cell wall perforation, the cytoplasm is drawn from all directions to the site of stylet penetration, where it accumulates into a large mass (Plate II, fig A). When lying nearby, the nucleus is drawn together with the cytoplasm to the feeding site. The nucleus swells on its passive migration and within less than 2 min appears optically empty (Fig. 3B). The nucleolus, whose size is not affected, then shows Brownian movement inside the apparently liquified nucleoplasm. After ingestion has been initiated by deep stylet thrusts (Plate II, fig. B), a large amount of the accumulated cytoplasm is sucked into the pharyngeal lumen with each stylet retraction. The nucleus is also ingested whenever the nuclear membrane is perforated by the tip of the stylet.

Reaction of the cell's protoplast after feeding. After the nematode's withdrawal, the feeding tube (Plate II, figs. C, D), anchored in the cell wall indicates the feeding site. Cyclosis is not resumed whenever parts of the accumulated cytoplasm are ingested.

Cytoplasm, left inside the cell, is still drawn to the feeding site where it coagulates. Coagulation is fast and spreads over the entire cell. In root hairs, fed upon at their base, the nucleus is sometimes pulled by fine cytoplasmic threads to the former feeding site. The nucleus swells as before on its migration and becomes disorganized within a few minutes.

In a few instances normal cytoplasmic streaming is resumed in cells that have been punctured by the nematode. This occurs, when the nematode abruptly abandons its feeding activities, soon after the stylet has perforated the cell wall, without ingesting any of the accumulated cytoplasm. When this happens, a small plug of gelled cytoplasm persists for some time at the feeding site, but the cytoplasm continues to stream past it.

Cytoplasmic streaming most likely plays an important part in the translocation of virus particles within plant cells, where the cell-to-cell spread is assumed to take place via the plasmodesmata. It is therefore conceivable, but probably difficult to prove, that a successful virus transmission is only possible under conditions of interrupted feeding.

It remains to be determined whether the reactions observed within parasitized cells result from injections of salivary secretions or simply from mechanical injury or, most probably, from a combination of both. It is known that in plant cells cytoplasmic streaming and the migration and changes in the volume of the nucleus can be influenced purely by mechanical stimuli.

Histopathology of Damaged Roots

Thin sections (0.5–1.9 μ) of seedling roots of *N. tabacum*, damaged by *T. similis*, show that the nucleus, when not removed by the nematode, remains hypertrophied in parasitized epidermal cells and root hairs. The irregular structures within the 'empty' nucleoplasm stain red with Feulgen stain and are thus probably coagulated nucleo-proteins. The volume of the nucleolus is not affected. Furthermore, sections reveal that the stimulus responsible for the disorganization of the nucleus can permeate the subepidermal cell layer, reaching cells that are not directly exposed to nematode attack. This pathogenic effect does not, however, spread deeper into cortical cells. This is probably prevented by pronounced wall thickening of the cells in vicinity of parasitized or indirectly affected cells. The sections also show that epidermal cells, as well as outer cortical cells, can collapse considerably under a massive attack by the nematodes. When this happens, feeding tubes sometimes extend into cortical cells to a depth of two to three cell layers. This indicates that, under the conditions of gregarious feeding, the nematodes are able to gain access to cortical cells without the necessity of root crack formation (Wyss, unpublished).

Sections of parasitized tobacco root tips neither show a definite meristem nor an elongation zone. As in tomato roots, parasitized by T. christiei (22), the differentiation of vessel elements is close to the apex of the root.

GENERA LONGIDORUS AND XIPHINEMA

Information on observation of feeding by Longidorus and Xiphinema on roots in agar culture is still scarce and incomplete. This is probably due to the difficulties of keeping members of the family Longidoridae under such conditions. It appears that Xiphinema spp. are more suitable as a subject for these investigations than Longidorus spp. as some relatively detailed descriptions of the feeding activities of X. bakeri (25), X. brevicolle (3), X. diversicaudatum (9), and X. index (3,5,9) are available. The feeding of only two Longidorus spp. has so far been observed in agar culture (3, 31). Knowledge on this subject is, however, still very fragmentary.

General Feeding Behaviour

Fisher and Raski (9) gave an interesting account on the feeding of X. index on seedling roots of Vitis vinifera, growing in agar culture. Newly hatched larvae fed on the outer cortical cells of the piliferous region. Shortly before moulting they attacked the root tip, where small galls were produced. The next stage larvae migrated back and forth from the piliferous region to the root tip, whereas the older larval stages and the females always fed there, adjacent to the meristematic cells. When X. bakeri was added to seedling roots of different coniferous plants, it at first fed at random along the roots. One week later, however, most of the nematodes had transferred their feeding activities to the root tip (25). A root tip preference was also shown by X. index on herbaceous plants, such as Brassica oleracea, Solanum dulcamara and Lycopersicon esculentum. Sometimes aggregations of nematodes, feed ing at the same root tip, were noticed (5). In contrast to these observations, X. brevicolle and X. index fed at various sites along the roots of seedlings of Vitis vinifera, Urtica urens and Bidens tripartita. The nematodes were only occasionally seen to feed near the tip (3). It is thus evident, that Xiphinema spp. are not necessarily purely root tip feeders.

Longidorus spp., however, show a distinct preference for the root tip and induce characteristic terminal galls on the roots of many herbaceous plants. In agar culture, L. africanus (3) and L. elongatus (31) both inserted their stylets into the tips of seedling roots of B. tripartita, resp. F. vesca var. semperflorens. In observation boxes, L. attenuatus was seen to feed on the tips of lateral roots of sugar beet, causing small galls (30).

Feeding Duration

Longidorus and Xiphinema spp. possess a long stylet resembling a hypodermic needle, which can be inserted deeply into the root tissue. When the diameter of the root is sufficiently small, the tip of the stylet is able to penetrate into the stele. According to Pitcher and Posnette (18), vascular feeding by X. diversicaudatum is only possible in fine feeder roots with a diameter of 250 μ or less. Deep stylet penetration into the root probably explains why Longidorus and Xiphinema spp. have much longer feeding periods at one site than Trichodorus spp. On S. dulcamara, X. index remained for 26 h at the same feeding site (5). Feeding periods of X. brevicolle lasted several hours and could extend to 3 days (3). In X. bakeri, oesophageal pulsations were seen to last 24 h in two extreme cases (25).

Feeding periods of L. africanus on B. tripartita and U. urens were relatively short and did not exceed 15 min (3). L. elongatus, however, fed longer on seedling roots of F. vesca var. semperflorens and when the stylet had been pushed deeply into the root tip, oesophageal bulb pulsations were observed to last up to 70 min (31). According to Fritzsche and Hofferek (11), who studied L. macrosoma feeding in soil on various plants, the nematodes remained at one feeding site for a period from a few minutes to as long as 4 days.

Feeding Phases

It is not yet possible to give a detailed description of the feeding phases of Longidorus and Xiphinema spp., as information is still scarce. Sutherland (25) saw X. bakeri move its stylet to and fro during exploration. The tip of the stylet, however, never projected beyond the oral aperture. Dorsoventral twisting was a characteristic feature of stylet movement during this phase. The wall of an epidermal cell was perforated by several short stylet thrusts, which again showed a slight amount of dorsoventral twisting; the stylet was then pushed steadily into the cell. After a short period of inactivity, pulsations in the oesophageal bulb became evident. These pulsations were not continuous, periods of activity alternated with periods of inactivity. When the nematode had stopped feeding on a given cell, the stylet then punctured the next deeper cell. Feeding was concentrated on a column of cells, progressing from the epidermis into deeper tissues. According to Fisher and Raski (9), the stylet of X. index is not limited merely to straight penetration. The tip of the stylet can also be deflected into adjacent cells.

Details of salivation do not seem to have been published, but it probably occurs after the cell wall has been perforated. Fisher and Raski (9) gave a detailed example of the feeding of X. diversicaudatum on rose roots. They noticed that intermittent contractions of the oesophageal bulb followed cell wall perforation. After a few minutes these intermittent contractions, which are probably characteristic of salivation, were followed by continuous contractions, probably indicating ingestion. During ingestion, muscular contractions of the oesophageal bulb in Xiphinema spp. appeared to perform a peristaltic action in order to force cell contents into the intestine (5, 25). The basal bulb of X. index elongated as its lumen was dilated (5). This was also observed in first stage larvae of L. elongatus inside the egg prior to hatching (36). Oesophageal pulsations were then continuous, although irregular and lasted up to 20 min. The rate of contractions varied between 75-95 per min. Observations at high power (X 1500) revealed the ingestion of fluids during these spells of pulsation. Film analysis showed that the bulb became stretched by the contraction of the bulb muscles when the oesophageal lumen was dilated. When the muscles relaxed, the bulb shortened again and the narrowing of the lumen, which was associated with the shortening of the bulb, ran in the direction to the intestine, thus pressing fluids through the oesophago-intestinal valve.

After feeding, X. brevicolle withdrew its stylet from the root tissue without retracting it. X. index, however, retracted its stylet while still in the feeding position (3).

Visible Damage and Histopathology of Parasitized Roots

Only a brief outline is given, as the pathogenicity of Longidorus and Xiphinema spp. is dealt with in detail by Cohn in these Proceedings.

Xiphinema spp. Darkening of the damaged areas and breakdown of cortical cells are commonly observed, when roots are parasitized by Xiphinema spp. along their length. Cross sections of grape roots, attacked by X. index, showed the formation of a layer of phellogen tissue, 3-5 cell layers beneath the directly affected cells (4). On Petunia roots the feeding site of X. diversicaudatum stained intensively red to violet with safranines, indicating that mucins were probably secreted by the nematode (10).

No symptoms were produced when newly-hatched larvae of X. index fed in the piliferous region of V. vinifera seedling roots, but when these larvae fed at the root tip, small galls were formed. Terminal galls on rose roots, induced by X. diversicaudatum, showed cortical hyperplasia. Furthermore hypertrophied cortical cells with thickened walls and granular cytoplasm were associated with feeding sites. In infected roots, meristematic activity was retarded and vascular differentiation extended far into the root tip (6).

Longidorus spp. Longidorus spp. are apparently genuine root tip feeders (4). Root tips of the herbaceous plant Bidens tripartita, if parasitized by L. africanus, became galled and showed cortical hyperplasia. The meristem was inactivated and vascular differentiation extended towards the root tip. Lateral roots were often initiated close to the tip. Cessation of root growth and swelling of the root tip became evident 24 h after the nematode stopped feeding (4). A similar reaction also occurred in F.vesca var. semperflorens, when a single pre-adult larva of L. elongatus fed near the meristem. An irreversible cessation of root growth and a swelling just behind the elongation zone became evident 24 h after inoculation (31). In celery, swollen root tips, resulting from the feeding by L. elongatus, did not exhibit pronounced cortical hyperplasia. Sections of these galls showed clusters of hypertrophied cells with dense cytoplasm and enlarged nuclei near necrotic cells, where feeding probably occurred (Wyss, unpublished).

Histochemical investigations of terminal root galls of B. tripartita, induced by L. africanus, gave evidence of an increase in simple sugars and inulin, which could be attributed to the injection of enzymes that enhanced the hydrolysis of polysaccharides and the biosynthesis of inulin (8).

GENERAL CONSIDERATIONS

Trichodorus spp. can acquire and inoculate the viruses they transmit, when they feed on epidermal cells, root hairs and, less frequently, also on the outer cortical cells. Longidorus and Xiphinema spp. seem to be able to perform this function in any root tissues, as their stylets are long enough to pierce the stele of small rootlets. It appears, however, that Longidorus spp. can acquire and transmit viruses mainly in undifferentiated cells of the root tip. Feeding in this region induces the formation of tip galls in a similar way as described for those nematode species of the suborder Tylenchina, which also possess a long needle-like stylet, for instance Hemicycliophora arenaria (15) or Dolichodorus heterocephalus (16). Tylenchina have, however, never been proved to transmit viruses.

Dropkin (7) mentioned that 'those nematodes proven to be virus vectors have relatively mild effects on host cells'. However, Trichodorus similis always kills the cells from which food is ingested. When the feeding of T. similis on tobacco is compared to that of the tylenchid epidermal feeder Paratylenchus projectus, the cell's reaction to feeding suggests that the latter species could, indeed, be an efficient virus vector. P. projectus feeds in epidermal cells of tobacco for many days without causing any disturbance to the cell's protoplast and cytoplasmic streaming continues in a normal fashion during and after feeding. The question why only members of the order Dorylaimida function as virus vectors still remains unanswered.

By means of electron microscopy, Taylor and Robertson (27, 28,

29) located virus particles within members of the three nematode genera known to be virus vectors. These particles are apparently selectively and specifically adsorbed on to certain surfaces of the nematode's stomodaeum. In L. elongatus, virus particles are associated with the inner surface of the odontostyle guiding sheath, whereas in Xiphinema and Trichodorus they are adsorbed to the surface of the cuticle lining the oesophageal lumen, and also the pharyngeal lumen in Trichodorus.

It would be worthwhile examining whether a similar process of virus adsorption might also occur in members of the suborder Tylenchina after they have fed on virus-infected roots. If this should prove to be the case, differences in the location of the oesophageal gland openings into the oesophageal lumen are probably decisive. In the suborder Tylenchina, which includes all the root parasitic Tylenchida, the duct of the dorsal gland empties into the oesophageal lumen just behind the stylet knobs. It appears that only secretions from this gland are injected into the food cell. In the root parasitic Dorylaimida, however, the ducts of all the glands open into the lumen of the basal bulb. In Trichodorus, Longidorus and Xiphinema, salivary secretions, which are injected into the plant cell, thus have to pass surfaces on which virus particles are adsorbed (27, 28, 29) whereas this seems **unlikely to be so** in the Tylenchina.

REFERENCES

1. Chen, T.A. & Mai, W.F. (1965). The feeding of Trichodorus christiei on individually isolated corn root cells. Phytopathology 55, 128.

2. Christie, J.R. & Perry, V.G. (1951). A root disease of plants caused by a nematode of the genus Trichodorus. Science 113 491-493.

3. Cohn, E. (1970). Observations on the feeding and symptomatology of Xiphinema and Longidorus on selected host roots. J. Nematol. 2, 167-173.

4. Cohn, E. & Orion, D.(1970). The pathological effect of representative Xiphinema and Longidorus species on selected host plants. Nematologica 16, 423-428.

5. Cotten, J. (1974). Feeding behaviour and reproduction of Xiphinema index on some herbaceous test plants. Nematologica 19, 516-520.

6. Davis, R.A. & Jenkins, W.R. (1960). Nematodes associated with roses and the root injury caused by Meloidogyne hapla Chitwood, 1949, Xiphinema diversicaudatum (Micoletzky, 1927) Thorne, 1939, and Helicotylenchus nannus Steiner, 1945. Bull. Md. agric. Exp. Stn. A-106, 16 pp.

7. Dropkin, V.H. (1969). Cellular responses of plants to nematode infections. Ann. Rev. Phytopath. 7, 101-122.

8. Epstein, E. (1972). Biochemical changes in terminal root galls caused by an ectoparasitic nematode, Longidorus africanus: Phenols, Carbohydrates and Cytokinins. J. Nematol. 4, 246-250.

9. Fisher, J.M. & Raski, D.J. (1967). Feeding of Xiphinema index and X. diversicaudatum. Proc. Helminthol. Soc. Wash. 34, 68-72.

10. Fritzsche, R. & Hofferek, H. (1969a). Beiträge zum Saugverhalten und Nährpflanzenkreis von Xiphinema diversicaudatum (Mikoletzky) Thorne. Arch. Pflanschutz 5, 111-118.

11. Fritzsche, R. & Hofferek, H. (1969b). Nahrungsaufnahme und Nahrpflanzenkreis von Longidorus macrosoma Hooper. Arch. Pflanzenschutz 5, 423-429.

12. Hewitt, W.B., Raski, D.J. & Goheen, A.C. (1958). Nematode vector of soil-borne fanleaf virus of grapevines. Phytopathology 48, 586-595.

13. Hügger, Ch. (1973). Preferred feeding site of Trichodorus christiei on tomato roots. J.Nematol. 5, 228-229.

14. Horner, C.H. & Jensen, H.J. (1954). Nematodes associated with mints in Oregon. Plant Dis. Rep. 38, 39-41.

15. McElroy, F.D. & van Gundy, S.D. (1968). Observations on the feeding processes of Hemicycliophora arenaria. Phytopathology 58, 1558-1565.

16. Paracer, S.M., Waseem, M. & Zuckerman, B.M. (1967). The biology and pathogenicity of the awl nematode Dolichodorus heterocephalus. Nematologica 13, 517-524.

17. Pitcher, R.S. (1967). The host-parasite relations and ecology of Trichodorus viruliferus on apple roots, as observed from an underground laboratory. Nematologica 13, 547-557.

18. Pitcher, R.S. & Posnette, A.F. (1963). Vascular feeding by X. diversicaudatum (Micol.). Nematologica 9, 301-302.

19. Pitcher, R.S. & McNamara, D.G. (1970). The effect of nutrition and season of year on the reproduction of Trichodorus viruliferus. Nematologica 16, 99-106.

20. Rhoades, H.L. (1965). Parasitism and pathogenicity of Trichodorus proximus to St. Augustine grass. Plant Dis. Rep. 49, 259-262.

21. Rhoades, H.L. & Linford, M.B. (1961). A study of the parasitic habit of Paratylenchus projectus and P. dianthus. Proc. Helminthol. Soc. Wash. 28, 185-190.

22. Rohde, R.A. & Jenkins, W.R. (1957). Host range of a species of Trichodorus and its host-parasite relationships on tomato. Phytopathology 47, 295-298.

23. Russell, C.C. & Perry, V.G. (1966). Parasitic habit of Trichodorus christiei on wheat. Phytopathology 56, 357-358.

24. Schindler, A.F. (1957). Parasitism and Pathogenicity of Xiphinema diversicaudatum, an ectoparasitic nematode. Nematologica 2, 25-31.

25. Sutherland, J.R. (1969). Feeding of Xiphinema bakeri. Phytopathology 59, 1963-1965.

26. Taylor, C.E. (1971). Nematodes as vectors of plant viruses. In: Plant parasitic nematodes Vo. 2, Ed. B.M. Zuckerman et al. New York and London: Academic Press, pp. 185-211.

27. Taylor, C.E. & Robertson, W.M. (1969). The location of raspberry ringspot and tomato black ring viruses in the nematode vector, Longidorus elongatus (de Man). Ann. appl. Biol. 64, 233-237.

28. Taylor, C.E. & Robertson, W.M. (1970). Sites of virus retention in the alimentary tract of the nematode vectors, Xiphinema diversicaudatum (Micol.) and X. index (Thorne and Allen). Ann. appl. Biol. 66, 375-380.

29. Taylor, C.E. & Robertson, W.M. (1970). Location of tobacco rattle virus in the nematode vector, Trichodorus pachydermus Seinhorst. J. gen. Virol. 6, 179-182.

30. Whitehead, A.G. (1967). Rep. Rothamsted exp. Stn for 1966, p. 147.

31. Wyss, U. (1970). Parasitierungsvorgang und Pathogenität wandernder Wurzelnematoden an Fragaria vesca var. semperflorens. Nematologica 16, 55-62.

32. Wyss, U. (1971a). Saugverhalten und Pathogenität von Trichodorus spp. in steriler Agarkultur. Nematologica 17, 501-507.

33. Wyss, U. (1971b). Der Mechanismus der Nahrungsaufnahme bei Trichodorus similis. Nematologica 17, 508-518.

34. Wyss, U. (1972). Trichodorus similis (Nematoda). Saugen an Wurzeln von Sämlingen (Rübsen). Film E 1763 der Enc. Cin. Göttingen, 12 pp.

35. Wyss, U. (1971a). Trichodorus similis (Nematoda). Reaktion der Protoplasten von Wurzelhaaren (Nicotiana tabacum) auf den Saugvorgang. Film E 2045 der Enc. Cin. Göttingen (In press).

36. Wyss, U. (1974b). Longidorus elongatus (Nematoda). Embryonalentwicklung. Film E 2046 der Enc. Cin. Göttingen (In press).

37. Zuckerman, B.M. (1961). Parasitism and pathogenesis of the cultivated cranberry by some nematodes. Nematologica 6, 135-143.

DISCUSSION

Commenting on several questions, Wyss explained that the feeding tube illustrated in his film on Trichodorus feeding is most probably formed from saliva only, as the tube is also formed when a dead cell is punctured. The perforation of the cell by Trichodorus appear to be accomplished mechanically and no leakage from the punctured cell has been observed. The difference between the reaction to the puncturing of a plant cell with a needle compared with Trichodorus, is that in the former case there is a certain amount of coagulation near the puncture but normal plasma streaming persists. Wyss considered that Trichodorus may react to some stimulus from the root cells which excites a feeding response. Roggen suggested that the fast retraction of the stylet shown in the film could be accounted for by stretched connective tissue in equilibrium with a contracted muscle and the sudden release of the muscle's tension; Robertson, however, doubted the feasibility of this hypothesis.

According to experimental work by Pitcher the life cycle of T. viruliferus on apple is 7-8 weeks. Wyss said he could not determine the duration of the life cycle of T. similis in his culture because contamination prevented the nematodes from completing one. Egg laying and moulting had been observed; eggs were produced only after feeding on root tips, although the observed reaction to any cell fed on appeared to be the same.

It was asked if tylenchids ingested virus particles and if so, what prevented them from acting as vectors. Taylor referred to van Hoof (H.A. van Hoof. Neth.J.Pl.Pathol., 73, 193-194 (1967) who had demonstrated the presence of tobacco rattle virus in *Pratylenchus penetrans* by inoculating susceptible test plants with ground up suspensions of the nematode. He supposed that in tylenchids there were no suitable retention sites to which viruses could become adsorbed and hence they could not act as vectors. Thomason suggested that tobacco rattle was more likely to be transmitted by *Trichodorus* to an unsuitable host than to one suited to *Trichodorus* feeding as in the latter case the cells are destroyed too quickly to allow infection to take place. However, Taylor pointed out that viruses multiply in microseconds and could spread to neighbouring cells before cytoplasmic movement had been stopped by the action of *Trichodorus* feeding.

SOME FEATURES OF NEMATODE-BORNE VIRUSES AND THEIR RELATIONSHIP WITH THE HOST PLANTS

G.P. MARTELLI

Instituto di Patologia Vegetale, Universita degli Studi,

Bari, Italy

INTRODUCTION

Among plant viruses that are moved from host to host by living organisms, those transmitted by nematodes constitute, as a whole, a heterogeneous collection of infectious entities, which can grossly be differentiated by virtue of particle morphology and relationship with vectors. All these viruses comprise two of the sixteen groups presently included in the virus classification scheme proposed by Harrison *et al*. (47), i.e. nepovirus and tobravirus groups.

Major differences between the two groups are that nepoviruses have polyhedral particles and are transmitted by species of *Xiphinema* and/or *Longidorus*, whereas tobravirus have rod-shaped rigid particles and are transmitted by species of *Paratrichodorus* and/or *Trichodorus*. However, both groups of viruses have some biological properties in common such as: (i) presence of at least two nucleoprotein components; (ii) multipartite genome constituted by single-stranded RNA; (iii) transmission through seed; (iv) transmission by inoculation of sap; (v) moderate to wide experimental host-range.

At present there are 15 nepoviruses: arabis mosaic (AMV), grapevine fanleaf (GFV), strawberry latent ringspot (SLRV), tomato black ring (TBRV), cocoa necrosis (CNV), grapevine chrome mosaic (GCMV), artichoke Italian latent (AILV), tobacco ringspot (TRSV), tomato top necrosis (TTNV), raspberry ringspot (RRSV), mulberry ringspot (MRSV), cherry rasp leaf (CRLV), tomato ringspot (TomRSV), peach rosette mosaic (PRMV), cherry leafroll (CLRV) and 4 putative ones: eunonymus ringspot (EuRSV), myrabolan latent ringspot (MLRV), grapevine Joannes-Seyve 26-205 and grapevine CM112 are reported in the literature. Tobacco ringspot virus is the type member of the group.

Only 2 tobraviruses, tobacco rattle (TRV) and pea early browning (PEBV), are known, the former being the type member of the group.

Although it has been claimed that viruses other than "nepo" and "tobra" can experimentally be transmitted by nematodes (10, 26, 27, 71, 99) none of these will be dealt with in this paper as, for most of them, the mechanism of transmission and its significance in nature remains to be established. In fact, it appears that a biologically unequivocal and economically important role of nematodes as vectors, has been ascertained, so far, only in relation to members of the nepo and tobravirus groups.

Rather than presenting a comprehensive review of the characteristics and properties of nematode-borne viruses and of the diseases they cause, the present article aims to illustrate briefly some of the features of these viruses and of their relationships with the host plants, which have been treated only in part in previous reviews (8, 49, 117, 118), and it is hoped this may prove useful to those who are engaged in virus transmission work.

MORPHOLOGICAL, STRUCTURAL AND CHEMICAL CHARACTERISTICS OF NEMATODE-BORNE VIRUSES.

Nepoviruses

All nepoviruses have isodiametric particles with icosahedral symmetry, 25-30 nm in diameter, and exhibiting angular outlines when seen in the electron microscope. Their chemical constitution is simple, for the particles are composed of protein and ribonucleic acid (RNA) only. Whereas the protein shell is always the same for any given nepovirus, its RNA content may vary, yielding particles with different physical and biological properties as discussed below.

The protein coat. Recent studies have demonstrated that the protein coat of several nepoviruses (55, 58, 76) is composed of 60 structural subunits each consisting of a single polypeptide molecule of mol. wt 5-6 x 10^4, presumably containing about 500 amino acid residues. Conceivably, a similar structural model applies to all members of the group. The forces that stabilize the capsid are, more than likely, protein-protein interactions (57). Since the contribution of RNA-protein linkages to the stability of the protein shell seems negligible, the presence of nucleic acid is not necessary to keep coat protein components assembled into a stable form. Hence, hollow particles devoid of RNA may exist. In fact, these are often found along with nucleoprotein components in virus preparations, and are usually referred to as top (T) component. T particles have sedimentation coefficients ranging from 51 to 55 **S** and a mol. wt of 3-3.4 x 10^6. That these are not an artifact

Table 1. Serological grouping and characteristics of nematode-borne viruses.

VIRUS GROUPS	Vector	No. of centrifugal components and sedimentation coefficients	Molecular weight of RNA species
Tomato black ring virus			
- TBRV English strain	*L. attenuatus*	Three components: 55 *S* (T), 92-101 *S* (M), 117-129 *S* (B).	RNA-1: 2.5×10^6
- TBRV Scottish strain	*L. elongatus*		RNA-2: 1.5×10^6
- Cocoa necrosis	Unknown		RNA-3: 0.5×10^6
- Grapevine chrome mosaic	Unknown (*X. vuittenezi*)		
- Artichoke Italian latent	*L. attenuatus*		
Arabis mosaic virus			
- AMV type strain	*X. diversicaudatum* *X. coxi* *L. caespiticola*	Three components: 53 *S* (T), 93 *S* (M), 126 *S* (B).	RNA-1: 2.4×10^6 RNA-2: 1.4×10^6
- AMV hop strain	*X. diversicaudatum*		
- Grapevine fanleaf	*X. index* *X. italiae*		
Strawberry latent ringspot			
- Several isolates closely related or serologically indistinguishable	*X. diversicaudatum* *X. coxi*	Possible three components although only one reported 134 *S* (B).	No information

Table 1. (cont..)

Raspberry ringspot virus			
– RRSV type strain	L. elongatus	Three components: 50 S (T), 91 S (M), 125 S (B).	RNA-1: 2.4×10^6 RNA-2: 1.4×10^6
– RRSV English strain	L. elongatus L. macrosoma L. caespiticola L. leptocephalus X. diversicaudatum		
– RRSV cherry strain	L. macrosoma X. diversicaudatum		
– RRSV grapevine strain	L. macrosoma (?)		
Tobacco ringspot virus			
– Several strains serologically distinguishable	X. americanum	Three components: 53 S (T), 91 S (M), 126 S ($\overline{B}$).	RNA-1: 2.2×10^6 RNA-2: 1.4×10^6
– Satellite virus	X. americanum		RNA: $1.15\text{–}1.25 \times 10^5$
Tomato ringspot virus			
– Several isolates closely related or serologically indistinguishable	X. americanum	Three components: 53 S (T), 119 S (M or B?), 127 S (B or B_2?).	RNA: 2.3×10^6
Peach rosette mosaic virus			
– Several strains serologically distinguishable	X. americanum	Three components: 52 S (T), 115 S (M), 134 S ($\overline{B}$).	No information

Table 1 (cont..)

Cherry leafroll virus			
- Cherry leafroll	<u>X. diversicaudatum</u> <u>X. coxi</u>	Two components: 115 <u>S</u> (M), 128 <u>S</u> (B).	RNA-1: 2.4×10^6 RNA-2: 2.1×10^6
- Elm mosaic			
- Golden elderberry virus			
Tobacco rattle virus			
- Several European and American strains serologically distinguishable	At least 12 <u>Trichodorus</u> species	Two components: 243 <u>S</u> (S), 300 <u>S</u> (L).	RNA S particles: $0.7\text{–}1.0 \times 10^6$ RNA L particles: 2.5×10^6
Pea early browning virus			
- Several European strains serologically distinguish-able	At least 5 <u>Trichodorus</u> species	Two components: 210 <u>S</u> (S), 286 <u>S</u> (L).	RNA S particles: 1.3×10^6 RNA L particles 2.5×10^6

arising during extraction and purification processes is also demonstrated by their direct visualization under the electron microscope in plant cells infected with strawberry latent ringspot virus (93). Furthermore, for tobacco ringspot virus, evidence has been provided that T component does not originate in vivo through leakage of degraded nucleic acid from nucleoprotein particles (104).

Extraction procedures and perhaps other factors linked with host plant and environmental conditions may have a bearing in determining the recovery of T particles from infected tissues. For instance, Kenten (58) reports that because of sensitivity to butanol, T component of cocoa necrosis virus is rarely found in preparations treated with this chemical, and Martelli and Quacquarelli (75) were unable to detect T particles in tomato black ring virus preparations purified with methods usually yielding them (40).

Besides serving as a protection for viral RNA, the protein coat has important biological functions among which a most critical one is the determination of specificity in transmission by vectors. As discussed in detail by Taylor and Robertson in these Proceedings evidence has been obtained for raspberry ringspot virus (48) that the protein surface of the virus particles has a major role in the specific retention and transmission of strains of the virus by Longidorus species. Furthermore, the protein shell is responsible for serological specificity, which often represents the sole basis for establishing relationships between different nepoviruses or for distinguishing between strains and/or serotypes of a same virus. Even though little is known on the role played by the coat protein of nepoviruses in the host's infection process, the results of recent work (22, 45, 85, 86) indicate that the protein material is not required in the initiation of infection as it occurs with other multicomponent plant viruses (5). In raspberry ringspot virus the coat protein does not seem to be involved in determining virulence nor in the specification of other characters such as infection of resistant Lloyd George raspberries or systemic invasion of French bean (48).

The nucleic acid. So far, all nepoviruses studied in detail proved to contain single-stranded RNA with the following molar percentages of nucleotides: G24:A23:C22:U31 (47). Rather than in a single molecule, the nucleic acid occurs as two functional species (RNA-1 and RNA-2) having different size and different biological functions. The suggestion that this condition was common to all members of the group (46) was based on work carried out with raspberry ringspot (45, 46, 76, 85), tobacco ringspot (22, 45, 76), arabis mosaic (76) and cherry leafroll (55) viruses, and was further supported by comparable findings with tomato black ring virus (86). In this connection, work in progress in my laboratory indicates that both grapevine fanleaf and artichoke Italian latent viruses also have two RNA species, thus fitting the above pattern.

All RNAs-1 have about the same dimensions (mol. wt 2.2-2.5 x 10^6) irrespective of the virus they come from, whereas RNAs-2 may differ in size depending on the virus of origin. A light RNA-2 (mol. wt 1.4-1.5 x 10^6) occurs in a large group of viruses comprising tobacco ringspot, raspberry ringspot, arabis mosaic, tomato black ring (22, 45, 76, 86) and possibly, grapevine fanleaf and artichoke Italian latent, whereas a heavier RNA-2 (mol. wt 2.1 x 10^6) has been reported for cherry leafroll virus (55).

It has been ascertained beyond doubt that RNA-2, rather than being a degradation by-product of the larger RNA-1, is synthesized independently so that both RNA species together account for the whole viral genome (22, 45, 46, 48, 55, 86). Such a condition which, incidentally, conforms to that known for several other multi-component virus systems (123), has intriguing biological implications because some characters are determined by one and some by the other RNA species. For instance, in raspberry ringspot virus, RNA-2 carries the genetic information for coat protein and therefore it controls specificity of transmission and serological behaviour (48; see also Taylor and Robertson in these Proceedings). This is a property common to several small-sized RNAs of other viruses with multipartite genomes (23, 63, 97). RNA-2 also possesses determinants for yellow mosaic symptoms (46, 48) which, remarkably, is also a characteristic of the smallest functional RNA of many split-genome viruses belonging to different groups (4, 23, 69). As reported by Harrison and co-workers (48), RNA-1 of raspberry ringspot virus has many genetic determinants, the most crucial of which is the replication gene, which renders this RNA indispensable for the onset of infection.

The two RNA species enter the constitution of nucleoprotein particles sedimenting at different rates, known as middle (M) and bottom (B) components. The slow-sedimenting M component contains always one RNA-2 molecule whereas B component particles contain either one molecule of RNA-1 or two molecules of RNA-2 (22, 45, 85).

Satellite RNAs. In addition to RNA-1 and RNA-2, accessory (satellite) nucleic acid molecules are present in some nepoviruses such as tobacco ringspot (100, 101) and tomato black ring (85) viruses. Although differing in size [mol. wt of TRSV S-RNA is 1.15-1.25 x 10^5 (112) and mol. wt of TBRV RNA-3 is 0.5 x 10^6 (86)], both satellite RNAs interact with the activator virus by inhibiting infectivity or reducing lesion size. Satellite RNAs are not able to multiply on their own, are incapsidated in a protein shell which is morphologically and serologically indistinguishable from that of the respective supporting virus (101), and can be eliminated from virus cultures without affecting their multiplication (86, 101).

The presence of satellite RNA in a virus culture may alter considerably its sedimentation profile because of the increased number of sedimenting classes of particles. This is particularly well illustrated by tobacco ringspot virus, whose preparation, when contaminated with S-RNA, may yield up to 14 types of nucleoproteins with different buoyant densities, sedimenting at different rates (105).

Although the biological significance of satellite RNAs is not understood, it is interesting that they can be transmitted by nematodes equally well as the activator virus. This was recently proved for encapsidated S-RNA of tobacco ringspot virus (77).

Tobraviruses

All tobraviruses have straight tubular particles of two or, more rarely, three predominant lengths: long (L), 180-210 nm and short (S), 45-115 nm, depending on the isolate. The particles have helical symmetry with a pitch of the helix of 2.5 nm and show a central canal about 5 nm in diameter (11, 87, 88). Their overall diameter is approximately 23 nm but variations up to 8% between strains of a same virus or different members of the group have been recorded (11, 42). The two ends of the particle differ in shape, one being slightly convex and the other flared (42).

The protein coat. The protein coat of tobraviruses is composed of subunits whose number and molecular weight has not yet been established with certainty. Values ranging from 1.9 to 2.9 x 10^4 have been reported by various authors for the mol. wt of single polypeptide chains composing a subunit. As discussed by Ghabrial and Lister (32) diversity in the method used for mol. wt determination may account for this discrepancy, even though the existence of variations in the mol. wt of protein subunits in different members of the group cannot be excluded (11). Also the reported number of aminoacid residues per subunit varies considerably (i.e. from 174 to 218) according to the author (78, 89, 107, 108).

Similarly to other rod-shaped viruses, tobacco rattle virus coat protein polymerizes into a stable disk form in absence of its nucleic acid. Addition to the latter is required for obtaining or restoring the helical construction (35, 79, 110).

The protein shell is responsible for serological specificity, but there is only slight evidence of its association with specificity of transmission by nematode vectors (see Taylor and Robertson in these Proceedings). Coat protein does not seem essential in the onset of infection nor does it interact with nucleic acids in hybridization experiments (28, 65, 67, 97).

The nucleic acid. Both members of the tobravirus group so far known, possess single-stranded RNA. As in the nepoviruses, the nucleic acid occurs as two distinct species each located in different particles, i.e. long particle RNA and short particle RNA. The former has a mol. wt of about 2.5×10^6, irrespective of the strain or the virus it comes from, whereas short particle RNA may vary in size and weight depending on the particle length. Thus, mol. wts ranging from 0.7 to 1.3×10^6 were recorded for two isolates of tobacco rattle virus and one of pea early browning virus (11). Since the virus genome is constituted by both species of RNA, their contemporary occurrence is essential for full symptom expression and production of complete virus particles. Hybridization experiments with different tobacco rattle virus strains (28, 65, 67, 69, 97, 109) have amply demonstrated that the longer RNA species is capable of independent replication but, as it is unable to code for coat protein, it yields unstable entities (i.e. coatless RNA) which are difficult to subculture. Short particle RNA contains genetic determinants for the specification of coat protein and of some pathogenic characters (33, 69). However, pathogenicity can also be controlled by genes located in the long particle's RNA (33). Interestingly, tobacco rattle virus hybrids are more readily obtained if the parent isolates are closely related serologically (33, 67, 98) so that, in nature, "compatible" combinations are required for new strains to arise.

GROUPING OF NEMATODE-BORNE VIRUSES

Grouping of nematode-borne viruses can be based on: (i) number of centrifugal components constituting the particle population and, (ii) serological relatedness. Certainly, this is not an attempt to recognize or propose taxonomic subgroups within the classification scheme set forth for these viruses (47) but it endeavours only to assemble them by taking into account differences evidenced by published work.

Grouping according to number of centrifugal components

As a result of the presence of two nucleic acid species, nepoviruses consist of a population of particles differing from one another in the percentage of RNA content and, therefore, yielding centrifugal classes which sediment at different rates in density gradient columns. Particle heterogeneity, diversity in RNA content and in buoyant density differentiate three subgroups:

1. Tobacco ringspot virus group. Members of this group may or may not show serological relationships among them and are transmitted by *Longidorus* or *Xiphinema*. The characterizing similarity is the occurrence of a slow-sedimenting T component (51-55 *S*) and at least two faster-sedimenting nucleoproteins, i.e. M (91-102 *S*) and B (117-129 *S*) components. This condition was experimentally verified in tobacco ringspot (114), arabis mosaic (80), raspberry ringspot (82), cocoa necrosis (58), tomato blackring (81), tomato top necrosis (3) and grapevine fanleaf (A. Quacquarelli, unpublished results) viruses,

whereas other possible members such as strawberry latent ringspot, mulberry ringspot, eunonymus ringspot (91), artichoke Italian latent and grapevine chrome mosaic viruses, are either still insufficiently known for an adequate evaluation, or are missing one component (75). In this connection it should be pointed out that the purification method (58), the host in which the virus was propagated, the time of year when the virus was extracted (62) and the age of infection (102) influence the recovery of different components and their relative ratio.

In this group, M particles contain about 30% RNA and, if properly separated from other components, should not be infective (22, 45). On the contrary, B particles are infective alone but their infectivity is increased by mixing with M particles (3, 45). The results of recent work (22, 45, 46, 85) suggest that within B component, two classes of particles (B_1 and B_2) should be resolved, depending on whether they contain one RNA-1 or two RNA-2 molecules. Indeed, this expectation has already been fulfilled with tobacco ringspot virus (85). Hence, B_1 and B_2 particles should be expected to contain a different RNA percentage (anyhow above 40%). Neither of them, alone, should be infective.

Most of the above information derives from studies on tobacco ringspot and raspberry ringspot viruses and it remains to be seen whether these notions apply also to other nepoviruses of the same type.

2. Tomato ringspot virus group. Tomato ringspot virus apparently differed from other nepoviruses in that it did not seem to possess a M component (115). Very recently, however, two nucleoprotein particles with a sedimentation coefficient of 119 S and 127 S, respectively were detected in purified preparations of four different isolates of the virus (106). The authors identify these components as B_1 and B_2 but, based on their sedimentation behaviour, B_1 appears lighter than most B particles of nepoviruses of the tobacco ringspot type and only slightly denser than M particles of cherry leafroll virus. Anyhow, it seems relevant that RNA molecules of different size, and perhaps having diverse biological function, are associated with the two types of particles. The smaller RNA component of tomato ringspot virus is similar to that of cherry leafroll (55, 129).

Peach rosette mosaic virus resembles tomato ringspot virus because of the high sedimentation coefficients (115 and 134 S, respectively) of its nucleoprotein components (19). However, until more detailed information is available it does not seem worthwhile attempting a comparison between these two entities.

3. Cherry leafroll virus group. One of the characterizing features of cherry leafroll virus is the absence of T component. So far, this

has held true for all strains studied (53, 55, 138). Virus preparations are composed of two nucleoprotein particles sedimenting at about 115 S and 128 S, respectively. The slow-sedimenting M particles contain about 40% RNA, i.e. a much higher quantity (i.e. a larger molecule) than the comparable component of nepoviruses of the tobacco ringspot type. The faster-sedimenting B particles have about 43% RNA (55). Neither of the components alone is considered to be infective, mutual complementation being required for full genetic expression.

Myrabolan latent ringspot (24) and grapevine Joannes-Seyve (26-205) (21) viruses apparently lack T component and exhibit a M component with a high sedimentation (115 S), comparable to that of cherry leafroll virus. No serological relationship between the above three viruses has been found. Even if additional investigations are perhaps needed to establish whether myrabolan latent ringspot and grapevine Joannes-Seyve (26-205) viruses constitutes a two- or three-component system, the point remains that, based on the reported high density of M particles, they seem to differ from the viruses of the tobacco ringspot type. Another putative nepovirus isolated from the grapevine, bearing no serological relatedness to many known members of the group, has a sedimentation profile showing only two very close-sedimenting nucleoprotein components (25). However, too little is known at the moment on the properties of this virus to allow for its placing in one of the above tentative groups.

As for tobraviruses, all members of the group have two or, more rarely, three nucleoprotein components all containing about 5% RNA (38, 39). Accessory particles made up of protein only (T type) do not occur, although the very short particles (45-55 nm) found in some isolates of tobacco rattle virus have been sometimes referred to as T component (107). In tobraviruses short and long particles are the equivalent of M and B particles of nepoviruses and like them interact biologically with one another.

Grouping according to serological relatedness

Although the available information is far from being complete, several broad groups comprising viruses that share antigenic determinants can be established within nepo and tobraviruses. In this connection, it should be noted that the existence of a distant serological relationship between two or more viruses sustains the retention of different names in some [e.g. arabis mosaic and grapevine fanleaf viruses (9)] but not in other cases [e.g. type and grapevine strain of raspberry ringspot virus (125)]. This explains why in Table 1, some of the groupings contain only one vernacular virus name whereas others include several differently named entities.

The tomato blackring virus (TBRV) group includes at least five members, not all of which are directly related serologically. For instance, grapevine chrome mosaic virus does not share antigens with any of the TBRV strains (75) but it is related, though distantly, with cocoa necrosis virus which, in turn, is related to TBRV (58). Artichoke Italian latent virus, although not showing any obvious serological relatedness with TBRV, has enough biological characters in common with it to favour the idea that it belongs in the same group.

Serological diversity is also found among members of the arabis mosaic virus group which includes a serotype causing nettlehead disease of hop and having the same nematode vector (*X. diversicaudatum*) as the type strain, and a serotype transmitted by different nematode species (*X. index* and *X. italiae*) inducing the fanleaf disease of the grapevine.

Several naturally occurring serological variants of raspberry ringspot (82, 125), tobacco ringspot (34, 95), peach rosette mosaic (20) and cherry leafroll (54, 128) viruses have been recorded. Conversely, in tomato ringspot virus most isolates, irrespective of the differences in the hosts they infect and in the diseases they cause, seem to be closely related or indistinguishable serologically (115, 121). A comparable situation occurs with strawberry latent ringspot virus (1, 64).

In the tobraviruses, strains of tobacco rattle and pea early browning constitute two rather well defined serological clusters, although distant relationships between members of the two groups have been reported (70).

RELATIONS WITH HOST TISSUES

Root tips, i.e. the site of acquisition and inoculation of nematode-borne viruses by the vectors, should represent a desirable material for studying virus-host interactions. Yet, very few investigations at the ultrastructural level have taken into account the root tip area of infected plants. These refer mostly to tobacco ringspot virus in artificially inoculated French beans in whose roots virus particles were visualized, even though with a certain amount of difficulty (14, 15). A similar difficulty was experienced by Gerola and co-workers (31) in the search for grapevine fanleaf virus in *Vitis vinifera* L. roots galled by *Xiphinema index*. Eventually, these authors were able to spot a few presumable virions, and were not much luckier when looking for particles in the leaves of several host plants. In my laboratory, many attempts to visualize grapevine chrome mosaic (74), grapevine fanleaf and tomato black ring viruses in host leaf tissues did not succeed, owing to the indistinct appearance of possible virus particles and to lack of structures, like crystalline aggregates, that would facilitate their identification. In fact, the arrangement of particles in crystalline or paracrystalline arrays seems uncommon in nepoviruses. Two remarkable exceptions refer to arabis mosaic virus, whose superimposed layers of particles in a spherical configuation are the characterizing feature of infected cells of Chenopodium amaranticolor Coste et Reyn. and Petunia hybrida Vilm. (29,30), and to tobacco ringspot virus which may occur in large crystalline aggregates in tobacco cells (94). Microcrystals of tomato ringspot virus particles were also seen in Datura stramonium L. cells (17).

Nepoviruses are very invasive, entering different tissues and organs of the infected host. Thus, for instance, cherry leafroll virus was detected in the apical meristem of axillary and terminal buds, root tips, pollen, ovules and mature seeds as well as in differentiated leaf cells (56, 127). Likewise, tobacco ringspot virus particles were abundantly found in phloem sieve tubes and companion cells (36), in apical initials of shoot meristems (94) and, as already mentioned, in root meristems (14, 15). Arabis mosaic virus, on the other hand, was seen in foliar parenchymas and phloem elements of C. amaranticolor and P. hybrida (29, 30). Virus particles usually occur in the cytoplasm and only with grapevine fanleaf virus were they occasionally observed in the nucleus (90).

The detection of virus particles in particular plant tissues or organelles does not necessarily mean that the particles were synthesized and/or assembled there, as they might have been translocated from the actual site of synthesis. For instance, Atchinson and Francki (2) have provided evidence that there is very little if any multiplication of tobacco ringspot virus in the root-tips of French bean, so that the accumulation of virus in that area is likely due to active transport from other parts of the plant.

As to the cytology of nepoviruses-infected plants it is evident that their cells are affected, though to a variable extent, by the invading pathogens. Severe modifications of chloroplast up to disruption were reported as induced by arabis mosaic virus, along with a general degeneration of cell constituents (29, 30), and peculiar cell-wall projections resulted from infection of Nicotiana clevelandii Gray with cherry leafroll virus (56) and of soybean with tobacco ringspot virus (36). However, two peculiar ultrastructural features seem to be common to nearly all nepoviruses so far studied i.e. tubular structures containing virus particles and membranous inclusion bodies. The tubular inclusions can be visualized in negative staining mounts by the simple "squash homogenate" technique (126), or in thin section. In negative staining mounts these appear as straight or slightly flexuous, seldom branched, single-walled tubules about 50 nm in diameter, containing numerous particles (up to 150) in linear arrangement (124, 126, 127). In thin sections, details of their organization and topological relationship to cell constituents are more easily observed. Thin section studies have also confirmed that a single row of particles is contained within each tubule, except for cherry leafroll virus where, on occasions, multiple rows of particles could be seen (56). In strawberry latent ringspot virus, the tubule wall consists of two layers, the outermost of which joins with a membrane that surrounds the whole structure like a sheath (93). In all other cases known, it appears that the tubule wall is a unit membrane. Also, there is no evidence of an outer sheath (14, 15, 56, 90, 94).

Tubular inclusions have been visualized in the cytoplasm, in the proximity (93) or not (56, 90, 94) of membranous inclusion bodies, or inside cell wall protrusions (36, 56), but the most intriguing finding is their consistent association with plasmodesmata. This has led to the suggestion that the tubules are primarily involved in the movement of virus from cell to cell (15, 17, 93). In this way they eventually gain entrance into sieve tubes, the starting point for long distance translocation and systemic infection of the host (36).

It is not known whether all virus particles are contained, at one time or another, within tubular structures but certainly a great proportion of them are. This condition may have a bearing in determining the configuration in which virus particles become disposed upon aggregation. A close examination of intracellular arrays of nepoviruses, reveals that they seldom show the cubic or pseudo-cubic packing of the crystalline structures, as it is customary for other small icosahedral plant viruses (96, 131). Instead the virus aggregates appear as if they were made out of stacked rows of particles in a straight or, more often, curved stratification (17, 90, 94, 124). It is therefore possible that nepovirus "crystals" originate from a close gathering of particle-containing tubules followed by dissolution of the tubule walls. This interpretation seems compatible even with the odd spherical aggregates of arabis mosaic virus (29, 30).

Cytoplasmic inclusion bodies represent another consistent ultra-structural feature of nepovirus infections. The inclusion bodies are aggregates of cellular material sometimes as large as the nucleus next to which they lie. They appear in the early stages of infection and contain ribosomes, endoplasmic reticulum and membranous vesicles. Such inclusions have been found in cells infected with arabis mosaic (30), cherry leafroll (56), grapevine fanleaf (31) and strawberry latent ringspot (93) viruses. Those of the last named virus contain internally masses of hollow structures resembling empty viral capsid and tubular inclusions at the periphery (93). The function of these bodies is not clear, but their early appearance before virus particles are visualized, and the presence of possible viral material within them, suggest that they may play a role in virus synthesis and/or assembly. It seems worth pointing out that some of the cytological abnormalities reported for nepoviruses, such as cell wall protrusions, virus-containing tubules and membranous inclusion bodies, have also been encountered in plant tissues infected with different members of the comovirus group (51, 52, 59, 60, 116, 122), confirming the alleged similarity between como and nepoviruses (129).

Insofar as tobraviruses are concerned, European, North and South American isolates of tobacco rattle virus have been studied at the fine-structure level. In infected plants, particles of the Brasilian pepper ringspot isolate were observed in every type of tissue, including shoot apical and root tip meristems, but not in xylem vessels (61). Both short and long particles could be identified in the cytoplasms of cells infected with some (16, 112) but not with other (43, 61) isolates, although short particles were present and could be extracted from infected leaf samples. A most striking characteristic of long particles of Brasilian isolates is their association with mitochondria in arrays perpendicularly oriented to the organelle's surface (43, 61, 112). This preferential association was explained on the basis of a difference in the mean surface charge of long and short particles (11) but it is not known whether it has any biological significance. Circumstantial evidence, however, suggests that mitochondria may be specifically involved in tobacco rattle virus replication. A similar particle-mitochondrion topological relation was not detected in cells infected with Californian isolates (16).

Tobacco rattle virus-infected cells can be severely altered, showing a progressive degeneration of major organelles that may lead to necrosis. Some strains of the virus produce non-crystalline inclusions (X-bodies) mostly composed of aggregated and highly modified mitochondria. Although the function of X-bodies is known, it was suggested that they may be sites of viral nucleic acid synthesis (44).

DISSEMINATION AND SURVIVAL OF NEMATODE-BORNE VIRUSES

Long-range spread

Dissemination of nematode-borne viruses over long distances cannot be achieved by natural means since the vectors and other biotic factors involved in their spread operate on a limited range. Therefore, long distance movement takes place primarily through infected propagating plant material. Over time this has affected the geographic distribution of these viruses although it is still possible to locate, for some of them, the geographic area of origin. Thus, for instance, it does not appear unsuitable to speak of European nepoviruses with reference to tomato black ring, raspberry ringspot, arabis mosaic and strawberry latent ringspot viruses all of which, along with the relative nematode vectors, are widely distributed in central European countries where they infect species of the native flora as well as cultivated crops. Not only are records sporadic for any of the above viruses outside of Europe, but in very few cases is there evidence that the viruses became established in the new environment.

An identical situation can be figured for American nepoviruses (i.e. tobacco and tomato ringspot viruses) which are largely confined to the North American continent and flora. Again, their occurrence in areas other than America is uncommon, and in no instance has an actively spreading outbreak of either of them been recorded elsewhere.

Two nepoviruses, namely cherry leafroll and grapevine fanleaf, may be regarded as cosmopolitan. Apart from Europe, where it is commonly encountered in cultivated and wild hosts (13), cherry leafroll virus is apparently established in North America, occurring in native plant species. Hence, it may not have been introduced from Europe. The situation, however, is puzzling for in America the presumed vectors of the virus are either absent (X. diversicaudatum) or rarely found (X. coxi). Other means of transmission (e.g. pollen) may therefore be operating. Grapevine fanleaf virus represents a primary example of long distance dissemination through commercial exchange of plant material. In fact, there is little doubt that the virus, which is now found in all the major vine-growing areas of the world, is probably of European origin and it was spread around with infected scions and rootstocks.

Owing to its world-wide distribution, tobacco rattle virus also deserves the appellative of cosmopolitan, whereas the other tobravirus, pea early browning, is confined to Europe (39).

Information on the geographic distribution of the lesser known nematode-transmitted viruses is inadequate. Nevertheless, it seems significant that some such entities, which as a group were hitherto reported only from temperate regions, are being discovered also in tropical areas (e.g. cocoa necrosis virus, a serotype of tomato black ring virus, in Ghana).

Short-range spread

As illustrated by Murant (83) in a paper reviewing the complex host-nematode-virus relationships, wild and cultivated plants function in the ecology of nematode-transmitted viruses as hosts of both vectors and pathogens, as well as agents of dissemination (through infected seed and propagating material) and persistence of the virus in the soil (through infected seed, roots and perennating organs). Therefore, the plants act mainly as reservoirs of viruses, thus ensuring their survival, and constitute a source of inoculum for the vector. The spread over short distance results from the combined action of viruliferous nematodes and infected seed.

Nematodes alone, because of their limited mobility, cannot be regarded as highly efficient agents of virus dispersal, the rate of spread of the diseases they incite seldom exceeding 2-3 feet a year (41, 50). Furthermore, they are unable to guarantee virus survival in absence of susceptible hosts. In fact, in the nematodes, infectivity persists from 8 weeks to no more than 11 months (for a review see 83) which is enough to cover the time lag between one susceptible crop and the next, but would not be sufficient to perpetuate the virus through prolonged fallow without a suitable host. However, nematodes play a unique epidemiological role in that they mediate entrance of viruses into plants. Hence, if a virus becomes separated from its nematode vector it has no way of infecting new hosts, nor has it the opportunity of getting firmly established in a new environment. Absence of X. americanum in Europe (73, 117, 132) might explain why tomato and tobacco ringspot viruses, which undoubtedly must have been introduced many times over with infected material in different European countries, are so rarely encountered and have never been recovered from wild plants. On the same basis, it is possible to give a reason for the rare occurrence of European nepoviruses outside of their area of origin and, at least in part, for the amazingly wide distribution of grapevine fanleaf virus. The latter has often been disseminated together with its major vector (X. index) which has now a comparable geographical distribution (111). Undoubtedly, the close association of the virus with X. index has favoured the establishment of active infection foci in areas far apart, which in turn have contributed to further virus dissemination through infected stocks. On this account, the discovery of arabis mosaic virus consistently associated with X. diversicaudatum (119) is, for New Zealand, far more relevant and threatening than the finding of another previously unrecorded nepovirus (tobacco ringspot virus) without its main vector (120).

Transmission through seed is a consistent feature of all nematode-borne viruses, although the percentage of infected seed varies in relation to the host and the parasite. Thus, tobraviruses have a low rate of transmission through seed (1-6%), notwithstanding the unusually high figure (37%) of a Dutch isolate of pea early browning virus (6), whereas nepoviruses can attain transmission levels as high as 100% (1, 37, 68, 84).

Seed-transmission is of little or no ecological value for vegetatively propagated woody species unless the seedlings are used as rootstocks, but it represents an effective means of virus dissemination and survival in herbaceous plants. Since many virus-transmitting nematode species (Longidorus and Trichodorus in particular) thrive on weeds that are also hosts of crop-damaging viruses, it becomes clear how these viruses establish themselves in arable land, inducing recurrent outbreaks in susceptible crops grown in rotation on the same field.

As a general trend, in the ecology of nematode-borne viruses the involvement of several different plant species serving as hosts for the virus and the vector can then be envisaged. Many such examples have been illustrated in Europe and North America mostly with reference to nepoviruses (see 83 for review). A remarkable exception to the above pattern is provided by grapevine fanleaf virus, which although artificially transmissible to a fairly wide range of herbaceous plants, has no natural hosts other than the grapevine. Moreover, this virus is not transmitted through grape seeds (7, 12), for it invades the endosperm and the seed coat but not the embryo (12) and it occurs only in a very low proportion of seeds of experimentally-infected herbaceous hosts (18). The latter notion, however, has no ecological importance, for none of the herbaceous hosts in which the virus is seed-borne is liable to be found infected in nature nor, as is often the case with the herbaceous plants and *Xiphinema*, do they seem able to support active growth of vector populations (116). On the other hand all the hosts of *X. index* so far known (72, 92, 111,130) are non hosts of the virus so that, in this particular instance, the spread and survival of grapevine fanleaf virus relies on a simpler ecological system, whose basic terms are a single host (grapevine) and the vector (*X. index* and/or *X. italiae*).

REFERENCES

1. Allen, W.R., Davidson, T.R. & Briscoe, M.R. (1970). Properties of a strain of strawberry latent ringspot virus isolated from sweet cherry growing in Ontario. *Phytopathology* **60**, 1262-1265.

2. Atchinson, B.A. & Francki, R.I.B. (1972). The source of tobacco ringspot virus in root-tip tissue of bean plants. *Physiol. Pl. Path.* **2**, 105-111.

3. Bancroft, J.B. (1968). Tomato top necrosis virus. *Phytopathology* **58**, 1360-1363.

4. Bancroft, J.B. & Lane, L.C. (1973). Genetic analysis of cowpea chlorotic mottle and brome mosaic viruses. *J. gen. Virol.* **19**, 381-389.

5. Bol, J.F., van Vloten-Doting, L. & Jaspars, E.M.J. (1971). A functional equivalence of top component *a* RNA and coat protein in the initiation of infection by alfalfa mosaic virus. *Virology* **46**, 73-85.

6. Bos, L. & van der Want, J.P.H. (1962). Early browning of pea, a disease caused by a soil- and seed-borne virus. *Net. J.Pl. Path.* **68**, 368-390.

7. Boubals, D. (1969). Observations sur la non-transmission par graine de deux viroses chez la vigne. (Vitis vinifera) Ann. Amel. Pl. 19, 213-219.

8. Cadman, C.H. (1963). Biology of soil-borne viruses. Ann. Rev. Phytopath. 1, 143-172.

9. Cadman, C.H., Dias, H.F. & Harrison, B.D. (1960). Sap-transmissible viruses associated with diseases of grapevine in Europe and North America. Nature, 187, 577-579.

10. Caveness, F.E., Gilmer, R.M. & Williams, R.J. (1974). Transmission of cowpea mosaic by Xiphinema basiri in Western Nigeria. In these Proceedings.

11. Cooper, J.I. & Mayo, M.A. (1972). Some properties of the particles of three tobravirus isolates. J. gen. Virol. 16, 285-297.

12. Cory, L. & Hewitt, W.B. (1968). Some grapevine viruses in pollen and seed. Phytopathology 58, 1316-1320.

13. Cropley, R. & Tomlinson, J.A. (1971). Cherry leafroll virus. C.M.I./A.A.B. Description of plant viruses 80, 4 pp.

14. Crowley, N.C., Davison, E.M., Francki, R.I.B. & Owusu, G.K. (1969). Infection of bean root-meristems by tobacco ringspot virus. Virology 39, 322-330.

15. Davison, E.M. (1969). Cell to cell movement of tobacco ringspot virus. Virology 37, 694-695.

16. De Zoeten, G.A. (1966). California tobacco rattle virus, its intracellular appearance and the cytology of infected cells. Phytopathology 56, 744-753.

17. De Zoeten, G.A. & Gaard, G. (1969). Possibilities for inter- and intracellular translocation of some icosahedral plant viruses. J. ultrastruct. Res. 40, 814-823.

18. Dias, H.F. (1963). Host range and properties of grapevine fanleaf and grapevine yellow mosaic viruses. Ann. appl. Biol. 51, 85-95.

19. Dias, H.F. (1972). Purification and some characteristics of peach rosette mosaic virus. Ann. Phytopath., h. ser., 97-103.

20. Dias, H.F. (1972). Strains of peach rosette mosaic virus differentiated by cross absorption and immunodiffusion tests. Ann. Phytopath., h.ser., 105-106.

21. Dias, H.F. (1973). A sap-transmissible virus associated with a severe disease of the hybrid Vitis sp. Joannes-Seyve (26-205). Riv. Pat. Veg. 9 suppl., 64-67.

22. Diener, T.O. & Schneider, I.R. (1966). The two components of tobacco ringspot virus: origin and properties. Virology 29, 100-105.

23. Dingjan-Versteegh, A., van Vloten-Doting, L. and Jaspars, E.M.J. (1972). Alfalfa mosaic virus hybrids constructed by exchanging nucleoprotein components. Virology 49, 716-722.

24. Dunez, J., Delbos, R., Desvignes, J.C., Marenxaud, C., Kuszala, J. & Vuittenez, A. (1971). Mise en evidence d'un virus de type ringspot sur Prunus cerasifera. Ann. Phytopath., h. ser., 117-128.

25. Ferreira, A.A. & De Sequeira, O.A. (1972). Preliminary studies on an undescribed grapevine virus. Ann. Phytopath., h. ser., 113-120.

26. Fritzsche, R. & Schmelzer, K. (1967). Übertragung des Nelken-rigfleken-Virus durch Nematoden. Naturwissenschaften 54, 498-499.

27. Fritzsche, R. & Kegler, H. (1968). Nematoden als Vektoren von Viruskrankheiten der Obstgewachse. Proc. VII European Symp. Virus Dis. Fruit Trees, Aschersleben, 289-295.

28. Frost, R.R., Harrison, B.D. & Woods, R.D. (1967). Apparent symbiotic interaction between particles of tobacco rattle virus. J. gen. Virol. 1, 57-70.

29. Gerola, F.M., Bassi, M. & Betto, E. (2965). Some observations on the shape and localization of different viruses in experimentally infected plants, and on the fine structure of the host cells. I. Arabis mosaic virus in Chenopodium amaranticolor. Caryologia 18, 353-375.

30. Gerola, F.M., Bassi, M. & Giussani-Belli, G. (1966). Some observations on the shape and localization of different viruses in experimentally infected plants, and on the fine structure of the host cells. IV. Arabis mosaic virus in Petunia hybrida Hort. Caryologia 19, 481-491.

31. Gerola, F.M., Bassi, M. & Belli, G. (1969). An electron microscope study of different plants infected with grapevine fanleaf virus. Giorn. bot. Ital. 103, 271-290.

32. Ghabrial, S.A. & Lister, L.M. (1973). Anomalies in molecular weight determinations of tobacco rattle virus protein by SDS-polyacrylamide gel electrophoresis. Virology 51, 485-488.

33. Ghabrial, S.A. & Lister, R.M. (1973). Coat protein and symptom specification in tobacco rattle virus. Virology 52, 1-12.

34. Gooding, G.V., Jr. (1970). Natural serological strains of tobacco ringspot virus. Phytopathology 60, 708-713.

35. Haidar, M.A., Pfeiffer, P., Fritsch, C. & Hirth, L. (1973). Sequential reconstruction of tobacco rattle virus. J. gen. Virol. 21, 83-97.

36. Halk, E.L. & McGuire, J.M. (1973). Translocation of tobacco ringspot virus in soybean. Phytopathology 63, 1291-1300.

37. Hansen, A.J., Nyland, G., MacElroy, F.D. & Stace-Smith, R. (1974). Origin, cause, host range and spread of cherry rasp leaf disease in North America. Phytopathology 64, 721-727.

38. Harrison, B.D. (1970). Tobacco rattle virus. C.M.I./A.A.B. Description of plant viruses 12, 4 pp.

39. Harrison, B.D. (1973). Pea early browning virus. C.M.I./A.A.B. Descriptions of plant viruses 120, 4 pp.

40. Harrison, B.D. & Nixon, H.L. (1960). Purification and electron microscopy of three soil-borne plant viruses. Virology 12, 104-117.

41. Harrison, B.D. & Winslow, R.D. (1961). Laboratory and field studies on the relation of arabis mosaic virus to its nematode vector Xiphinema diversicaudatum (Micoletzky). Ann. appl. Biol. 49, 631-633.

42. Harrison, B.D. & Wood, R.D. (1966). Serotypes and particle dimensions of tobacco rattle viruses from Europe and America. Virology 28, 610-620.

43. Harrison, B.D. & Roberts, I.R. (1967). Association of tobacco rattle virus with mitochondria. J. gen. Virol. 3, 121-124.

44. Harrison, B.D., Stefanac, Z. & Roberts, I.M. (1969). Role of mitochondria in the formation of X-bodies in cells of Nicotiana clevelandii infected with tobacco rattle virus. J. gen. Virol. 6, 127-140.

45. Harrison, B.D., Murant, A.F. & Mayo, M.A. (1972). Two properties of raspberry ringspot virus determinated by its smaller RNA. J. gen. Virol. 17, 137-141.

46. Harrison, B.D., Murant, A.F. & Mayo, M.A. (1972). Evidence for two functional RNA species in raspberry ringspot virus. J. gen. Virol. 16, 339-348.

47. Harrison, B.D., Finch, J.I., Gibbs, A.J., Hollings, M., Shepherd, R.J., Valenta, V. & Wetter, C. (1971). Sixteen groups of plant viruses. Virology 45, 356-363.

48. Harrison, B.D., Murant, A.F., Mayo, M.A. & Roberts, I.M. (1974). Distribution of determinants for symptom production, host range and nematode transmissibility, between the two RNA components of raspberry ringspot virus. J. gen. Virol. 22, 233-247.

49. Hewitt, W.B. & Grogan, R.G. (1967). Unusual vectors of plant viruses. Ann. Rev. Microbiol. 21, 205-224.

50. Hewitt, W.B., Goheen, A.G., Raski, D.J. & Gooding, G.V., Jr. (1962). Studies on virus diseases of the grapevine in California. Vitis 3, 57-83.

51. Honda, Y. & Matsui, C. (1972). Electron microscopy of intracellular radish mosaic virus. Phytopathology 62, 448-452.

52. Hooper, G.R., Spink, G.C. & Myers, R.L. (1972). Electron microscopy of leaf enations on Chinese white winter radish infected with radish mosaic virus. Virology 47, 833-837.

53. Jones, A.J. (1973). A comparison of some properties of four strains of cherry leaf roll virus. Ann. appl. Biol. 74, 211-217.

54. Jones, A.T. & Murant, A.F. (1971). Serological relationship between cherry leafroll, elm mosaic and golden elderberry viruses. Ann. appl. Biol. 69, 11-15.

55. Jones, A.T. & Mayo, M.A. (1972). The two nucleoprotein particles of cherry leaf roll virus. J. gen. Virol. 16, 349-358.

56. Jones, A.T., Kinninmonth, A.M. & Roberts, I.M. (1973). Ultrastructural changes in differentiated leaf cells infected with cherry leaf roll virus. J. gen. Virol. 18, 61-64.

57. Kaper, J.M. (1973). Arrangement and identification of single isometric viruses according to their dominating stabilizing interactions. Virology 55, 299-304.

58. Kenten, R.H. (1972). The purification and some properties of cocoa necrosis virus, a serotype of tomato black ring virus. Ann. appl. Biol. 71, 119-126.

59. Kim, K.S. & Fulton, J.P. (1971). Tubules with viruslike particles in leaf cells infected with bean pod mottle virus. Virology 43, 329-337.

60. Kim, K.S. & Fulton, J.P. (1972). Fine structure of plant cells infected with bean pod mottle virus. Virology 49, 112-121.

61. Kitajima, E.W. & Costa, A.S. (1969). Association of pepper ringspot virus (Brasilian tobacco rattle virus) and host cell mitochondria. J. gen. Virol. 4, 177-181.

62. Ladipo, J.L. & de Zoeten, G.A. (1972). Influence of host and seasonal variation on the components of the tobacco ringspot virus. Phytopathology 62, 195-201.

63. Lane, L.C. & Kaesberg, P. (1971). Multiple genetic components in bromegrass mosaic virus. Nature 232, 40-43.

64. Lister, R.M. (1964). Strawberry latent ringspot: a new nematode-borne virus. Ann. appl. Biol. 54, 167-176.

65. Lister, R.M. (1966). Possible relationships of virus-specific products of tobacco rattle virus infections. Virology 28, 350-353.

66. Lister, R.M. (1968). Functional relationships between virus-specific products of infection by viruses of the tobacco rattle type. J. gen. Virol. 2, 43-58.

67. Lister, R.M. (1969). Tobacco rattle NETU viruses in relation to functional heterogeneity plant viruses. Fed. Proc. ed. Amer. Soc. Exp. Biol. 28, 1875-1889.

68. Lister, R.M. & Murant, A.F. (1967). Seed-transmission of nematode-borne viruses. Ann. appl.Biol. 59. 49-62.

69. Lister, R.M. & Bracker, C.E. (1969). Defectiveness and dependence in three related strains of tobacco rattle virus. Virology 37, 262-275.

70. Maat, D.Z. (1963). Pea early browning virus and tobacco rattle virus: two different but serologically related viruses. Neth. J. Pl. Path. 69, 287-293.

71. Mali, V.R. & Hooper, D.J. (1973). Observations on Longidorus euonymus n.sp. and Xiphinema vuittenezi Luc et al. 1964 (Nematoda: Dorylaimida) associated with spindle trees infected with euonymus mosaic irus in Czechoslovakia. Nematologica 19, 459-467.

72. Martelli, G.P. & Raski, D.J. (1963). Osservazioni su Xiphinema index Thorne et Allen, fico e degenerazione infettiva della vite. Inftore Fitopatol. 13, 4 6-420.

73. Martelli, G.P. & Lamberti, F. (1967). Le specie di Xiphinema Cobb 1913 trovate in Italia e commenti sulla presenza di Xiphinema americanum Cobb (Nematoda: Dorylaimoidea). Phytopath. medit. 6, 65-85.

74. Martelli, G.P. & Castellano, M.A. (1971). A brief account of the use of the uranyl soak method for the visualization of some viruses in plant tissue. Phytopath. medit. 10, 76-81.

75. Martelli, G.P. & Quacquarelli, A. (1972). Hungarian chrome mosaic and tomato black ring: two similar but unrelated plant viruses. Ann. Phytopath., h. ser., 123-141.

76. Mayo, M.A., Murant, A.F. & Harrison, B.D. (1971). New evidence on the structure of nepoviruses. J. gen. Virol. 12, 175-178.

77. McGuire, J.M. & Schneider, I.R. (1973). Transmission of satellite of tobacco ringspot virus by Xiphinema americanum. Phytopathology 63, 1429-1430.

78. Miki, T. & Okada, Y. (1970). Comparative studies on some strains of tobacco rattle virus. Virology 42, 993-998.

79. Morris, I.J. & Semancik, J.S. (1973). In vitro protein polymerisation and nucleoprotein reconstitution of tobacco rattle virus. Virology 53, 215-224.

80. Murant, A.F. (1970). Arabis mosaic virus. C.M.I./A.A.B. Descriptions of plant viruses 16, 4 pp.

81. Murant, A.F. (1970). Tomato black ring virus. C.M.I./A.A.B. Descriptions of plant viruses 38, 4 pp.

82. Murant, A.F. (1970). Raspberry ringspot virus. C.M.I./A.A.B. Descriptions of plant viruses 4, 4 pp.

83. Murant, A.F. (1970). The importance of wild plants in the ecology of nematode-transmitted plant viruses. Outlook Agric. 6, 114-121.

84. Murant, A.F. & Lister, R.M. (1967). Seed-transmission in the ecology of nematode-borne viruses. Ann. appl. Biol. 59,

85. Murant, A.F., Mayo, M.A., Harrison, B.D. & Goold, R.A. (1972). Properties of virus and RNA components of raspberry ringspot virus. J. gen. Virol. 16, 327-338.

86. Murant, A.F., Mayo, M.A., Harrison, B.D. & Goold, R.A. (1973). Evidence of two functional RNA species and a "satellite" RNA in tomato black ring virus. J.gen. Virol. 19, 275-278.

87. Nixon, H.L. & Harrison, B.D. (1959). Electron microscopic evidence on the structure of the particles of tobacco rattle virus. J. gen. Virol. 21, 582-585.

88. Offord, R.E. (1966). Electron microscopic observations on the structure of tobacco rattle virus. J.Mol.Biol. 17, 370-375.

89. Offord, R.E. & Harris, J.I. (1965). The protein subunit of tobacco rattle virus. Fed. Europ. Biochem. Soc. Abstr. 2ns Meeting, Vienna, 216.

90. Pena-Iglesias, A. & Rubio-Huertos, M. (1971). Ultrastructura de hojas de Chenopodium quinoa Willd. enfectadas con el virus entrenudo corto infeccioso de la vid. Microbiol. Espan. 24, 183-192.

91. Puffinberger, C.W. & Corbett, M.K. (1973). Characterization of a NEPO virus isolated from euonymus. Phytopathology 63, 805.

92. Radewald, J.D. & Raski, D.J. (1962). Studies on the host range and pathogenicity of Xiphinema index. Phytopathology 52, 748-749.

93. Roberts, I.M. & Harrison, B.D. (1970). Inclusion bodies tubular structures in Chenopodium amaranticolor plants infected with strawberry latent ringspot virus. J.gen. Virol. 7, 47-54.

94. Roberts, D.A., Christie, R.G. & Ascher, M.C. Jr. (1970). Infection of apical initial in tobacco shoot meristems by tobacco ringspot virus. Virology 42, 217-220.

95. Rush, M.C. & Gooding, G.V. Jr. (1970). The occurrence of tobacco ringspot virus strains and tomato ringspot virus in host indigenous to North Carolina. Phytopathology 60, 1756-1760.

96. Russo, M., Martelli, G.P. & Quacquarelli, A. (1968). Studies on the agent of artichoke mottled crinkle. IV. Intracellular localization of the virus. Virology 34, 679-693.

97. Sänger, H.L. (1968). Characteristics of tobacco rattle virus. I. Evidence that its two particles are functionally defective and mutually complementing. Mol. Gen. Genet. 101, 346-367.

98. Sänger, H.L. (1969). Functions of the two particles of tobacco rattle virus. J. Virol. 3, 304-312.

99. Schmidt, H.B., Fritzsche, R. & Lehemann, W. (1963). Die Übertragung des Weidelgrasmosaic-virus durch Nematoden. Naturwisserschaften 50, 136-137.

100. Schneider, I.R. (1969). Satellite-like particles of tobacco ringspot virus that resembles tobacco ringspot virus. Science 166, 1627-1629.

101. Schneider, I.R. (1971). Characteristics of a satellite-like virus of tobacco ringspot virus. Virology 45, 108-122.

102. Schneider, I.R. & Diener, T.O. (1965). The ratio of infectious to non-infectious nucleoprotein in purified preparations from tobacco ringspot virus-infected plants as related to the age of infection. Phytopathology 55, 1075.

103. Schneider, I.R. & Diener, T.O. (1966). The correlation between the proportions of virus-related products and the infectious component during synthesis of tobacco ringspot virus. Virology 29, 92-99.

104. Schneider, I.R. & Diener, T.O. (1967). Nucleic acid and specific infectivity changes in bottom component of tobacco ringspot virus. Phytopathology 57, 829.

105. Schneider, I.R., Hull, R. & Markham, R. (1972). Multidense satellite of tobacco ringspot virus: a regular series of components of different densities. Virology 47, 320-330.

106. Schneider, I.R., White, R.M. & Civerolo, E.L. (1974). Two nucleic acid-containing components of tomato ringspot virus. Virology 57, 139-146.

107. Semancik, J.S. (1966). Purification and properties of two isolates of tobacco rattle virus from pepper in California. Phytopathology 56, 1190-1193.

108. Semancik, J.S. (1970). Identity of structural protein from two isolates of TRV with different length of associated short particles. Virology 40, 618-623.

109. Semancik, J.S. & Kajiyama, M.R. (1967). Properties and relationships among the RNA species from tobacco rattle virus. Virology 33, 523-532.

110. Semancik, J.S. & Reynolds, D.A. (1969). Assembly of protein and nucleoprotein particles from extracted tobacco rattle virus protein and RNA. Science 164, 559-560.

111. Siddiqi, M.R. (1974). Xiphinema index. C.I.H. Descriptions of plant-parasitic nematodes 45, 4 pp.

112. Silberschmidt, K., Weigl, D.R. & Salomao, T.A. (1970). Some electron microscopic observations on tobacco cells invaded by the Brazilian tomato yellow band disease virus. Phytopath. Z. 68, 210-220.

113. Sogo, J.M., Schneider, I.R. & Koller, Th. (1974). Size determination by electron microscopy of the RNA of tobacco ringspot satellite virus. Virology 57, 459-466.

114. Stace-Smith, R. (1970). Tobacco ringspot virus. C.M.I./A.A.B. Descriptions of plant viruses 17, 4 pp.

115. Stace-Smith, R. (1970). Tomato ringspot virus. C.M.I./A.A.B. Descriptions of plant viruses 18, 4 pp.

116. Stefanac, Z. & Ljubesic, N. (1971). Inclusion bodies in cells infected with radish mosaic virus. J. gen. Virol. 13, 51-57.

117. Taylor, C.E. (1971). Nematodes as vectors of plant viruses. In Plant parasitic nematodes, Eds. B.M. Zuckerman, W.F. Mai and R.A. Rhode, Acad. Press, London, vol. II, 185-211.

118. Taylor, C.E. (1972). Nematode transmission of plant viruses. Pans 18, 269-282.

119. Thomas, W. & Procter, C.H. (1972). Arabis mosaic virus in Cyphomandra betacea Sendt. N.Z. J. Agric. Res. 15, 395-404.

120. Thomas, W. & Procter, C.H. (1973). Tobacco ringspot virus in horse radish. N.Z. J. Agric. Res. 16, 233-237.

121. Uyemoto, J.K. (1970). Symptomatologically distinct strains of tomato ringspot virus isolated from grape and elderberry. Phytopathology 60, 1838-1841.

122. Van der Scheer, C. & Groenewegen, J. (1971). Structure in cells of Vigna unguiculata infected with cowpea mosaic virus. Virology 46, 493-497.

123. Van Kammen, A. (1972). Plant viruses with a divided genome. Ann. Rev. Phytopathol. 10, 125-150.

124. Vovlas, C., Martelli, G.P. & Quacquarelli, A. (1971). Le virosi delle piante ortensi in Puglia. VI. Il complesso delle maculature anulari della Cicoria. Phytopath. medit. 10, 244, 254.

125. Vuittenez, A., Kuszala, J., Rüdel, M. & Brückbauer, H. (1970). Détection et étude sérologique du virus latent des taches du annulaires du fraisier (strawberry latent ringspot), du virus des anneaux moir de la tomate (tomato black ring) et du virus des taches annulaires du fromboisier (raspberry ringspot) chez des vignes du Palatinat. Ann. Phytopath. 2, 279-327.

126. Walkey, D.G.A. & Webb, M. J.W. (1968). Virus in plant apical meristems. J. gen. Virol. 3, 311-313.

127. Walkey, D.G.A. & Webb, M.J.W. (1970). Tubular inclusion bodies in plants infected with viruses of the NEPO type. J. gen. Virol. 7, 159-166.

128. Walkey, D.G.A., Stace-Smith, R. & Tremaine, J.H. (1973). Serological, physical and chemical properties of strains of cherry leaf roll virus. Phytopathology 63, 566-571.

129. Walters, H.J. (1969). Beetle transmission of plant viruses. Adv. Virus. Res. 15, 339-363.

130. Weiner, A. & Raski, D.J. (1966). New host records for Xiphinema index Thorne et Allen. Pl. Dis. Reptr. 50, 27-28.

131. Weintraub, M. & Ragetli, H.W.J. (1970). Electron microscopy of the bean and cowpea strains of southern bean mosaic virus within leaf cells. J. ultrastruct. Res. 32, 167-189.

132. Weischer, B. (1974). Xiphinema-Arten europaischer Weinberge. Weinb. Kell. 21, 61-76.

DISCUSSION

The use of Tobra in preference to Netu for those viruses trans-mitted by *Trichodorus* species was queried and also the relationship of brome mosaic virus to the Nepo virus group. Martelli replied, with verification from Taylor, that a sub-committee of the International Committee on Nomenclature of Viruses had published (B.D. Harrison *et al*, *Virology* 45, 356-363 (1971)) a list of 16 distinct virus groups of which Tobra and Nepo had, amongst other characteristics, nematode vectors. The Bromo and Como virus groups were easily differentiated from each other and other virus groups and these groups might also have some nematode vectors.

The importance of the tubules observed in infected plant cells in virus taxonomy, and their composition, was raised by Estey, to which Martelli replied that little was known about tubules therefore they could not, as yet, be used in virus taxonomy; he speculated that they were probably lipoproteins. Vuittenez mentioned reports in the literature of tubular inclusions observed in cells of the root tips. Martelli affirmed that he had seen two reports of virus aggregates, possibly tubules, in root cells; it was not known if nematodes could acquire these tubules.

In response to questions about the importance of seed and pollen transmission of Nepo and Tobra viruses, Martelli said that this is important with the Nepo viruses as the infected seeds act as virus reservoirs and enable the viruses to persist in soil over-winter; the resultant seedlings do not show the typical virus symptoms. With the exception of pea early browning virus, which has few hosts, seed transmission is not thought to be as important in the Tobra virus group. Cooper stated that little was known about pollen transmission within this virus group and went on to say that pollen transmission was not important in the Nepo virus group, with the exception of cherry leaf roll virus where pollen transmission may be more important than nematode transmission.

The movement of Nepo and Tobra viruses from the roots to upper parts of the plants was queried by Yassin who also asked if aphids could acquire these viruses from infected leaves. As cell to cell movement was known to be relatively slow, Martelli suggested that some other movement of the virus within the plant takes place. Taylor said that viruses in leaves indeed were available to leaf feeders as shown by van Hoof (H.A. van Hoof, Nematologica 10, (1964)) experimentally transmitted tobacco rattle virus when he infected Trichodorus on the leaves of plants. He also said that satisfactory bait tests could be made by placing the leaves of healthy indicator plants into soil containing trichodorid nematodes and Tobra viruses.

Methods of virus eradication by meristem tip culture and heat therapy were discussed by Martelli and Pitcher and Thomason who remarked that plants could outgrow virus under optimum growing conditions. Virus and vector specificity were discussed in relation to a strain of arabis mosaic virus infecting grapevines in Hungary, France and Germany. This virus, which is closely related to the type strain (AB10) of arabis mosaic virus, is transmitted, according to Martelli and verified by Vuittenez, by a population of Xiphinema diversicaudatum from vine in France but not by a population of X. diversicaudatum from peach in Poland.

Pitcher suggested that failure to transmit the virus by an expected vector nematode may be due to the technique used, and pointed to the length of time it had taken to show hop AMV was transmitted by X. diversicaudatum because of problems with techniques. Cohn asked if X. index had been tried as a vector for AMV from grapevine but Dalmasso said the species had not been tried. In further discussions on specificity of transmission Cohn disclosed that he had not tried X. italiae as a vector of grapevine fanleaf virus other than as published (E. Cohn, E. Tanne and F.E. Nitzany. Phytopathology 60, 181-182 (1970)) and Martelli reported that he had only undertaken one experiment in which he had obtained a single transmission with X. italiae. The transmission of peach rosette virus by some populations of X. americanum was also quoted by Martelli in support of the complexities of virus and vector specificity.

ACQUISITION, RETENTION AND TRANSMISSION OF VIRUSES BY NEMATODES

C.E. TAYLOR and W.M. ROBERTSON

Scottish Horticultural Research Institute

Invergowrie, Dundee, Scotland.

INTRODUCTION

Nematode-transmitted viruses associated with disease outbreaks in the field fall into two groups called nepoviruses, with isometric particles about 30 nm in diameter and transmitted by species of Xiphinema and Longidorus, and tobraviruses, which have straight tubular particles and are transmitted by Trichodorus and Paratrichodorus species (see Martelli in these proceedings). At present 5 Xiphinema have been implicated in the transmission of 6 nepoviruses and 4 Longidorus species in the transmission of 3 nepoviruses (Table 1); 12 trichodorids have been shown to transmit the two known tobraviruses, tobacco rattle and pea early browning viruses, and it seems likely that many other species of Trichodorus or Paratrichodorus could transmit either or both of these viruses. Brome mosaic, carnation ringspot and prunus necrotic ringspot viruses have isometric particles and have been transmitted either by Longidorus macrosoma or Xiphinema diversicaudatum in laboratory experiments but it is not known whether these species are involved as vectors in nature.

The appearance of a patch or patches of diseased plants in a crop is often the first indication of nematode-transmitted infection. In perennial crops such as grapevines or raspberries, the disease may not make its appearance until two or three years after planting and spread of infection from the periphery of the patches rarely exceeds a few metres each year. A further characteristic is the persistence of infection in the soil for long periods, even during a fallow or when disease resistant crops are grown.

Since the initial discovery of X. index as a vector of

grapevine fanleaf virus (29) the problem of associating a vector with the causative virus has been investigated with a variety of techniques under different experimental conditions and, consequently, the various published experimental data on the transmission processes are not strictly comparable. Transmission may be affected by the way in which nematodes are extracted from the soil (63), by the potting mixture used in test pots (2), by the size of the test pots (11; 66), or by the age and kind of plants exposed to nematodes in bait tests (5). Such factors may be a reflection of the way in which nematodes feed intermittently or the variation in time taken to find suitable feeding places on the roots.

ACQUISITION AND RETENTION

Transmission efficiency

Adult and juvenile stages of the vector species appear to be capable of transmitting the associated viruses equally well. Grapevine fanleaf is transmitted by all stages of *Xiphinema index* (44), and tomato ringspot virus by all stages of *X. americanum* (65). Arabis mosaic and strawberry latent ringspot viruses are transmitted by adults and juveniles of *X. diversicaudatum* (38; 20); males appear to transmit strawberry latent ringspot virus less frequently than females (20; 63) although this may be due to differences in the frequency of feeding rather than inherent differences in efficiency of transmission. In early experiments it was reported that only juveniles of *Longidorus elongatus* transmitted tomato black ring virus (23) but later, under different experimental conditions, both adults and juveniles were shown to transmit the virus (71; 57). Tobacco rattle virus is also transmitted by males, females and larvae of *Paratrichodorus pachydermus* (49; 17) and by adults and larvae of *P. teres* (31). In experiments using a single nematode in each transmission test, Ayala and Allen (2) showed that a single specimen of any stage of *P. allius*, except the first stage which was not tested, transmit tobacco rattle virus; Teliz (64) transmitted tobacco ringspot virus by single *X. americanum*; and Harrison (20) simultaneously transmitted strawberry latent ringspot and arabis mosaic with single *X. diversicaudatum*.

Acquisition

The minimum periods published for nematodes to acquire viruses from infected plants and then to transmit them to other plants, refer to access periods and not to feeding periods. In general, increasing the times of access to both infected sources and healthy bait plants, results in increased transmission efficiency, although in experiments in which several nematodes are used in each test pot this undoubtedly points to the increasing proportion of nematodes that have fed. In some early experiments, *X. diversicaudatum* and arabis mosaic virus(38) and X. index and grapevine fanleaf virus, (28) transmissions were obtained after an access period on infected plants of 24 h, the minimum time tried. More recently, it has been shown that nepoviruses and tobraviruses can be acquired within an hour or less under suitable experimental conditions. Teliz et al (65) found that 1 h was sufficient for X. americanum to acquire tomato ringspot virus and to transmit to cucumber bait plants, but the rate of transmission increased when access was increased to a day or more. Tobacco rattle virus has also been transmitted by P. allius with exposures of 1 h to infector and bait plants, but again efficiency increased with the duration of the access periods. Nematode vectors probably can acquire sufficient virus for subsequent inoculation during a single feed on an infected plant which could be as brief as a few seconds.

Likewise, inoculation of a bait plant can probably occur in a single, brief feed. Inoculation access periods of less than 24 h have been established experimentally for several nematode vectors, and as with acquisition of virus, frequency of transmission increased with time of access. Teliz et al (65) obtained a few transmissions of tomato ringspot virus to cucumber seedlings after exposure to X. americanum for 1 h but obtained 100% transmission with an access period of 4 days. Similarly, efficiency of transmission of tobacco rattle virus by T. allius increased as the access time increased from 1 h up to 48 h (1).

Retention

The usual procedure for investigating the period for which nematodes can retain virus is to obtain nematodes from a highly infective population and add these in batches to pots of sterilized compost in which bait plants are planted at intervals to assay the efficiency of transmission. Alternatively, samples of natural soil can be planted with bait plants after intervals of fallow, although this method is subject to the possible error of germinating weed seeds acting as sources of virus infection (41; 63). Laboratory tests done in these ways show that viruses can persist in their vectors for several weeks or more, although the numbers of nematodes in a population able to transmit decreases with the period of starvation (55). X. americanum transmitted tobacco ringspot virus after 49 weeks of starvation at 10°C (3) and X. diversicaudatum transmitted strawberry latent ringspot virus after 84 days and arabis mosaic virus after 112 days (63). Viruses may persist in Xiphinema vectors even when allowed to feed on virus-immune hosts. X. index retained grapevine fanleaf virus for 12 wk when allowed to feed on virus-immune fig (8), and X. diversicaudatum on Malling Jewel raspberry which is immune to arabis mosaic virus apparently transmitted the virus for 8 months (27). Raspberry ringspot and tomato black ring viruses persist in L. elongatus for a maximum of about 8 wk at 20°C (41; 52; 54; 55) and it seems likely that Longidorus retains

Table 1. Viruses and their nematode vectors

Virus	Vector	Examples of Crops infected
	NEPOVIRUSES	
Arabis mosaic		
Type strain	*Xiphinema diversicaudatum* (122;37)	Cherry, cucumber, grapevine, raspberry, rhubarb
	X. coxi (10)	
Hop strain	*X. diversicaudatum* (67)	Hop
Cherry leaf roll	*X. diversicaudatum* (13) *X. coxi* (10) *X. vuittenezi* (9)	Cherry, blackberry, elm, rhubarb
Grapevine fanleaf	*X. index* (29) *X. italiae* (6)	Grapevine
Mulberry ringspot	*Longidorus martini* (70)	Mulberry
Raspberry ringspot		
Type (Scottish) strain	*L. elongatus* (55)	Blackberry, raspberry, redcurrant strawberry
English strain	*L. macrosoma* (19) *L. elongatus* (57)	
Cherry strain	*L. macrosoma* (14) *X. diversicaudatum* (14)	Cherry

Table 1 (cont..)

Strawberry latent ringspot	X. diversicaudatum (40) X. coxi (43)	Black currant, cherry, celery, peach, plum, raspberry, rose, strawberry
Tobacco ringspot	X. americanum (16) X. coxi (34)	Bean, blueberry, gladiolus, grapevine, tobacco
Tomato black ring		
English strain	L. attenuatus (23)	Celery, globe artichoke, lettuce, peach, potato, raspberry, strawberry, sugar beet, tomato
Scottish strain	L. elongatus (23)	
Tomato ringspot	X. americanum (4;35)	Blackberry, cherry, grapevine, peach, raspberry, tobacco
	TOBRAVIRUSES	
Pea early browning		
Dutch isolates	P. pachydermus (30) P. teres (30)	Pea, lucerne
English isolates	P. anemones (21) T. primitivus (21) T. viruliferus (17)	Pea, lucerne

Table 1 (cont..)

Tobacco rattle		
European isolates	P. anemones (32) P. nanus (32) P. pachydermus (50) P. teres (32) T. cylindricus (32) T. minor (39) T. primitivus (18;46) T. similis (7) T. viruliferus (32)	Potato, bulbous ornamentals, tobacco
American isolates	P. allius (36) P. christiei (68) P. porosus (2)	Potato, globe artichoke, lettuce, pepper, tobacco
	OTHER VIRUSES	
Brome mosaic	X. diversicaudatum (48) L. macrosoma (12)	Grasses, cereals
Carnation ringspot	X. diversicaudatum (15)	Carnation
Prunus necrotic ringspot	L. macrosoma (14)	Plum

virus for lesser periods than *Xiphinema* or trichodorids. Van Hoof (33) reported transmission of tobacco rattle virus by *P. pachydermus* in soil stored for about 2 yr; Ayala and Allen (2) found that the Californian strain of tobacco rattle virus persisted in *T. allius* for 20 wk in soil without host plants, but for 27 wk when virus-immune sweet pea was present.

Despite the long retention of viruses in their nematode vectors, the little experimental evidence available suggests that nematode-transmitted viruses do not persist through a moult, and do not pass through nematode eggs (27; 58; 2). By analogy with aphids, this suggests that the viruses are retained in a non-persistent manner (68) and probably extra-cellularly attached to some part of the feeding apparatus which is shed during a moult. This hypothesis has been examined by electron microscopy of ultra-thin sections to locate sites of retention (see below).

SPECIFICITY OF TRANSMISSION

So far nematode vectors have been shown to occur only in the genera *Xiphinema*, *Longidorus*, *Trichodorus* and *Paratrichodorus*, although it is known that nematodes in other genera can ingest virus particles from infected plants. This has been shown by cutting open nematodes that have been allowed to feed on infected plants and demonstrating the presence of virus by electron microscopy or by rubbing suspensions of the nematodes on to the leaves of suitable host plants in which the viruses cause symptoms. Tobacco rattle virus was shown to occur in the vector *Trichodorus* sp. by this technique (43) and Taylor (53) demonstrated that *L. elongatus* can ingest arabis mosaic and strawberry latent ringspot viruses although it does not transmit them. The viruses detected in this way are probably from the nematode's intestines, and because of the "one-way" action of the oesophago-intestinal valve they are unlikely to be involved in the transmission process.

Nevertheless, despite the ease with which nematodes can acquire viruses from plants there are relatively few vectors, particularly in the genera *Longidorus* and *Xiphinema*, and a puzzling feature is why a given vector can transmit one virus and not another and how a given virus can be transmitted by one species of vector though not by another species that may be closely related to it taxonomically. Thus *X. diversicaudatum* can transmit arabis mosaic and strawberry latent ringspot viruses but does not transmit grapevine fanleaf virus, which is distantly serologically related to arabis mosaic virus but transmitted by *X. index* (Table 1). *L. elongatus* transmits the Scottish strains of raspberry ringspot and tomato black ring viruses but *L. macrosoma* transmits only the English strain of raspberry ringspot and *L. attenuatus* transmits only the English strain

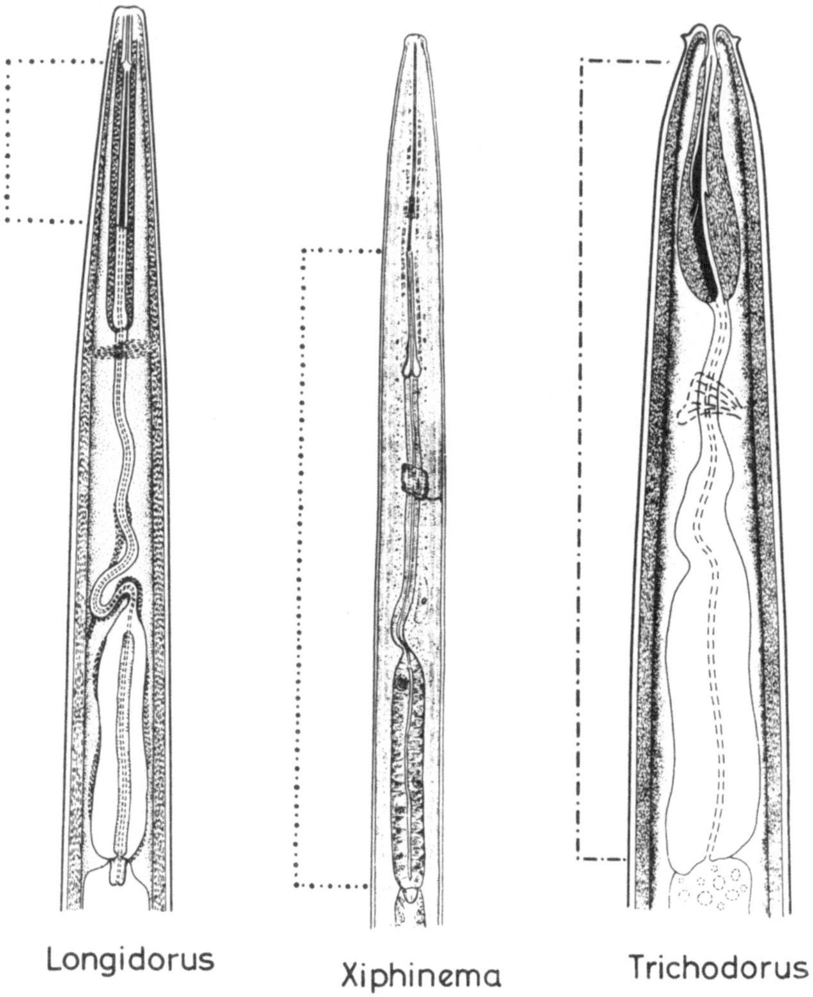

Fig. 1. Sites of specific retention of viruses in vector nematodes. The virus is associated with the cuticle of the alimentary tract in the areas indicated.

of tomato black ring virus. In trichodorids such vector specificity seems somewhat less well developed than in Xiphinema and Longidorus and several trichodorid species can transmit tobacco rattle virus or pea early browning virus (Table 1). Nevertheless, there is some evidence of specificity, because P. anemones will transmit a British but not a Dutch strain of pea early browning virus (21); and van Hoof (32) reported that P. pachydermus transmitted only one of the five Dutch isolates of tobacco rattle virus that were tested and concluded that the nematode becomes viruliferous only when a virus isolate is available that "suits" it.

Table 2. Virus acquisition, retention and transmission

Sequence	Process	Interaction
Plant	Multiplication	
Plant/nematode	Feeding	P/V
Nematode	Retention	V/N
Nematode/plant	Inoculation	P/N/V
Plant	Infection	V/P

P = plant N = nematode V = virus

Specificity is a complication of the already involved process of transmission. Nematodes acquire and transmit viruses when they feed at the root tip of the host plant. However, there is a sequence of complex events involving the biological properties and interactions of virus, nematode and plant at several stages (Table 2). The probability and efficiency of transmission depends on these events being successfully completed at each stage; failure at any one stage will cause failure to transmit. Retention of virus is an interaction between virus and nematode.

RETENTION SITES AND MECHANISM OF TRANSMISSION

Sites of retention

The sites of retention of viruses in nematodes have been discovered by means of electron microscopy of ultrathin sections of virus-carrying nematodes. Taylor and Robertson (59) found that in Longidorus elongatus particles of raspberry ringspot or tomato black ring viruses accumulate as a single layer on the inner surface of the guiding sheath (Plate I, fig. 1; Text-fig. 1). In Xiphinema

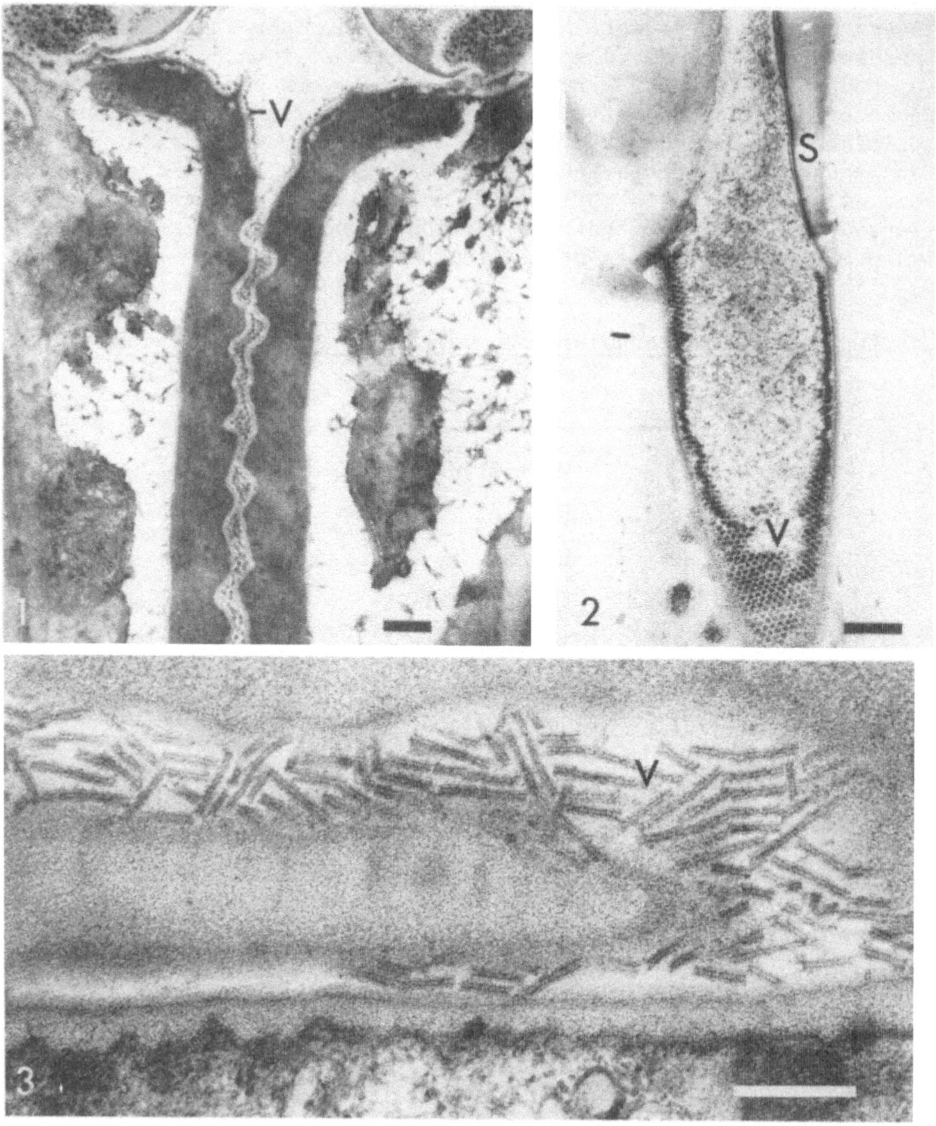

PLATE I. Virus at sites of retention in *Longidorus*, *Xiphinema* and *Trichodorus*.

1. Raspberry ringspot virus adsorbed to the guiding sheath in *L. elongatus*. 2. Arabis mosaic virus adsorbed to the odontophore in *X. diversicaudatum*. Note absence of particles in association with stylet (S). 3. Tobacco rattle virus in the oesophagus of *T. pachydermus*. Bars represent 200 nm.

diversicaudatum carrying arabis mosaic virus and X. index carrying grapevine fanleaf they found virus particles only in association with the cuticular lining of the oesophageal lumen, occurring as a monolayer from the anterior part of the odontophore to the posterior end of oesophageal basal bulb (60) (Plate I; fig. 2); similar observations have been made on X. americanum carrying tobacco ring-spot virus (42) and Raski et al (45) have confirmed the behaviour of strains of grapevine fanleaf virus in X. index. There is some evidence that virus particles are adsorbed more readily to the lining of the odontophore than to that in the remainder of the oesophagus. Tobacco rattle virus is found in a similar position to that of the viruses in Xiphinema spp, namely in association with the cuticular lining of the lumen of the pharynx and the oesophagus (61) (Plate I, fig. 3). Sites of retention in Longidorus on the one hand, and Xiphinema and trichodorids on the other, thus parallel differences in the periods of persistence of viruses in the vector.

Specificity of retention

It is thought that the virus particles are selectively and specifically adsorbed on the sites of retention as the virus-containing plant sap passes from the plant to the nematode's intestine. Other viruses present in the sap but not transmitted by the nematode are not retained at these sites. For example, particles of arabis mosaic virus were not found in L. elongatus on the guiding sheath or any other cuticular surface in the oesophageal region after the nematode had been allowed to feed on infected plants (59). Specificity therefore seems to reflect the specific association between the protein surface of the virus and the cuticular surface in the nematode at the site where viruses are retained. From the evidence of electron microscopy it seems inevitable that the surface of virus particles are important in specific retention, although the factors involved at the interface between virus particle and nematode cuticle remain a matter of speculation for the present. From a comparison of the relative transmissibility of pairs of serologically related viruses by the same nematode vector species (Table 3) it would seem that vector specificity is the most strongly developed where the viruses are the most distantly related serologically, and decreases with increasing antigenic similarity.

Further evidence can be derived from recent investigations on the structure of the ribonucleic acid (RNA) of raspberry ringspot virus and the production of hybrids between strains of the virus (24; 25). The RNA of the virus is composed of two molecules of different sizes, the larger RNA-1 and the smaller RNA-2, which can be manipulated separately thus enabling different properties of the virus to be assigned to one piece or the other; for example the protein coat of the virus is determined by RNA-2. Hybrids were made between the Scottish strain of raspberry ringspot virus transmitted by L. elongatus and the English strain of the virus trans-

Table 3. Nematode transmissibility and serological relationship of viruses

NEMATODE	VIRUSES COMPARED* A	B	TRANSMISSION A	B	% CROSS-REACTING ANTIBODY
Trichodorus anemones	PEBV-B	PEBV-D	Frequent	None	1
Xiphinema index	GFV	AMV	Frequent	None	8
X. diversicaudatum	AMV	GFV	Frequent	None	8
Longidorus elongatus	TBRV-S	TBRV-E	Frequent	Trace	12
L. macrosoma	RRV-E	RRV-S	Frequent	Trace	25
L. elongatus	RRV-S	RRV-E	Frequent	Less Frequent	25
L. elongatus	RRV-S	RRV-LG	Frequent	Frequent	90

*PEBV = pea early browning virus
GFV = grapevine fanleaf virus
AMV = arabis mosaic virus
TBRV = tomato black ring virus
RRV = raspberry ringspot virus

The suffixes B, D, E, LG and S refer to particular strains of the viruses.

(Information in Tables 3, 4 and 5 from Harrison, Robertson and Taylor, 1974a (26)).

mitted by L. macrosoma; the differences in the transmission of the two strains is shown in Table 4. The relative transmissibility of isolates of the parental and hybrid types is shown in Table 5, where it is seen that transmissibility by L. elongatus is determined by RNA-2; and the source of RNA-1 has no effect on transmission. The RNA-2 which determines the protein coat of the virus also determines transmissibility, and this suggests that transmissibility by the vector depends critically on the nature of the protein surface of the virus particles.

It may seem paradoxical that although serologically distinctive strains of a virus may have different vectors (e.g. English and Scottish strains of tomato black ring virus transmitted by L. macrosoma and L. elongatus respectively) serologically unrelated viruses may have the same vector (e.g. arabis mosaic and strawberry latent ringspot virus transmitted by X. diversicaudatum). It is suggested that this can happen because the features of the particle

Table 4. Specific transmission by Longidorus elongatus and L. macrosoma of strains E (English) and S (Scottish) of raspberry ringspot virus and their serological affinity.

Virus strain	Reciprocal of antiserum titre		Transmission by nematodes	
	RRV-E	RRV-S	L. elongatus	L. macrosoma
RRV-E	512	256	Less frequent	Frequent
RRV-S	138	1024	Frequent	Trace

Table 5. Transmission by Longidorus elongatus and serological specificity of hybrids between strains E and S of raspberry ringspot virus.

	RNA constitution of isolates*			
	RNA-1(S) RNA-2(S)	RNA-1(E) RNA-1(S)	RNA-1(S) RNA-2(E)	RNA-1(E) RNA-2(E)
Transmission (%) by L. elongatus	84	84	28	24
Serological specificity of isolates	S	S	E	E

*Hybrid isolates were made by using mixtures of purified samples of RNA-1 and RNA-2 as inoculum and culturing isolates from single lesions (25).

surface that are involved in serological specificity may not be identical with those that determine specific retention, so that a major change occurring in one type of feature often but not always results in a change in the other type (26). If surface charge density is the governing factor determining the adsorption of virus particles at the sites of retention, this may be similar in two serologically unrelated viruses; but also, a major change in an antigenic (serological) determinant might well lead to an alteration in the surface charge density of a virus.

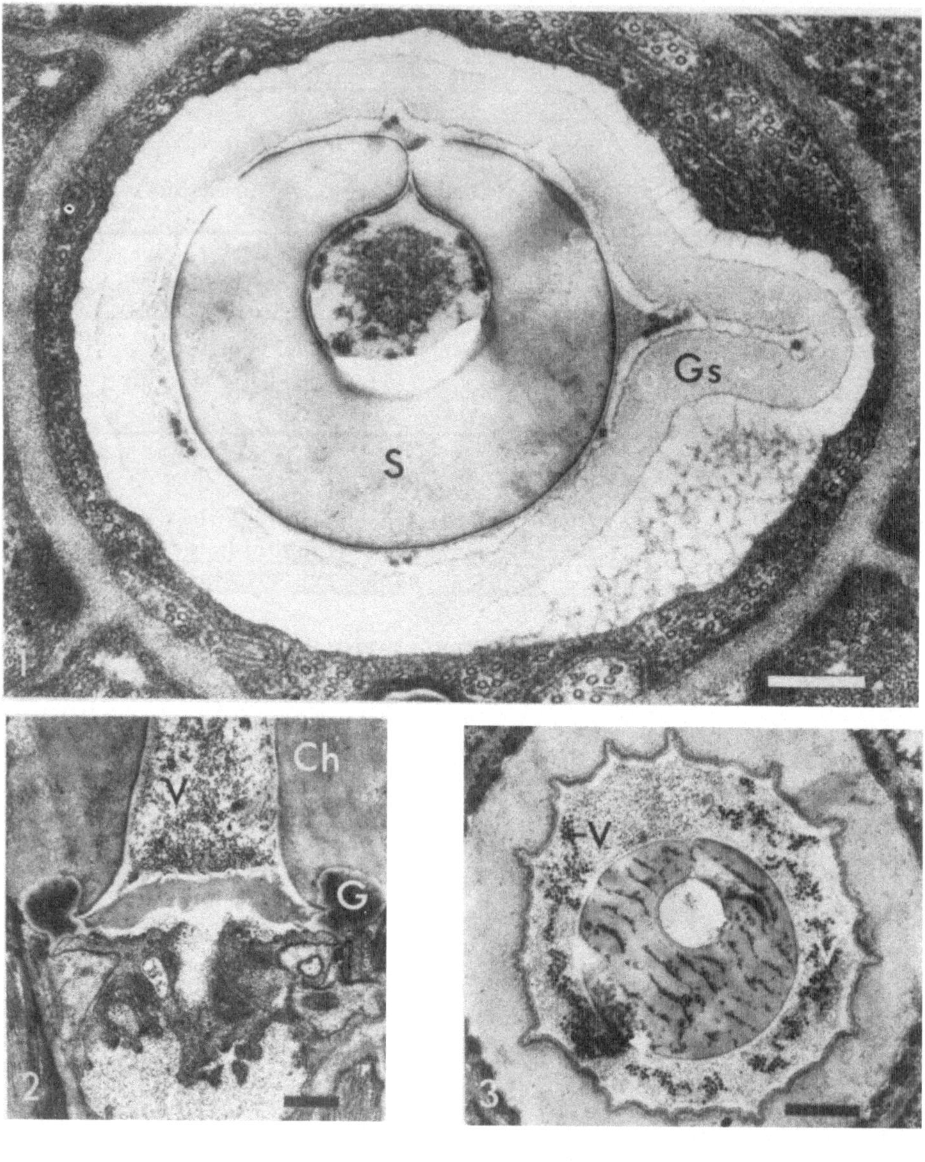

PLATE II.

1. T.S. of stylet (S) and guiding sheath (Gs) in L. macrosoma with particles of raspberry ringspot virus (English strain). 2. L.S. in area of guide ring (G) showing virus particles non-specifically associated in the lumen of the cheilostome (Ch). 3. T.S. of stylet in cheilostome of X. diversicaudatum; virus particles (V) in cheilostome lumen.
Bars represent 500 nm.

Mechanism of dissociation and release

For virus to be transmitted, virus particles must first associate with, and later dissociate from the site of retention in the nematode. Dissociation seems most likely to occur when saliva passes from the nematode into the plant cell during the initial phases of feeding and may be brought about by the pH or ionic conditions they change, or because of some enzymic effect of the saliva either on the surfaces of nematode or virus particles, or on materials which may be involved in binding the virus particles to the cuticular surface in the nematode. However it occurs the process of dissociation is likely to take place relatively slowly because viruses can persist for many weeks in their vectors and some can be transmitted serially to several plants.

Taylor and Robertson (62) have recently produced electron microscopy evidence indicating that some degree of specificity may also be involved in the dissociation of viruses within the nematode. They exposed *L. macrosoma* to plants infected with the English strain (E) or the Scottish strain (S) of raspberry ringspot virus and, as expected, obtained transmissions with strain E but not strain S. However, examination of ultrathin sections showed virus particles adsorbed to the cuticular guiding sheath and spear and with the same pattern of distribution both in *L. macrosoma* exposed to strain S as well as strain E (Plate II, fig. 1). A possible interpretation of these results is that strain S of raspberry ringspot virus becomes adsorbed to the site of retention in *L. macrosoma* but does not become specifically dissociated, as does strain E, and hence is not transmitted.

Effect of moulting

The available evidence suggests that viruses are not retained through the moult (27; 58). This is readily reconciled with the site of virus retention in *Longidorus* because the lining of the stoma, the stylet and the guiding sheath are shed with the nematode's outer cuticle. In *Xiphinema* and trichodorid nematodes the cuticular lining of the oesophagus is not cast with the outer cuticle. However, Taylor and Robertson (60 and unpublished) have shown by electron microscopy that in *Xiphinema*, as well as in *Longidorus*, considerable structural changes occur during the moult when the lining of the oesophagus is sloughed off and passes posteriorly into the intestine together with any virus particles that may have been present. After each moult, therefore, *Xiphinema* has a new, virus-free lining to its food canal and so is unable to transmit until it has again acquired virus particles from infected plants.

Non-specific transmission

Laboratory experiments have produced a few anomalous situations in which viruses are transmitted apparently in a non-specific manner. For example, Valdez (66) transmitted raspberry ringspot virus (E) with *X. diversicaudatum* and also with *L. caespiticola* and *L. leptocephalus*, which have not been associated with virus infection in field situations (55); the "other viruses" referred to in Table 1 have also been transmitted only in laboratory experiments. It has been long suspected that viruses may be retained for short periods in the lumen of the stoma in *Xiphinema* and *Longidorus* and transmissions occur because the nematodes are transferred quickly from infectors to bait plants which are in small quantities of soil to facilitate the immediate access of the nematode to the plant roots. Taylor and Robertson (59) published electron micrographs of virus-like particles in the cheilostome of *L. elongatus* fed on arabis mosaic virus-infected plants, although the virus was not transmitted. Since then evidence has accumulated to show that during feeding of *Xiphinema* and *Longidorus* clumps of virus particles may lodge in the triangular section of the lumen just anterior to the guide ring (Plate II, figs 2,3) and may be present in sufficient quantity to provide an infectious inoculum when the nematodes next feed.

The possible ways in which retention and transmission can occur are summarised in Table 6.

Table 6. Specific and non-specific retention and transmission

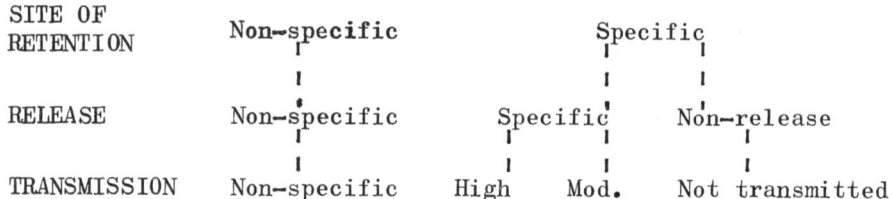

SITE OF RETENTION	Non-specific		Specific	
RELEASE	Non-specific	Specific		Non-release
TRANSMISSION	Non-specific	High	Mod.	Not transmitted

Inoculation and infection of plants

The last phase in the transmission process is the inoculation of the virus particles into plant cells. Successful transmissions can occur only when infective virus particles are inoculated into cells that are not so damaged or otherwise affected by the nematode that virus is unable either to replicate in them or to pass into adjacent cells. Inoculation presumably occurs when the virus particles are carried with the nematode's saliva into a plant cell and this presents another interface at which specificity of transmission could be determined, although there is no evidence for this.

REFERENCES

1. Ayala, A. & Allen, M.W. (1966). Transmission of the Californian tobacco virus by three species of the nematode genus Trichodorus. Nematologica 12, 87 (Abstr.)

2. Ayala, A. & Allen, M.W. (1968). Transmission of the Californian tobacco virus by three species of the nematode genus Trichodorus. J. agric. Univ. P.R. 52, 101-125.

3. Bergeson, G.B., Athow, K.L., Laviolette, F.A. & Thomasine, Sister M. (1964). Vector relationships of tobacco ringspot virus (TRSV) and Xiphinema americanum and the importance of this vector in TRSV infection of soybean. Phytopathology 54, 723-728.

4. Breece, J.R. & Hart, W.H. (1959). A possible association of nematodes with the spread of peach yellow bud mosaic virus. Plant Dis. Rep. 43, 989-990.

5. Cadman, C.H. & Harrison, B.D. (1960). Studies on the behaviour in soils of tomato black ring, raspberry ringspot and arabis mosaic viruses. Virology 11, 1-20.

6. Cohn, E., Tanne, E. & Nitzany, F.E. (1970). Xiphinema italiae, a new vector of grapevine fanleaf virus. Phytopathology 60, 181-182.

7. Cremer, M.G. & Kooistra, G. (1964). Investigations on notched leaf ('Kartelblad') of Gladiolus and its relation to tobacco rattle virus. Nematologica 10, 69-70 (Abst.)

8. Das, S. & Raski, D.J. (1968). Vector efficiency of Xiphinema index in the transmission of grapevine fanleaf virus. Nematologica 14, 55-62.

9. Flegg, J.J.M. (1969). Tests with potential nematode vectors of cherry leaf-roll virus. Rpt. E. Mall. Res. Sta. 1968, 155-157.

10. Fritzsche, R. (1964). Untersuchungen über dei virusübertragung durch Nematoden. Wiss. Z. Univ. Rostock, Malh.-Naturwiss. Reihe 13, 343-347.

11. Fritzsche, R. (1967). Methoden der Uhertragung pflanzen pathogener Viren durch Nematoden. Biol. Zentralbl. 86, 753-759.

12. Fritzsche, R. (1970). Beitrage zum Übertragungsmechanismus Planzen-Pathogener Viren durch Nematoden. Proc. 9th Int. Nematol-Symp., Warsaw, 1967 pp. 293-300.

13. Fritzsche, R. & Kegler, H. (1964). Die Ubertragung des Blattrolvirus der Kirsche (cherry leaf-roll virus) durch Nematoden. Naturwissenschaften 51, 299.

14. Fritzsche, R. & Kegler, H. (1968). Nematoden als Vektoren von virus krankheiten der Obstgewachse. TagBer. dt. Akad. Landw-Wiss.Berl. 97, 2890295.

15. Fritzsche, R. & Schmelzer, K. (1967). Übertragharkeit Nelkenringflecken-virus durch Nematoden. Naturwissenschaften 54, 498-499.

16. Fulton, J.P. (1962). Transmission of tobacco ringspot virus by Xiphinema americanum. Phytopathology 52, 375.

17. Gibbs, A.J. & Harrison, B.D. (1964). A form of pea early-browning virus found in Great Britain. Ann. appl. Biol. 54, 1-11.

18. Harrison, B.D. (1961). Soil-borne viruses - tobacco rattle virus. Rep. Rothamsted Exp. Sta. 1960, p. 118.

19. Harrison, B.D. (1964). Specific nematode vectors for serolog -ically distinctive forms of raspberry ringspot and tomato black ring viruses. Virology 22, 544-550.

20. Harrison, B.D. (1967a). The transmission of strawberry latent ringspot virus by Xiphinema diversicaudatum (Nematoda) Ann. appl. Biol. 60, 405-409.

21. Harrison, B.D. (1967b). Pea early-browning virus (PEBV). Rep. Rothamsted Exp. Sta. 1966, p. 115.

22. Harrison, B.D. & Cadman, C.H. (1959). Role of a dagger nematode (Xiphinema sp.) in outbreaks of plant disease caused by arabis mosaic virus. Nature (London) 184, 1624-1626.

23. Harrison, B.D., Mowat, W.P. & Taylor, C.E. (1961). Transmission of determinants for symptom production, host range and nematode transmissibility between the two RNA components of raspberry ringspot virus. J. Gen. Virol. 17, 137-141.

24. Harrison, B.D. Murant, A.F. & Mayo, M.A. (1972). Two properties of raspberry ringspot virus determined by its smaller RNA. J. Gen. Virol. 17, 137-141.

25. Harrison, B.D., Murant, A.F., Mayo, M.A. & Roberts, I.M. (1974a). Distribution of determinants for symptom production, host range and nematode transmissibility between the two RNA components of raspberry ringspot virus. J. Gen. Virol. 17, 137-141.

26. Harrison, B.D., Robertson, W.M. & Taylor, C.E. (1974b). Specificity of retention and transmission of viruses by nematodes. J. Nematology (in press).

27. Harrison, B.D. & Winslow, R.D. (1961). Laboratory and field studies on the relation of Arabis mosaic virus to its nematode vector Xiphinema diversicaudatum (Micoletzky). Ann. appl. Biol. 49, 621-633.

28. Hewitt, W.B. & Raski, D.J. (1962). Nematode vectors of plant viruses. In K. Maramorosch (ed.) Biological transmission of disease agents, Academic Press, New York. pp. 63-72.

29. Hewitt, W.B., Raski, D.J. & Goheen, A.C. (1958). Nematode vector of soil-borne fanleaf virus of grapevines. Phytopathology 48, 586-595.

30. Hoof, H.A. van (1962). Trichodorus pachydermus and T. teres, vectors of the early browning virus of peas. Tijdschr. Plantenziekten 68, 391-396.

31. Hoof, H.A. van (1964). Serial transmission of rattle virus by a single male of Trichodorus pachydermus Seinhorst. Nematologica 10, 141-144.

32. Hoof, H.A. van (1968). Transmission of tobacco rattle by Trichodorus species. Nematologica 14, 20-24.

33. Hoof, H.A. van (1970). Some observations on retention of tobacco rattle virus in nematodes. Neth. J. Pl. Pathol. 76, 329-330.

34. Hoof, H.A. van (1971). Virus transmitted by Xiphinema species in the Netherlands. Neth. J. Pl. Path. 77, 30-31.

35. Iwaki, M. & Komuro, Y. (1971). Viruses isolated from narcissus (Narcissus spp.) in Japan. II. Tomato ringspot virus and its transmission by Xiphinema americanum. Ann. Phytopath. Soc. Japan, 37, 108-116.

36. Jensen, H.J. & Allen, T.C. (1964). Transmission of tobacco rattle virus by a stubby-root nematode, Trichodorus allius. Pl. Dis. Reptr. 48, 333-334.

37. Jha, A. & Posnette, A.F. (1959). Transmission of a virus to strawberry plants by a nematode. Nature (London) 184, 962-963.

38. Jha, A. & Posnette, A.F. (1961). Transmission of Arabis mosaic virus by the nematode Xiphinema diversicaudatum (Micol.). Virology 13, 119-123.

39. Komuro, Y., Yoshino, M. & Ichinohe, M. (1970). Tobacco rattle virus isolated from Aster showing ringspot syndrome and its transmission by Trichodorus minor Colbran. Ann. Phytopath. Soc. Japan, 36, 17-26.

40. Lister, R.M. (1964). Strawberry latent ringspot: a new nematode-borne virus. Ann. appl. Biol. 54, 167-176.

41. Lister, R.M. & Murant, A.F. (1967). Seed-transmission of nematode-borne viruses. Ann. appl. Biol. 59, 49-62.

42. McGuire, J.M., Kim, J.S. & Dou Thit, L.B. (1970). Tobacco ringspot in the nematode Xiphinema americanum. Virology 42, 212-216.

43. Putz, C. & Stocky, G. (1970). Premieres observations sur une souche de strawberry latent ringspot virus transmise par Xiphinema coxi Tarjan et associee a une maladie du framboise en Alsace. Ann. Phytopath. 2, 329-347.

44. Raski, D.J. & Hewitt, W.B. (1963). Plant parasitic nematodes as vectors of plant viruses. Phytopathology 53, 39-47.

45. Raski, D.J., Maggenti, A.R. & Jones, N.O. (1973). Location of grapevine fanleaf and yellow mosaic virus particles in Xiphinema index. J. Nematology, 5, 208-211.

46. Sänger, H.L. (1961). Untersuchungen über schwer übertragbare Formen des Rattlevirus. Proc. 4th Conf. Potato Virus. Dis., Braunschweig, Germany 1960 pp. 22-28.

47. Sanger, H.L., Allen, M.W. & Gold, A.H. (1962). Direct recovery of tobacco rattle virus from its nematode vector. Phytopathology 52, 750 (Abstr.)

48. Schmidt, H.B., Fritzsche, R. & Lehmann, W. (1963). Die Übertragung des Weidelgras mosaik-virus durch Nematoden. Naturwissenschaften 50, 386.

49. Sol, H.H. (1963). Some data on the occurrence of rattle virus. Netherlands J. Pl. Pathology, 69. 208-214.

50. Sol, H.H., van Heuven, J.C. & Seinhorst, J.W. (1960). Tijdschrift Plantenziekten 66, 228-231.

51. Taylor, C.E. (1962). Transmission of raspberry ringspot virus by Longidorus elongatus (de Man), (Nematoda: Dorylaimoidea). Virology 17, 493-494.

52. Taylor, C.E. (1968a). The association of ringspot viruses with their nematode vectors. C.R. 8th Symp. Int. Nematol., Antibes, 1965, pp.109-110.

53. Taylor, C.E. (1968b). Nematology - association of virus and vectors. Rep. Scot. Hort. Res. Inst. 14, 66.

54. Taylor, C.E. (1970). The association of Longidorus elongatus with raspberry ringspot and tomato black ring viruses. Proc. 9th Int. Nematol. Symp. Warsaw 1967, pp.283-289.

55. Taylor, C.E. (1972). Nematode transmission of plant viruses. PANS 18, 269-282.

56. Taylor, C.E. & Brown, A.J.F. (1973). Nematology - NATO survey of Longidorus and Xiphinema. Rep. Scot. Hort. Res. Inst., 19, 74-75.

57. Taylor, C.E. & Murant, A.F. (1969). Transmission of strains of raspberry ringspot and tomato black ring virus by Longidorus elongatus (de Man). Ann. appl. Biol. 64, 43-48.

58. Taylor, C.E. & Raski, D.J. (1964). On transmission of grape fanleaf by Xiphinema index. Nematologica 10, 489-495.

59. Taylor, C.E. & Robertson, W.M. (1969). The location of raspberry ringspot and tomato black ring viruses in the nematode vector, Longidorus elongatus (de Man). Ann. appl. Biol. 64, 233-237.

60. Taylor, C.E. & Robertson, W.M. (1970a). Sites of virus retention in the alimentary tract of the nematode vectors, Xiphinema diversicaudatum (Micol.) and X. index Thorne and Allen. Ann. appl. Biol. 66, 375-380.

61. Taylor, C.E. & Robertson, W.M. (1970b). Location of tobacco rattle virus in the nematode vector, Trichodorus pachydermus Seinhorst. J. Gen. Virol. 6, 179-182.

62. Taylor, C.E. & Robertson, W.M. (1973). Nematology - Electron microscopy. Rep. Scot. Hort. Res. Inst. 19, 77.

63. Taylor, C.E. & Thomas, P.R. (1968). The association of Xiphinema diversicaudatum (Micoletzky) with strawberry latent ringspot and arabis mosaic viruses in a raspberry plantation. Ann. appl. Biol. 62, 147-157.

64. Teliz, D. (1967). Effects of nematode extraction method, soil mixture and nematode numbers on the transmission of tobacco ringspot virus by Xiphinema americanum. Nematologica 13, 177-185.

65. Teliz, D., Grogan, R.G. & Lownsbery, B.F. (1966). Transmission of tomato ringspot, peach yellow bud mosaic and grape yellow vein viruses by Xiphinema americanum. Phytopathology 56, 658-663.

66. Valdez, R.B. (1972). Transmission of raspberry ringspot virus by Longidorus caespiticola, L. leptocephalus and Xiphinema diversicaudatum and of arabis mosaic virus by L. caespiticola and X. diversicaudatum. Ann. appl. Biol. 71, 229-234.

67. Valdez, R.B., McNamara, D.G., Ormerod, P.J., Pitcher, R.S. and Thresh, J.M. (1974). Transmission of the hop strain of arabis mosaic virus by Xiphinema diversicaudatum. Ann. appl. Biol. 76, 113-122.

68. Walkinshaw, C.H., Griffin, G.D. and Larson, R.H. (1961). Trichodorus christiei as a vector of potato corky ringspot (tobacco rattle) virus. Phytopathology 51, 806-808.

69. Watson, M.A. (1972). Transmission of plant viruses by aphids. In C.I. Kado and H.Q. Agrawal (ed.) Principles and techniques in plant virology, van Nostrand Reinhold: New York, pp. 131-167.

70. Yagita, H. & Komuro, Y. (1972). Transmission of mulberry ringspot virus by Longidorus martini Merny. Ann. Phytopath. Soc. Japan 38, 275-283.

71. Yassin, A.M. (1968). Transmission of viruses by Longidorus elongatus. Nematologica 14, 419-428.

DISCUSSION

In reply to a question Taylor stated that a number of chemicals, some of which may be present in the nematode's saliva, could alter the ionic charge in the virus particle's surface and thus release the virus from its site of retention within the nematode; however, it was not thought that the fixation process could cause this. In reply to another question he pointed out that it was not possible to prove that the virus particles, as observed at the site of retention within the nematode vector, could or would be transmitted; there was, however, strong circumstantial evidence from experimental results to support the view that it was the virus particles retained at the site of retention which were eventually transmitted. Pitcher supported Taylor's view by mentioning that tobacco rattle virus particles from plant sap measured identically with those particles found at the site of retention within the Trichodorus vector which had fed on the original plant source for the virus.

Information was requested by Sienhorst on the length of time a virus could be retained by its nematode vector and he quoted H.A. van Hoof's results where infected Trichodorus kept in fallow soil at $4^{\circ}C$ for 5 years were still able to transmit tobacco rattle virus. Taylor replied that enzymes break down the virus in plants but virus could be stored in a frozen sap suspension for many years; it appeared reasonable, therefore, to suppose that virus could persist for long periods within a nematode where it presumably was protected within a favourable environment. There was no evidence of virus replication at the sites of retention as virus requires a living cell in which to replicate.

The importance, as a diagnostic feature, of the quantities of virus retained by nematodes was queried by Martelli, who suggested that this feature of virus retention might have been due to differing virus concentrations in the plants the nematodes fed upon. Taylor, however, stressed that in Xiphinema diversicaudatum arabis mosaic virus was retained as a closely packed monolayer whereas strawberry latent ringspot virus adsorbed at the same site showed spaces between the particles. This spatial arrangement of virus particles at the site of retention was therefore a more important diagnostic feature than actual quantities of virus at the site of retention. He also suggested that this spatial arrangement of virus particles at the site of retention would in part answer Estey's question as to whether simultaneous transmission of two serologically distinct viruses depended on which virus the nematode first acquired. The full answer to this question, however, was unknown at the present time.

Information was requested by Smith on any work where species had been extensively tested and proved to be non-vectors of NEPO viruses. Taylor stressed the importance of field observations linking virus with its nematode vector. Species which were not known virus vectors in the field such as *Longidorus leptocephalus* and *L. caespiticola* were reported to transmit certain NEPO viruses. However, he felt that such transmissions were of a non-specific association between virus and vector.

Short Report

SOME NON-ROSACEOUS TREES AND SHRUBS AS HOSTS TO NEMATODE-BORNE VIRUSES

J.I. Cooper (Unit of Invertebrate Virology, Commonwealth Forestry Institute, Oxford, England.)

Incidental to field observations on poplar plantations arabis mosaic virus (AMV), tomato black ring virus (TBRV) and tobacco rattle virus (TRV) were detected in roots but not leaves of *Populus* x *euramericana* cultivars. However, Nepo viruses were transmitted from leaves of other woody plants in which they were associated with symptoms of disease.

At a site in south-west England where the soil was infested with AMV and *X. diversicaudatum* (fewer than 5/200 g) one tree of *Fraxinus excelsior* showed chlorotic chevron and vein banding patterns on some but not all leaves. AMV was transmitted from leaves with this symptom and from symptomless leaves on the same tree. Adjacent individuals of *F. excelsior* showed a mild mottle symptom in some leaves and also yielded AMV. In partial fulfilment of Koch's postulates the leaves of *F. omus* seedlings were inoculated with an isolate of AMV from *F. excelsior*; chlorotic local lesions formed in the inoculated leaves and systemic infection was accompanied by mild mottle symptoms.

Infection with AMV was detected in leaves of certain other members of the Oleaceae. Indeed, AMV was the most frequently detected sap-transmissible virus in *Ligustrum* spp., its prevalence being judged from a survey made in April-May of *Ligustrum ovalifolium* and *L. vulgare* growing in gardens in Oxford. When leafy shoots taken at random from 287 bushes in 173 gardens were examined, 280 showed chlorotic chevron or ring markings. AMV was detected in all 55 symptom-bearing samples tested serologically and none of these yielded viruses which reacted with sera to TBRV, raspberry ringspot (RRV), cherry leaf roll (CLRV) or strawberry latent ringspot viruses (SLRV). The soil from the gardens was not examined for nematodes but AMV was detected in only 4 of the 55 samples tested.

Yellow line, ring or blotch symptoms in leaves of *Jasminium* officinale and *Syringa vulgaris* were also associated with systemic AMV-infection. AMV was detected in leaves of an individual of *Hedera helix* having yellow blotch and chevron patterns and RRV was transmitted from leaves of *Laburnum anagyroides* showing pale chlorotic chevron symptoms in some but not all leaves.

CLRV was isolated from roots and leaves of Betula verrucosa growing at two sites in England; at one site 31 of 47 trees yielded virus. In 1973, virus-infected trees tended to have leaves of normal area only at the distal ends of branches. Thus, the mean area of 527 leaves from stem bases was 16 sq mm whereas that of 268 leaves from stem tips was 60 sq mm. None of the known virus-vector genera were observed in soil to a depth of 150 cm and CLRV was not detected in the roots of bait plants (Chenopodium amaranticolor, C. quinoa and Cucumis sativus) grown in similar soil for a four week period.

Short Report

DISTRIBUTION OF ARTICHOKE ITALIAN LATENT VIRUS AND ITS NEMATODE VECTOR IN APULIA

F. Roca, G.P. Martelli and G.L. Rana (Laboratorio di Nematologia agraria del C.N.R., Bari and Istituto di Patologia vegetale, Universita di Bari, Italy.)

A survey was carried out to investigate the relative distribution of artichoke Italian latent virus (AILV) (1), and its nematode vector, *Longidorus attenuatus* Hooper, in Apulia, southern Italy. In this region artichoke is cultivated on about 15,000 ha in three major areas, *viz.* in the surroundings of Bari, where the crop was first established, and in the provinces of Brindisi and Foggia, where it spread subsequently. In these areas a total of over 180 fields of artichokes were inspected during autumn and winter, 1973-74. Soil samples from the rhizosphere of plants chosen at random (4-6/field) were checked for the presence of *L. attenuatus* and the leaves of the same plants were used for sap-inoculating French bean, a suitable host for assaying AILV (2). The results of this survey show that AILV prevails in the Bari area, occurring in more than 70% of the sampled fields (Table 1). The virus is less widespread in Brindisi and Foggia provinces where it was recovered from 23 and 10% of the samples, respectively.

L. attenuatus has a comparable distribution pattern, although it was encountered less frequently than AILV. Only 42, 15 and 4% of soil samples contained *L. attenuatus* in Bari, Brindisi and Foggia, respectively (Roca and Lamberti, pers. comm.). Most of the nematode populations in Bari were associated with AILV but in Brindisi and Foggia a large proportion AILV-infected fields was without the vector (Table 1). This suggests that AILV has been distributed primarily in infected propagating material. Based on present knowledge it is difficult to establish whether or not *L. attenuatus* pre-existed the introduction of artichoke in the provinces of Brindisi and Foggia, but the likelihood exists that the nematode, at least in Foggia, was introduced with artichoke sprouts used for propagation.

The efficiency of *L. attenuatus* as a vector of AILV was tested in glasshouse transmission experiments in which different numbers of hand-picked nematodes from the rhizosphere of naturally infected artichokes were added to tomato seedlings grown in 50 ml plastic containers filled with steam-sterilized sandy loam. Although after 2 months the nematode populations decreased by an average of 50% (Table 2), there was a reasonable level of transmission which tended to increase with the number of nematodes used. These results, which confirm and extend those obtained in previous trials(2)indicate that L. attenuatus is an efficient vector of AILV and provides an explanation for the consistent association of the nematode and AILV-infected artichokes found in nature.

REFERENCES

1. Majorana, G. and Rana, G.L. (1970). Un nuovo virus latente isolato da Carciofo in Puglia. Phytopath. medit. 9, 193-196.

2. Rana, G.L. and Roca, F. (1973). Transmissione con nematodi del virus latente italiano del Carciofo (AILV). Atti II Congresso Internazionale Carciofo, Bari, Nov. 1974 (in the press)

Table 1. Distribution of AILV and L. attenuatus in three artichoke-growing districts of Apulia.

L. attenuatus	AILV	Bari (92 samples)	Brindisi (39 samples)	Foggia (50 samples)
+	+	40%	3%	2%
+	−	2%	13%	2%
−	+	33%	21%	8%
−	−	25%	64%	88%

+ presence of virus or nematode
− absence of virus or nematode

Table 2. Transmission of AILV by L. attenuatus

Number of nematodes inoculated	Number (minimum and maximum) of nematodes recovered after 2 months	Percentage of transmission to tomato seedlings
5	1-3	42
10	2-4	25
20	5-9	67
30	0-15	58

Short Report

NETTLEHEAD AND RELATED HOP DISEASES ASSOCIATED WITH THE HOP STRAIN OF ARABIS MOSAIC VIRUS AND ITS VECTOR XIPHINEMA DIVERSICAUDATUM

R.S. Pitcher (East Malling Research Station, Maidstone, England)

The hop strain of arabis mosaic virus (AMV-H) differs considerably from the 'type' strains (AMV-T) found in most other hosts (e.g. strawberry, raspberry & many wild hosts) both in its host range and the symptoms it produces on herbaceous indicator plants. Like AMV-T, it is transmitted very effectively by *X. diversicaudatum* (2). The epidemiology of AMV on hops (*Humulus lupulus*), which are large, long-lived perennials, differs from that on other, smaller and shorter-lived crops and poses special problems to plant pathologists and hop growers (1). Each individual hop root system occupies at least 2 m^3 of soil, so that a site containing only one *X. diversicaudatum*/200 ml (i.e. the smallest population detectable in the usual soil sampling unit) exposes each hop plant to c 10,000 vector nematodes, each of which is capable of infecting it. The population of *X. diversicaudatum* which can be tolerated in a hop field when AMV-H is also present is therefore very low. Any foci of AMV-H infection will, eventually, extend to all adjacent soil containing *X. diversicaudatum*, the virus moving faster within the plant (2m per plant) than *via* migrating nematodes, which need to move little to transmit the virus from plant to plant wherever the root systems interlace.

Nettlehead is the most severe of three symptom types commonly found in hop fields, but yield losses of 20-30% can be caused, even by the mildest infections. Since the soil is the primary source of infection, effective control involves grubbing the entire infected crop and replacing it with healthy stock after the vector and/or the virus has been virtually eliminated from the field (see also in these Proceedings report by Pitcher on chemical control).

REFERENCES

1. Thresh, J.M., Pitcher, R.S., McNamara, D.G. & Ormerod, P.J. (1972). The spread and control of nettlehead and related diseases of hop. Rep. E. Malling Res. Stn for 1971, 155-162.

2. Valdez, R.B., McNamara, D.G., Ormerod, P.J., Pitcher, R.S. & Thresh, J.M. (1974). Transmission of the hop strain of arabis mosaic virus by *X. diversicaudatum.* Ann. appl. Biol., 76, 113-122.

Short Report

VIRUS VECTOR SPECIES OF XIPHINEMA AND LONGIDORUS IN RELATION TO CERTIFICATION SCHEMES FOR FRUIT AND HOPS IN ENGLAND

J. Cotten (M.A.F.F., Plant Pathology Laboratory, Harpenden, England.)

The biology and control of virus vector species of *Xiphinema* and *Longidorus* are being investigated at the Plant Pathology Laboratory to assist in the selection of sites for propagation of virus tested nursery stock of strawberry, blackcurrant, top fruit and hop. The highest grades of stock can be propagated only on sites which have been soil sampled and on which the appropriate vector species of *Xiphinema* and *Longidorus* have not been detected.

A survey of the distribution of *Xiphinema* and *Longidorus* species by the Scottish Horticultural Research Institute, Invergowrie (see Brown in these Proceedings), has shown that some vector species, notably *X. diversicaudatum* and *L. elongatus*, occur commonly in England on a range of soil types and under a range of crops. *L. macrosoma* is more restricted in its distribution being confined largely to southern and central England. *L. attenuatus* occurs on light sandy soils and is mainly confined to East Anglia.

Despite the prevalence of vector nematodes, the incidence of Nepo viruses other than the hop strain of arabis mosaic virus (AMV) is very low, particularly in eastern England. The type strain of AMV rarely occurs in fruiting strawberry crops, and in recent years growing season inspections of the lower grades of certified strawberry runners derived from virus tested stocks have never revealed AMV infections, although these crops are produced on sites which are not sampled for vector nematodes. The low incidence of virus in vector populations has been confirmed by the Scottish Horticultural Research Institute's survey. The hop strain of AMV is probably common in hop gardens in England (see Pitcher in these Proceedings), but this is probably the result of using uncertified non-virus-tested stock rather than of dissemination by vector nematodes.

Microplot experiments at the Plant Pathology Laboratory to investigate the host range of *X. diversicaudatum* show clear differences in host efficiency. Of those perennial crops tested, strawberry supported the greatest density (over 7,000 nematodes/l of soil at the maximum); raspberry and ryegrass were also efficient hosts (over 2,500/l at the maximum). In contrast hop was a poor host, never supporting more than 500 nematodes/l of soil. Arable crops including spring barley, winter wheat, winter bean, potato, sugar beet and cabbage were also generally poor hosts, rarely supporting more than 1,000 nematodes/1 of soil, and arable fields would seem to be the best choice of sites for certifiable crops.

Field and laboratory studies of mortality in fallow soil suggest that cultural control of the vectors is not practicable. In laboratory tests, 43% of L. elongatus and 7% of X. diversicaudatum adults survived for 43 wk at 20°C, although juveniles survived less well. At 5°C there was no measurable mortality of either species (Fig. 1).

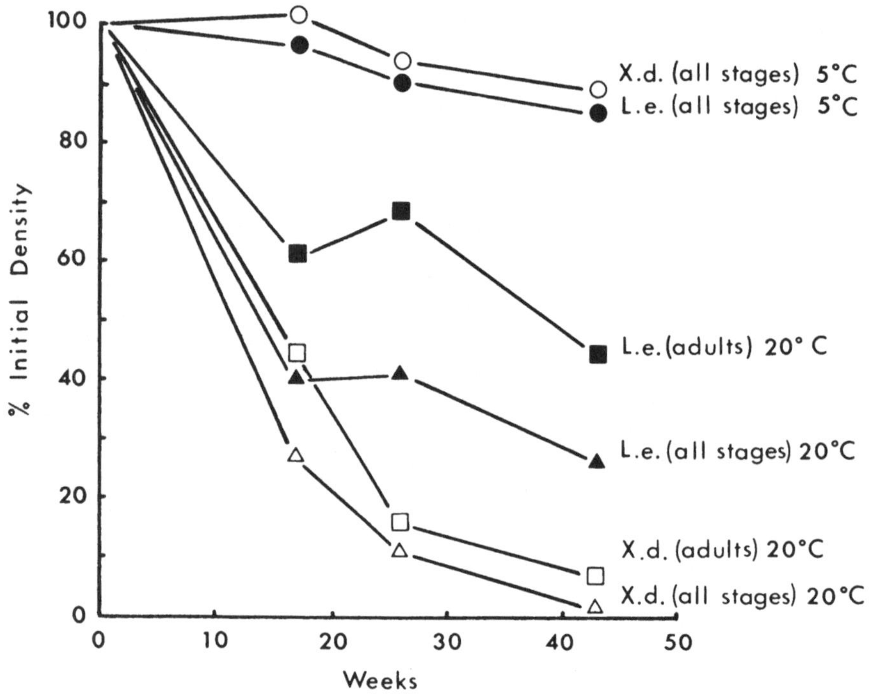

Fig. 1 Mortality of Xiphinema diversicaudatum (X.d) and Longidorus elongatus (L.e) in fallow soil at 20°C and 5°C

Vertical distribution of the nematodes, particularly in association with perennial crops, makes control with soil fumigant nematicides difficult. More than 800 L. macrosoma/1 of soil have been recovered at depths of 60 cm in association with raspberry and smaller numbers from considerably greater depths. L. elongatus and X. diversicaudatum have similarly been recovered from below 60 cm in association with a ryegrass ley. Under these conditions, attempts to control the vectors with normal doses and placement of nematicide are less than 100% successful. Table 1 shows the effect of dichloropropene (Telone) at 386 kg active ingredient/ha (322 l/ha) on L. macrosoma in a fine sandy loam with placement down to 15 cm. Control in the top 20 cm was 85%, but was only 73% from 20-40 cm and at 40-60 cm there was no measurable effect. Work is in progress to investigate the effect of increased dosage and deeper placement.

Table 1. Control of Longidorus macrosoma in a fine sandy loam soil with dichloropropene at 386 kg a.i./ha (322 l/ha)

Depth (cm)	L. macrosoma/500 ml soil		S.E. of difference of control and dichloropropene means
	Control	dichloropropene	
0-20	191.6	30.0**	
20-40	235.0	64.0**	±43.25
40-60	60.1	66.4	

**Significantly different from control at 1% level.

Short Report

NEMATODE TRANSMITTED VIRUSES IN BRITISH COLUMBIA, CANADA

F.D. McElroy (Canada Agriculture, 6660 NW Marine Drive, Vancouver, B.C., Canada.)

To date, two nematode transmitted viruses and their vectors have been found in British Columbia. Both are Nepo viruses and both are transmitted by *Xiphinema americanum*. However, the two viruses have limited distribution and they are currently of only minor economic importance.

Tomato ringspot virus (TmRSV) in raspberry. The distribution of TmRSV in raspberry is limited at least in part by the distribution of *X. americanum*. In British Columbia raspberries are grown mainly in three types of soil but the virus and vector occur only in clay loam soils in an area encompassing about 50 square miles. Even within this area the virus has limited distribution but where it does occur with the nematode vector it causes a serious problem. Raspberry plantings so infected begin to decline about three years after planting. While there is a decrease in the amount of vegetative growth each year the most serious problem to the grower is the effect on fruit quality. The virus causes a crumbly condition in the fruit which results in poor quality and difficulty in harvest. Replanting into such a field without adequate nematode control results in a reoccurrence of the problem and a necessity to replant within another 3-4 yr.

Other raspberry growing areas on loam and sandy loam soil types have neither the virus nor the vector. A different *Xiphinema* species, *X. bakeri*, predominates here. Repeated attempts under controlled conditions to transmit TmRSV with this species have failed. Further, no virus has been found naturally associated with this species. The same is true for *Longidorus elongatus* which has been found in raspberry but has a very limited distribution.

Cherry rasp leaf virus (CRLV) in cherry. The cherry rasp leaf disease occurs in the western United States and in British Columbia, Canada (1). It has been shown that the disease is caused by the sap-transmissible CRLV (3) and therefore differs from the English rasp leaf disease which is induced by the synergistic action of two viruses, prune dwarf virus and raspberry ringspot virus or arabis mosaic virus (2). Enation symptoms caused by CRLV are very severe and typically restricted to the lower part of the tree.

A survey was conducted in British Columbia to determine the involvement of nematodes in CRLV transmission in the field.

X. americanum was recovered from 43 of 64 samples collected and CRLV from 12 of those containing this species. No virus was recovered from nematodes taken from under indexed virus-free young peach trees growing near virus infected cherries nor from apparently healthy cherry trees. Where they occurred, viruliferous nematodes were present from the soil surface layers to a depth of at least 60 cm. CRLV was recovered once using bait cucumber seedlings to which only 10 nematodes had been added.

In British Columbia the disease is presently known to occur in only a few widely separated orchards involving single trees or at most less than twelve. However, there is a potential for spread or for infection of a newly planted orchard. Several common orchard weeds such as dandelion (Taraxacum officinale), plantain (Plantago major), and balsamroot (Balsamorhizer sagittata) are virus hosts and may serve as reservoirs from which nematodes can transmit the virus to cherry. In fact such a case has been partially documented (3). Indexed, virus-free cherry seedlings were planted into an orchard site which had been in apples since its first cultivation 43 yr previously. Two years after planting the cherry seedlings, six trees in a 4 x 6 m area displayed severe symptoms of rasp leaf. The virus was isolated from the affected parts of the trees, from symptomless dandelion and plantain growing among the affected trees and from X. americanum recovered from the soil beneath the trees.

REFERENCES

1. Bodine, E.W., E.D. Blodgett & T.B. Lott (1951). Rasp leaf. Pages 71-80 in U.S. Dep. Agric. Handb. 10.

2. Cropley, R. (1964). Further studies on European rasp leaf and leafroll diseases of cherry trees. Ann.appl.Biol. 53: 333-371.

3. Mansen, A.J., G. Nyland, F.D. McElroy & R. Stace-Smith (1974). Origin, cause, host range and spread of cherry rasp leaf disease in North America. Phytopathology 64: 721-727

Short Report

TRANSMISSION OF COWPEA MOSAIC BY *XIPHINEMA BASIRI* IN WESTERN NIGERIA

F.E. Caveness, R.M. Gilmer and R.J. Williams (International Institute of Tropical Agriculture, Ibadan, Nigeria.)

Cowpea mosaic is a serious virus disease of cowpea in the Ibadan area. Disease incidence may reach 80% by harvest time. The disease also occurs in soybean, sword bean, Mexican yam bean, and lima bean. Two well-defined disease syndromes occur in cowpea 1) a bright yellow fleck or mosaic and 2) a green blister with leaf distortion and puckering. Green blister type isolates are generally more injurious than the yellow fleck type isolates, but the two types are closely related serologically to each other and to a CPMV isolate from Arkansas, U.S.A. and El Salvador. Physical properties of the Nigerian CPMV isolates are: isodiametric particles about 27-29 nm, stable for several days when moist, inactivated by dryness, thermal inactivation point at 65-70°C, dilution end-point about 10^{-4}.

Economic injury to cowpea depends on three factors 1) virus isolate, 2) tolerance of the infected cowpea cultivar, and most important, 3) age of the host plant at time of infection. Early infections (7 days after emergence) reduce yields by 40-60% but late infections (after flowering) cause reduction of 5-10%.

The virus is transmitted through seeds from infected plants (1-5%) and such infections appear to supply inoculum for secondary spread in the field. In addition to the known vector, *Ootheca mutabilis* several new vectors have recently been identified: two beetles, *Luperodes lineata* and *Nematocerus acerbus*, the grasshoppers *Zonocerus variegatus* and *Catantops spissus*, and the thrips *Sericothrips occipitalis* and *Taeniothrips sjostedti*.

Soil containing a natural infestation of *Xiphinema basiri* Siddiqi, 1959 (= *X. ifacolum* Luc, 1961)[1] was planted to virus free cowpea seed in 500 cm^3 pots at 97 *X. basiri* per pot. Ten pots of cowpea were inoculated with CPMV six days after emergence. Ten uninoculated pots of cowpea served as controls. After 83 days all plants were pulled up and the pots replanted with virus free cowpea seed. At the end of 51 days 35 of 78 plants were infected with CPMV. None of the control plants was infected. The soil also contained populations of *Helicotylenchus pseudorobustus*, *Scutellonema clathricaudatum*, miscellaneous tylenchs, and saprozic forms.

A second trial using 100 hand picked X. basiri per pot in sterilized soil was run as above. Using 20 pots with four plants each seven of 80 plants were infected with CPMV 42 days after replanting. None of the control plants was infected. Investigations continue.

REFERENCE

Cohn, E. and S.A. Sher (1972). A contribution to the taxonomy of the genus Xiphinema. J. Nematology 4: 36-65.

ECOLOGY OF *XIPHINEMA* AND *LONGIDORUS*

BERNHARD WEISCHER

Biologische Bundesanstalt

D-44 Munster, Federal Republic of Germany

INTRODUCTION

In nematological literature there are few publications devoted entirely to ecology but on the other hand there are equally few which do not contain some information of an ecological nature. The majority of available information on the ecology of the approximately 70 *Xiphinema* and 40 *Longidorus* species is represented by short remarks connected with descriptions, surveys, population studies and so on, and thus in the literature the subject matter consists mainly of innumerable fragments without giving a comprehensive and clear picture. Climate, soil and plants are the main elements on which nematodes depend, but these are complexes each consisting of many interrelated and interacting physical, chemical and biological factors. The subject matter of this paper will be treated according to the most important of these environmental variables: temperature, water, soil, climate and season, plants, and other biotic factors. Preference is given to experimental data though they are almost nothing in number as compared to useless remarks like: "This species seems to prefer fairly dry soil".

ABIOTIC FACTORS

Temperature

According to their influence on nematodes five different temperature phases can be distinguished:

lethal low temperatures
minimum temperatures for activity
optimum temperatures

maximum temperatures for activity
lethal high temperatures

The actual figures for these temperatures vary from species to species and are in addition greatly influenced by other factors like time of exposure, moisture, host plant, stage of development and physiological activity of the nematodes. For *Xiphinema* and *Longidorus* few reliable data are available. Egg hatch in *X. americanum* occurs at all temperatures above freezing up to 32°C (22) and this may be true for most longidorids. Larval development is, however, more closely related to temperature. *X. americanum* matures more quickly in host-free soil as temperatures increase from 5° to 20°C; at 28°, however, the nematodes mature more slowly than at 20° (22). In *X. brevicolle*, *X. index*, *X. italiae* and *L. brevicaudatus* (now *L. sidiqii*) development is very slow below 16° and the duration of the life cycles of these nematodes are shorter with increasing temperature levels from 16° to 28°C (5). In *L. elongatus* no development occurs below 10° and the speed of development increases with temperatures increasing to 30° where the life cycle is completed within 9 weeks (58). In general *Longidorus* species have a shorter life cycle than *Xiphinema* species. At 20°-23°C on grapevine *X. index* completes its life cycle in 7-9 months, *X. brevicolle* in 4-7 months and *L. africanus* within less than 4 months (4). Under field conditions with fluctuating temperatures development is usually longer. *X. americanum* has its optimum temperature for reproduction at 20° (22), *X. index* at 24° (41), *L. elongatus* at 25° (58) and *L. africanus* at 30°C (28); no reproduction occurs below 24° with *L. africanus*. These data indicate remarkable differences in the optimum temperature for reproduction but it must, however, be considered that reproduction is not a clearly defined process. Most authors use it in the sense of "population development in time", and at the so called optimum temperature the population reaches the highest level within a given period. The quicker population increase with rising temperatures can be due to various factors such as progressively shorter life cycles, increase in egg production per female, or increase in number of gravid females. Hence, optimum temperatures for reproduction are useful in relation to culturing nematodes but they do not always give precise information about biology and ecology.

Populations of *X. americanum* cannot survive in frozen soil and only eggs and a few larvae overwinter (21, 22, 36), but in contrast most European *Xiphinema* species are not adversely affected by such temperatures. All stages of *X. diversicaudatum* and *X. vuittenezi* can be found in the soil throughout the year and no remarkable decline is observed during winter (15, 16, 47). With *L. elongatus* temperatures fluctuating around freezing are less damaging than constant exposure to frost and females are less sensitive than larvae (58); this is the reverse with *X. americanum*. In host-free soil, survival of X. americanum is greatest at temperatures between 5° and 20°C although populations decline during storage at any temperature (13). X. bakeri is less sensitive to storage. These nematodes increase in number during the first month of storage at 5° to 30° in naturally infested soil (52) which may be due to the hatching of eggs laid prior to storage. After the first month total numbers declined, especially at higher temperatures, and the populations comprised later development stages thus proving their greater tolerance (52).

Optimum temperature for migration of X. diversicaudatum, X. coxi and L. macrosoma is 20°-22°C (19). Short exposures to high temperatures does not affect their mobility, at least not in X. americanum. These nematodes pass as rapidly through a filter at 16° as at 29°C (29). Few data only are available on the influence of temperature on virus transmission but high temperatures are likely to be unfavourable. L. macrosoma transmits raspberry ringspot more efficiently at 20° than at 25° and no transmission occurs at 30°C (11).

Water

Though nematodes are active only in free water, species of Xiphinema and Longidorus show greatly differing relationships to humidity. Some species like X. hygrophilus and X. bergeri live in flooded soils, while others like X. neovuittenezi are confined to sandy soil with less than 500 mm annual rainfall. Some die in soils with less than 20% of water, while others like X. mediterraneum survive periods of desiccation by remaining quiescent and are reactivated by increasing humidity. Consequently one can expect specific reactions to soil water and there are indications that the occurrence of some species is more closely related to water content than to any other soil character. Unfortunately very few relevant data are available and are even fewer than data on temperature, probably because the latter is much easier to determine than soil moisture. Numbers of X. americanum follow changes in soil moisture but not in soil temperature. Populations increase at moisture levels of 13, 18 and 24% by weight (30). These nematodes do not survive outside the moisture range of 20-90% field capacity.

Many observations and experimental data show that fluctuations in soil moisture are more inimical to longidorids than constant levels. This high sensitivity probably is the main reason why these nematodes are found mostly at greater depths and why they are difficult to maintain in small pots. The optimum moisture for migration of X. diversicaudatum, X. coxi and L. macrosoma is 18-20% (21). Whether Xiphinema and Longidorus actively migrate downwards with increasing desiccation of upper soil layers is still an open question. Sampling in the field gave no clear answer. However, in experiments an undescribed Longidorus species (close to

L. elongatus) quickly moved downwards in a drying soil column reaching 75-100 cm within 2 months (45).

Soil water also influences tolerance to adverse temperatures. High soil moisture lessens the tolerance of L. elongatus to freezing temperatures (58). Most longidorids are sensitive to desiccation and cannot survive in really dry soils but there are species with a high tolerance e.g. X. mediterraneum becomes quiescent during summer when soils are drying out and is reactivated when moisture levels increase such as by a heavy rainfall (9). Still more impressive is X. bergeri which lives actively for four months per year in flooded paddy fields and survives the following desiccation period of eight months in a soil completely dry and as hard as concrete (31). It is interesting that another typical paddy nematode, Hirschmanniella spinicaudata, survives much better in dry soil than in a soil kept humid and this behaviour might be a common feature in species from habitats with extreme fluctuations in soil moisture. It is not known, however, whether longidorids in dry soils coil themselves up and secrete a protective sheath like other nematodes do.

Soil

In soil nematodes move through the system of pore spaces the size and distribution of which depend on the crumb size (see Pitcher in these Proceedings). Soil structure is the most important character for nematodes particularly because it is related to aeration and water. Because soil structure has very rarely been determined reference is made here to soil type although this is not correct.

Several reports connect the occurrence of a species with the soil type. X. brevicolle, X. italiae and X. neovuittenezi are mainly found in sandy soils near the coastline (3, 9). X. diversicaudatum is more common in heavy soils but occurs also in sandy soils and has even been found in peat soils in southern Britain (23). However, in my own studies on the occurrence of plant parasitic nematodes in swamps in northwestern Germany longidorids have never been found in original undisturbed swamps nor in drained moorland unless the process of mineralisation was already well advanced. In southern Germany X. vuittenezi is more frequent in loess than in sand whereas X. index occurs mainly in lighter soils (47). However, X. index can live in a variety of soils so that soil type is not the limiting factor for its geographical distribution (56, 57). Similarly X. americanum and L. elongatus are not very sensitive to soil type and occur in light soil and in heavy clay soil. Recently an interesting explanation (44) was given for such a tolerance: mobility is greatly influenced by the soil structure and since movement through soil is more essential for amphimictic species (in which males search for the females) than for parthenogenetic species, the latter can be more tolerant of soil type.

The majority of surveys have been carried out in cultivated soils. The occurrence of a species in arable land indicates its ability to exist here but whether it is a residue of the original fauna or was recently introduced cannot be decided. This can be demonstrated by two examples: X. index is widespread in vineyards in most vine growing countries but there is only one report from natural habitats, namely from woodlands in Iran (Sturhan, pers. com.). On the contrary, X. coxi is common in several types of natural habitats on diluvial soils but disappears when the soil is cultivated. It has been found under strawberry and in some vineyards but only when bordering old hedgerows or woodland (57). There -fore, studies on the distribution of longidorids in natural habitats are urgently needed. Recently a study was initiated on the occurrence of virus transmitting nematodes in natural habitats in northwestern Germany (see Rau in these Proceedings). Fig. 1 shows how species can be strictly limited to certain habitats under natural conditions. Similarly X. diversicaudatum is common in natural habitats on soils derived from Cretaceous chalk but was never found on Jurassic limestone. Where both soils are neighbouring there is a strict limitation and no individuals are found on the Jurassic part although suitable hosts may be present.

In Israel X. mediterraneum and other small species are predominant in clayey soils whereas the larger longidorids prevail in lighter soils (3). This observation relates soil structure directly with body size. Pore size, size of necks between soil particles and the length of the mean free path (= the mean distance a nematode can go before encountering an impossible neck or a blind end) are obviously more important for longer than for shorter nematode species (24). However, large species are found in heavy, compact clay soils and in most instances secondary spaces like cracks, clefts, traces of roots etc. are certainly more important than the primary pores between soil particles.

Population density and its fluctuations can be remarkably influenced by soil type. In Sardinian vineyards higher populations of X. index are found in light soils as opposed to heavy soils (40) and fluctuations in numbers of this nematode are more pronounced in heavy clay than in medium soil (2). L. elongatus builds up high populations on strawberry in light soil but in heavy soil the populations decline under this host (50). The population increase of X. brevicolle, X. diversicaudatum, X. index and X. italiae is quicker in heavy than in light soil which probably can be attributed to the more constant moisture (5). Depth distribution and stratification of Xiphinema and Longidorus are, of course, influenced by soil structure although the distribution of roots play a more important role as will be shown later. In relation to total

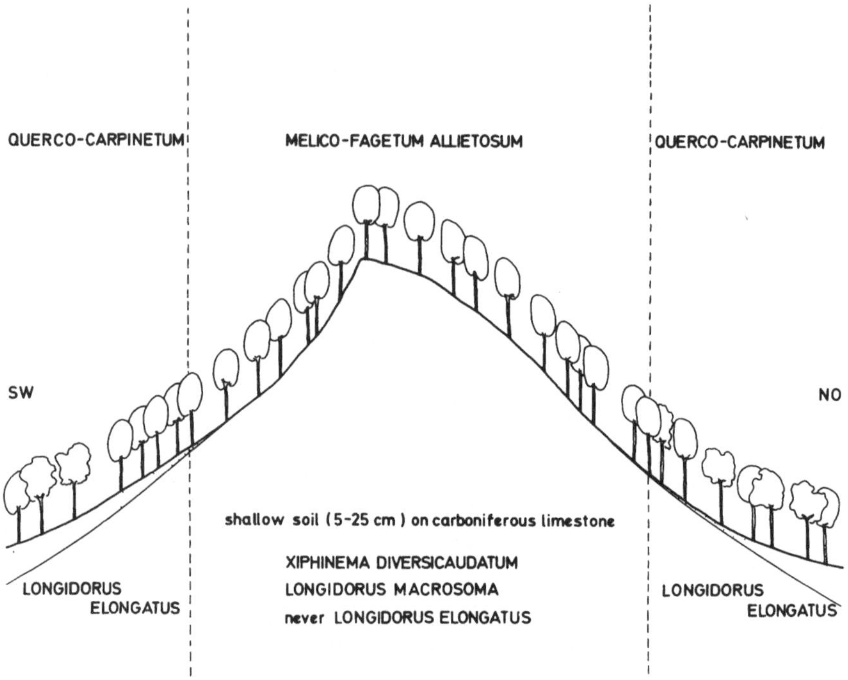

Fig. 1. Distribution of longidorids in a natural habitat on a mountain ridge in Germany. Longidorus elongatus is strictly limited to the deeper soils at the base. It has never been found in the shallow soil covering the slopes and the top, where L. macrosoma and Xiphinema diversicaudatum are common. The plant community on slopes and top is characterized by beech (Fagus silvatica) and Melica uniflora, a grass, whereas at the base oak (Quercus robur) and hornbeam (Carpinus betulus) are prevalent. (After Rau, 43).

numbers, most longidorids are most numerous at 10-30 cm depth with a rapid decline at more than 50 cm depth. Single individuals can, however, be found more than 2 m deep as long as living roots are present as found with X. index on grapes. Contrary to the majority of species, L. macrosoma and L. profundorum increase in number with increasing depth with a maximum at 70-80 cm (6, 17). The reasons for this pattern are unknown but these species may be particularly sensitive to fluctuations of environmental conditions. In many longidorids the percentage of adults increases with increasing depth (51).

The influence of soil structure on movement has already been mentioned. X. diversicaudatum covers 1.8 mm per day in heavy clay but 0.8 mm only in light sand. Similarly L. macrosoma migrates 1.4 mm in heavy clay and 0.4 mm in light sand. On the contrary the smaller X. coxi migrates 0.8 mm only in heavy clay but 2 mm in light sand (19). The best particle sizes for migration of X. americanum are from 150 to 700 µm but the optimum soil for movement is not always the best soil for other activities (39). Survival of X. americanum is greater in all soil types sieved to increase the pore space than in undisturbed soils. This is probably due to better aeration because survival is directly correlated with oxygen diffusion rates. Application of 7% oxygen to the atmosphere still resulted in a decline of X. americanum within 10 days whereas Meloidogyne incognita and Tylenchulus semipenetrans only decreased with a complete lack of oxygen (55) thus demonstrating a much greater tolerance.

Besides soil structure some chemical properties can influence longidorids. Mostly nematodes are highly tolerant of pH except for the extremes but populations of X. index are larger in soils with pH values between 6.5 and 7.5 as opposed to a pH of less than 6.5 (40). X. americanum has a similar range with maximum populations at pH 7 whereas X. chambersi has maximum populations at pH 4.5 (37). Concentration of nitrate ions and of phosphorous ions, and the cation exchange capacity (CEC-value) of a soil affect behaviour and survival of X. americanum, but other soil factors do not (49). A well known phenomenon affecting the handling of longidorids is their sensitivity to copper ions. Mortality of X. diversicaudatum in a concentration of 0.1 ppm copper ions is more than 50% compared to 1.3% of Aphelenchoides ritzemabosi (38).

Climate and Season

Climate and season are often treated as single environmental factors though in fact they are not. Temperature, humidity, light and their changes influence nematodes directly or indirectly via the plant hosts. Therefore their influence is always complex. Bearing this in mind some examples for climatic and seasonal influences on longidorids will be discussed. General occurrence of species is governed by climate. X. vuittenezi is a mainly continental species which does not occur in atlantic climate; X. italiae is confined to hot climate; L. elongatus is most common and most widely distributed in temperate climates whereas L. africanus is a typical inhabitant of tropical and subtropical regions. Similarly X. mediterraneum is very common in the Mediterranean region but is very rare in Germany (57). This species shows seasonal changes in activity (9) which obviously also occur in X. insigne in Japan (48). Studies on the influence of geographic location, latitude, mean annual precipitation and temperature on the morphology of 75 populations of the X. americanum-group revealed that shorter nematodes with smaller "a" ratios, smaller stylet lengths, and lower V-percentage are correlated with localities closer to the equator that have high amounts of precipitation and high temperatures, thus demonstrating a connection between climate and morphometrics (53).

In temperate climates reproduction i.e. gravid females and egg laying, occurs in spring and early summer only. This applies both to bisexual and parthenogenetic species of longidorids and is probably due to the presence of fresh young roots at this time of the year. Egg hatching occurs shortly after the eggs are laid but thereafter larval development does not show a clear relation to season. All stages can be found throughout the year and the relative proportions of larval stages and adults show no annual cycle (9, 16). X. americanum and L. africanus do not appear to follow this general pattern; all adults and second stage larvae die during winter and only some third and fourth stage larvae and the eggs overwinter (27, 28, 36). According to the local climate the period of reproduction is shifted towards the summer or towards the winter. Hence X. diversicaudatum reproduces from March to June in the west of France but in southern Italy gravid females occur as early as January (9). In temperate climates most species of Xiphinema and Longidorus show a seasonal fluctuation of population density, the extent of which depends mainly on the degree of climatic change (2, 7, 9, 18, 20, 21, 33, 36, 49).

BIOTIC FACTORS

Plants

The interrelationships between nematodes and plants are discussed elsewhere (see Cohn in these Proceedings). Here some more general aspects will be discussed. Xiphinema index provides a good example of the influence of a plant on the geographic distribution of a nematode species. About 90% of the reports on its occurrence are from grapevine and it has been found in most vine growing countries. In comparison with this host, other ecological factors like soil properties are of minor importance in relation to distribution and biology.

Recently an interesting relation between climate, nematodes and host plants was pointed out: in the centre of the area occupied by a nematode species the number of attacked host plants is much higher than at the periphery (9). So X. mediterraneum attacks many different plants in the Mediterranean area where it is endemic but in France at its northern border of distribution it is found on grapevines and fruit trees only, although other potential hosts are present. X. diversicaudatum shows a similar behaviour in that it attacks many plants in the Atlantic region whereas in Provence it is limited to roses and some fruit trees. It would be interesting to study whether these limitations are genetically fixed or caused by purely physiological reasons.

The local distribution of nematodes depend greatly on plants. X. bergeri, which is a typical nematode of rice paddies, can be found under tomatoes but only when these are grown on former rice-land (31). In Belgian strawberry fields X. diversicaudatum occurs mainly where fruit trees have previously been (12). Hedgerows often act as reservoirs of this nematode. In southern Britain they were estimated to have spread from an old hedgerow into former arable land later reverted to woodland at an average rate of 30 cm per year over a 75 years period (23).

With reference to vertical distribution I found more longidorids in the upper layer in natural habitats with a variety of annual and woody plants than in neighbouring cultivated fields where the majority occurred at depths greater than 20 cm. In vineyards most Longidorus and Xiphinema are found between 25 and 50 cm deep according to the distribution of suitable roots. In irrigated vineyards where most grape roots are in the upper soil layer the majority of X. index were in the same region, whereas in non-irrigated soils both roots and nematodes were more numerous at greater depths (C.E. Taylor pers. com.).

The influence of the host on nematode physiology and morphology can be demonstrated in X. americanum and X. bergeri. A population of X. americanum from blue spruce matured more quickly as temperatures increased from 5° to 20°C but not above that limit whereas a population from jack pine matured more quickly from 5° to 32°C thus showing a greater tolerance of larval development to high temperatures (22). Typical populations of X. bergeri from paddy fields have an average V-value of 32.2% whereas in some populations from woody plants the vulva is clearly more posterior (V = 35-39%) although all other characters fit perfectly (31).

The host ranges of the majority of Xiphinema and Longidorus species are incompletely known. Most species are polyphagous at least under greenhouse conditions. Generally they multiply on relatively more woody perennials than on herbaceous plants but more and more annual crops or weeds proved to be very efficient hosts. Higher populations are found under established woody plants than under arable cropping but this is probably due more to their sensitivity to alterations in the environment caused by agricultural practice than to the unsuitability of annual crops. The same applies to population dynamics. X. vuittenezi shows little fluctuation in the rhizosphere of grapevines but under annual crops the populations increase and decrease rapidly during the vegetative period particularly in the upper 20 cm i.e. above ploughing level (47). Regarding the influence of plants on population dynamics of longidorids it must, however, be considered that - due to the long life cycle of these nematodes - host plant effects are retarded. Thus the effect of the non-host maize on X. americanum is more pronounced in the second year because most individuals in the first growing season come from eggs laid in the previous year when a normal rate of reproduction occurred (14).

Many species of Xiphinema and Longidorus maintain themselves for long periods on several plants without multiplication. So, although the numbers of L. africanus drop below the initial density under a variety of plants they still are significantly greater than in fallow (27). Nutrition can change the host status of a plant as shown in X. americanum whose numbers are negatively related with the nitrogen supply (25, 35). Little is known on the existence of resistance to longidorids in general but there is a specific resistance to X. index in several species of Vitis e.g. V. candicans, V. solonis and V. rotundifolia. This resistance is not related to root-knot resistance (26). Some plants have a direct adverse effect such as raspberry which contains water soluble substances toxic to L. elongatus (54). X. bakeri, however, reproduces well on raspberry (32).

Other biotic factors

It is not unusual to find mixed populations of several species of Xiphinema and Longidorus. In vineyards, old pastures and woodlands I observed up to 4 species in one sample but in places with really high populations only one species was present and I think this is due to competition.

Microorganisms have been observed particularly in X. americanum and X. mediterraneum (1,38) where up to 50% of the females can be attacked. The presence of such parasites can well account for the well known difficulties in culturing these nematodes. The influence of viruses on their longidorid vectors has rarely been studied. Grapevine fanleaf virus causes slight physiological and anatomical changes and an increased rate of nematode survival occurred with X. index allowed to feed on grapevines infected with grapevine fanleaf virus (46, 10).

ECOLOGICAL FACTORS AND CONTROL

Some of the experimental data and observations on ecology described and discussed indicate that ecological factors can be used to control longidorids and virus spread but in the long term only. Due to the long individual life span and to their ability to survive long periods without feeding ecological control of these nematodes requires much time. Non-host rotations of 6-7 years are needed to control X. index in vineyards (8). Careful removal or killing of roots and keeping the soil strictly fallowed may even be more efficient. X. index persists in fallow soil following removal of grapes diminishing in number in accordance with the depletion of surviving roots (42). The selection of suitable fields can help to avoid difficulties since arable land with annual cropping usually contains fewer longidorids than old pastures, woodland or orchards. For direct control drying out of a soil can be a convenient practice e.g. in irrigated areas, but in other instances inundating for a moderate period may be more efficient. In such general terms ecological control looks promising. However, in practice the lack of detailed information becomes evident. Non-host rotations of 6-7 years may have a good effect but what are effective non-hosts of X. italiae, X. diversicaudatum or L. attenuatus to mention just 3 out of the 12 or more known virus vectors? Practically nothing is known about the other 100 or so species of Xiphinema and Longidorus. Thus, some pest management practices can be suggested but much more detailed and reliable information on the ecological requirements of these nematodes is needed.

REFERENCES

1. Adams, R.E. & Eichenmüller, J.J. (1964). Studies with Xiphinema americanum (Abstr.) Nematologica. 10, 70.

2. Amici, A. (1967). Fluttuazione della popolazione di Xiphinema index in una zona viticola italiana. Riv. Patol. Veg. 3, 99-104.

3. Cohn, E. (1969). The occurrence and distribution of species of Xiphinema and Longidorus in Israel. Nematologica 15, 179-192.

4. Cohn, E. & Mordechai, M. (1969). Investigations on the life-cycles and host preference of some species of Xiphinema and Longidorus under controlled conditions.

5. Cohn, E. & Mordechai, M. (1970). The influence of some environmental and cultural conditions on rearing populations of Xiphinema and Longidorus. Nematologica 16, 85-93.

6. Coolen, W.A. & D'Herde, C.J. (1970). Nematodes associated with glasshouse roses. Zesz. Probl.Postep.Nauk Roln. 92, 259-266.

7. Cotten, J. & Tuppen, C. (1972). Population studies with Longidorus macrosoma under annual and short and long-term perennial cropping regimes. Abstr. 11th Int. Symp. Nematol. Reading, 13-14.

8. Dalmasso, A. (1969). Méthodes de lutte contre un nématode vecteur du court-noué de la vigne. Phytoma 21, 20-23.

9. Dalmasso, A. (1970). Influence directe de quelques facteurs écologiques sur l'activité biologique et la distribution des espèces françaises de la famille des Longidoridae (Nematoda-Dorylaimida). Ann.Zool.Ecol.Anim. 2, 163-200.

10. Das, S. & Raski, D.J. (1969). Effect of grapevine fanleaf virus on the reproduction and survival of its nematode vector, Xiphinema index Thorne and Allen. J. Nematol. 1, 107-110.

11. Debrot, E.A. (1964). Studies on a strain of raspberry ringspot virus occurring in England. Ann.appl.Biol. 54, 183-191.

12. D'Herde, C.J. & van den Brande, J. (1964). Distribution of Xiphinema and Longidorus spp. in strawberry fields in Belgium and a method for their quantitative extraction. Nematologica 10, 454-458.

13. Elmiligy, I.A. (1971). Recovery and survival of some plant-parasitic nematodes as influenced by temperature, storage time and extraction technique. Meded. Fac. Landbouwwetensch. Gent 36, 1333-1339.

14. Ferris, V.R. & Bernard, R.L. (1971). Crop rotation effects on population densities of ectoparasitic nematodes. J. Nematol. 3, 119-122.

15. Flegg, J.J.M. (1966). Once-yearly reproduction in Xiphinema vuittenezi. Nature, London 212, 741.

16. Flegg, J.J.M. (1968). Life-cycle studies of some Xiphinema and Longidorus species in south-eastern England. Nematologica 14, 197-210.

17. Flegg, J.J.M. (1968). The occurrence and depth distribution of Xiphinema and Longidorus species in south-eastern England. Nematologica 14, 189-196.

18. Fritzsche, R. (1966). Beitrag zur Oekologie von Xiphinema diversicaudatum (Micoletzky) Thorne. Nachrichtenbl. dt. Pfl. schutzd. Berlin 20, 8-11.

19. Fritzsche, R. (1968). Beitrag zum Wanderungsverhalten von Xiphinema diversicaudatum (Micoletzky) Thorne, X. coxi Tarjan und Longidorus macrosoma Hooper sowie der Ausbreitung des Rhabarbermosaik-Virus im Feldbestand. Biol.Zbl. 87, 481-488.

20. Fritzsche, R. (1968). Oekologie und Vektoreignung von Longidorus macrosoma Hooper. Biol.Zbl. 87, 139-146.

21. Griffin, G.D. & Darling, H.M. (1964). An ecological study of Xiphinema americanum Cobb in an ornamental spruce nursery. Nematologica 10, 471-479.

22. Griffin, G.D. & Barker, K.R. (1966). Effects of soil temperature and moisture on the survival and activity of Xiphinema americanum. Proc. Helm Soc.Wash. 33, 126-130.

23. Harrison, B.D. & Winslow, R.D. (1961). Laboratory and field studies on the relation of arabis mosaic virus to its nematode vector Xiphinema diversicaudatum (Micoletzky). Ann.appl.Biol. 49, 621-633.

24. Jones, F.G.W., Larbey, D.W. & Parrot, D.M. (1969). The influence of soil structure and moisture on nematodes, especially Xiphinema, Longidorus, Trichodorus and Heterodera spp. Soil Biol. Biochem. 1, 153-165.

25. Kirkpatrick, J.D., Mai, W.F., Fisher, E.G. & Parker, K.G.(1959). Relation of nematode populations to nutrition of sour cherries (Abstr.) Phytopathology 49, 543.

26. Kunde, R.M., Lider, L.A. & Schmitt, R.V. (1968). A test of Vitis resistance to Xiphinema index. Amer. J. Enol. Viticult. 19, 30-36.

27. Lamberti, F. (1968). The effect of cropping on the population levels of Longidorus africanus. Plant Dis. Reptr. 52, 748-750.

28. Lamberti, F. (1969). Effect of temperature on the reproduction rate of Longidorus africanus. Plant Dis. Reptr. 53, 559.

29. Lownsbery, B.F. (1964). Some temperature reactions of Xiphinema americanum and implications to virus transmission. Plant Dis.Reptr. 48, 222-224.

30. Lownsbery, B.F. & Maggenti, A.R. (1963). Some effects of soil temperature and soil moisture on population levels of Xiphinema americanum. Phytopathology 53, 667-668.

31. Luc, M. (1973). Redescription de Xiphinema hallei Luc 1958 et description de six nouvelles especes de Xiphinema Cobb 1893 (Nematoda:Dorylaimoidea). Cah.ORSTOM ser. Biol. 21, 45-63.

32. McElroy, F.D. (1972). Studies on the host range of Xiphinema bakeri and its pathogenicity to raspberry. J. Nematol. 4, 16-22.

33. Malek, R.B. (1969). Population fluctuations and observations of the life cycle of Xiphinema americanum associated with cotton wood (Populus deltoides) in South Dakota. Proc. Helm.Soc.Wash. 36, 270-274.

34. Morone di Lucia, M.R. & Grimaldi de Zeo, S. (1973). Presenza di Microsporidi in gonadi di Xiphinema mediterraneum. Nematologia Mediterr. 1, 66-68.

35. Nigh, E.L. (1964). The influence of host nutrition on the development of Xiphinema americanum. Diss. Abstr. 24, 4340-4341.

36. Norton, D.C. (1963). Population fluctuations of Xiphinema americanum in Iowa. Phytopathology 53, 66-68.

37. Norton, D.C. & Hoffmann, J.K. (1974). Distribution of selected plant parasitic nematodes relative to vegetation and edaphic factors. J. Nematol. 6, 81-86.

38. Pitcher, R.S. & McNamara, D.G. (1972). The toxicity of low concentrations of silver and cupric ions to three species of plant parasitic nematodes. Nematologica 18, 385-390.

39. Ponchillia, P.E. (1972). Xiphinema americanum as affected by soil organic matter and porosity. J. Nematol. 4, 189-193.

40. Prota, U. (1970). Sull'influenza di alcune caratteristiche del suolo e dell'eta delle viti sulla distribuzione di Xiphinema index Thorne et Allen in Sardegna. Studi Sassar. Sez.III. 18, 1-12.

41. Radewald, J.D. & Raski, D.J. (1962). Studies on the host range and pathogenicity of Xiphinema index (Abstr.) Phytopathology 52, 748-749.

42. Raski, D.J., Hewitt, W.B., Goheen, A.C., Taylor, C.E. & Taylor, R.H. (1965). Survival of Xiphinema index and reservoirs of fanleaf virus in fallowed vineyard soil. Nematologica 11, 349-352.

43. Rau, J. (1973). Das Vorkommen von virusübertragenden Nematoden und der mit ihnen vergesellschafteten Viren in ungestörten Biotopen der landschaftlichen Großräume Niedersachsens. Mitt.Biol.Bundesanst. Dahlem 151, 302.

44. Robbins, R.T. & Barker, K.R. (1974). The effects of soil type particle size, temperature, and moisture on reproduction of Belonolaimus longicaudatus. J. Nematol. 6, 1-6.

45. Rossner, J. (1972). Vertikalverteilung wandernder Wurzeinematoden im Boden in Abhängigkeit von Wassergehalt und Durchwurzelung. Nematologica 18, 360-372.

46. Roggen, D.R. (1966). On the morphology of Xiphinema index reared on grape fanleaf virus infected grapes. Nematologica 12, 287-296.

47. Ruedel, M. (1971). Vorkommen einiger Arten der Gattung Xiphinema (Nematoda:Dorylaimidae) in Pfalz und Rheinhessen. Weinberg und Keller 18, 505-520.

48. Saisuga, T. & Yamamoto, Y. (1973). [Distribution and host plants of Xiphinema insigne Loos detected from exported lily bulbs]. Res. Bull. Plant Prot.Serv. Japan 9, 27-38. (Japanese with Engl. sum.)

49. Schmitt, D.P. (1973). Soil property influences on Xiphinema americanum populations as related to maturity of loess-derived soils. J. Nematol. 5, 234-240.

50. Sharma, R.D. (1965). Direct damage to strawberry by Longidorus elongatus (de Man 1876) Thorne and Swanger 1936. Meded. Landbouwhogesch. Gent 30, 1437-1443.

51. Sturhan, D. (1963). Der pflanzenparasitische Nematode Longidorus maximus, seine Biologie und Oekologie, mit Untersuchungen an L. elongatus und Xiphinema diversicaudatum. Z. angew.Zool. 50, 129-193.

52. Sutherland, J.R. & Ross, D.A. (1971). Temperature effects on survival of Xiphinema bakeri in fallow soil. J. Nematol. 3, 276-279.

53. Tarjan, A.C. (1969). Variation within the Xiphinema americanum-group (Nematoda:Longidoridae). Nematologica 15, 241-252.

54. Taylor, C.E. & Murant, A.F. (1967). Nematicidal activity of aqueous extracts from raspberry canes and roots. Nematologica 12, 488-494.

55. Van Gundy, S.D., Stolzy, L.H., Szuszkiewicz, T.E. & Rackham, R.L. (1962). Influence of oxygen supply on survival of plant parasitic nematodes in soil. Phytopathology 52, 628-632.

56. Weischer, B. (1966). Ein Beitrag zur geographischen Verbreitung und Oekologie von Arten der Gattung Xiphinema und Longidorus. Mitt.Biol. Bundeanst.Dahlem 118, 100-106.

57. Weischer, B. (1974). Xiphinema-Arten europäischer Weinberge. Weinberg und Keller 21, 61-76.

58. Wyss, U. (1970). Untersuchungen zur Populationsdynamik von Longidorus elongatus. Nematologica 16, 74-84.

DISCUSSION

In his talk Weischer contended that Xiphinema and Longidorus are not found in pure peat soils. Cotten and Winfield said that they had recovered Xiphinema and Longidorus from peaty soils in Britain, but Weischer observed that if the soils were cultivated then there would be some degree of mineralisation. Seinhorst, however, said that he had found X. diversicaudatum in sphagnum peat used for potting compost in the Netherlands, and hence he concluded that it was not safe to assume that peat was nematode-free and it should therefore always be sterilized before using it in experiments.

The possibility of ecological control was illustrated by Cohn by reference to L. vineacola infesting onion, carrot and celery crops in Israel. Nematode populations increase during December and decline during March; by sowing the crops early in the season the susceptible seedling stage is reached before the peak nematode population in December.

The importance or otherwise of soil factors such as temperature and soil moisture were discussed and examples quoted. Martelli observed that most Xiphinema species in Europe are not affected by soil temperatures, but he queried whether nematodes in the upper soil layers die out. Weischer suggested that in frozen soil nematodes are killed mostly by drying out. Winfield indicated that soil moisture could influence the vertical distribution of different nematode species differentially. He quoted experiments of K. Evans who found that in sandy soils Meloidogyne hapla infested the upper layers whereas L. leptocephalus was found in the lower layers. Potato crops grew poorly in a drought season in these soils and this was believed to be due to L. leptocephalus attacking the roots. Coomans queried whether factors like temperature and soil moisture are really important in natural habitats, and asked if nematode stratification had been studied in soils containing several species. Weischer said that

X. attorodorum provided an example of distribution in relation to soil type; in a transect most were found in silt, followed by clay and then coarse sand. Luc observed that sampling for X. attorodorum at constant depth does not eliminate variation due to root density along a transect. X. vuittenezi is apparently distributed in relation to the distribution of host plant roots but L. profundorum, which is found to a depth of 80 cm, is not. In South Africa X. elongatum is found in the top layers of soil, while X. hallei is always found below 40 cm. Lippens said that a survey of Belgian orchards had shown a constant stratification of longidorids but there were differences from plant to plant. Seinhorst observed that some nematodes e.g. L. elongatus prefer certain layers independent of root systems.

Interspecific competition was also quoted as another factor affecting nematode distribution. Lamberti referred to pot experiments in the glasshouse in which X. index populations had become infested with Meloidogyne sp. and this had led to a sharp decline in the numbers of X. index. Miller observed that in peanut crops in Virginia, U.S.A., Dolichodorus haplocaudata had been found to suppress reproduction of Meloidogyne hapla.

Coomans pointed out that in extreme soil conditions e.g. paddy soils, special attention should be given to the possibility of nematodes being in a quiescent phase. He referred to the cocoon-like structures present in Trichotylenchus and queried whether similar structures might be associated with some Xiphinema and Longidorus species. Luc said that in Senegal, peanut soils may have an 8-9 month dry season and special extraction methods were being developed for such situations. Thomason indicated that the biochemistry of quiescent nematodes is being studied in his laboratories.

THE ECOLOGY OF TRICHODORUS

A.L. WINFIELD AND D.A. COOKE

Agricultural Development and Advisory Service, Brooklands Avenue, Cambridge and Broom's Barn Experimental Station, Higham, Bury St. Edmunds, Suffolk.

INTRODUCTION

Members of the genus *Trichodorus* (21) are widespread in North America and Europe and have been recorded from all the main continental land masses of the world and from many islands, especially in the West Indies and in Australasia. Their economic significance was first demonstrated when Christie and Perry (20) showed that a species later described as *T. christiei* was responsible for damage to crop plants in the United States. They called it the stubby root nematode as a result of the symptoms it produced on host plants; this sobriquet was at first restricted to *T. christiei* but because similar symptoms are known to be caused by other species it is now customary to apply it to all members of the genus *Trichodorus*. Interest in the genus was further increased when it was shown that *Trichodorus* species could transmit tobacco rattle virus (84) and pea early browning virus (40). In 1973 Siddiqi (82) proposed the division of the genus *Trichodorus*, *sensu lato* into two genera, *Paratrichodorus* and *Trichodorus*. In this paper we use the generic name *Trichodorus* to encompass both *Paratrichodorus* and *Trichodorus*, *sensu stricto*, as there seems at present to be little evidence of biological or behavioural differences between the two new genera; specific names, however, are given according to Siddiqi (82). This may prove to be a false assumption (39) if future work shows differences in habit for *Trichodorus* spp. similar to those described for seven species of *Tylenchorhynchus* (12), each of which seemed to have preferred feeding sites on the roots of ryegrass. In this paper we follow the definition of the terms host and host-status given by Seinhorst (81). General accounts of the role of these nematodes both as causal organisms of direct damage to plants and as vectors of plant virus diseases are given by Taylor (90) and Hooper (46).

Electron microscopy of the somatic musculature of P. porosus (7) and P. christiei (34, 35) indicated a meromyarian arrangement of platymyarian muscle cells. This system restricts Trichodorus to simple and sluggish movements and is less advanced than the shallow coelomyarian type found, for example, in Longidorus elongatus which permits more complicated and lively movements. These observations might have a fundamental bearing on the biology and ecology of the genus, and are almost certainly important in its systematics (82).

DIRECT DAMAGE TO PLANTS

Paratrichodorus christiei was named by Allen (2) after J.R. Christie who in 1951, together with V.G. Perry, reported damage by this species to celery (Apium dulce), sugar beet (Beta vulgaris) and maize (Zea mays) in Florida (20). This widespread and polyphagous species has since been recorded causing damage to several crop plants in the U.S.A. (19, 70).

Stubby root injury or loss of root weight caused by P. christiei has been reported on wheat (Triticum sp.) (75), soybean (Glycine max) (77), grapefruit (Citrus paradisi) (86), tobacco (58), lucerne (Medicago sativa) (92) and cranberry (Oxycoccus macrocarpus) (104). On blueberry (Vaccinium corymbosum) pathogenicity appeared to be dependent on the stage of host development with cuttings suffering more damage than older plants (105). Christie (19) reported that cabbage and cauliflower (Brassica oleracea), chayote (Sechium edule), fig (Ficus carica) and various grasses were also damaged by this species. Later, it was found that although 2,900 P. christiei/l of soil caused root injury to cotton (Gossypium hirsutum) seedlings this was not of sufficient magnitude to evaluate quantitatively (13); conversely Alhassan and Hollis (1) could find no root necrosis on cotton but found that 533 P. christiei/l significantly decreased seedling weight and considered that even smaller numbers (160/l) might significantly damage young cotton seedlings in the field. Ruehle (74) found no root necrosis on roots of pine seedlings (Pinus spp.) inoculated with 500 or 5,000 P. christiei each but that there were significant decreases in seedling height and top and root weight after 6 months.

Hoff (36) showed that the stubby root symptoms caused by P. christiei on onion (Allium cepa) were the result of an increase in the size of the cortex both radially and longitudinally and that the early phases of this damage are a result of cell maturation nearer the apical meristem than would normally be the case, and not solely from a cessation of cell division as previously thought. Hoff and Mai (37) illustrated stubby root injury on onions; they also demonstrated that fumigation with 'Telone' (dichloropropene) greatly improved yield of onions grown in a soil naturally infested with

P. christiei and that adding nematodes to sterilised soil gave the same yields as those from infested soil, indicating that there was no pest/disease interaction (Table 1).

Table 1. Root and top weights of onions grown in soil naturally infested with Paratrichodorus christiei, in the same soil fumigated with Telone, and in another organic soil, steam sterilised then infested with 50 P. christiei/pot (After Hoff and Mai, (37)).

	Average fresh weight g/plant	
	Tops	Roots
Naturally infested soil + Telone	2.47	5.99
" " " unfumigated	0.13	1.18
Steam treated soil with 50 P. christiei added to each pot	0.11	1.02

Small numbers of P. christiei therefore had a drastic effect on onion but an initial level of 500 per plant had no influence on the growth and development of the root system of tomato (Lycopersicon esculentum) grown in glass tubes filled with vermiculite; at 750 per plant there were significantly more secondary roots shorter than 1 cm (38). Rohde and Jenkins (71) found accelerated cell maturation and no root cap or region of elongation in tomato roots parasitised by P. christiei.

Trichodorus sp. was associated with sugar cane (Saccharum officinarum) roots in Louisiana and P. christiei and P. porosus with sugar cane roots in Hawaii (50) and Apt and Koike (4) tested the effect of adding different numbers of P. christiei to three week old sugar cane seedlings grown in fumigated soil in metal containers. Roots of seedlings grown in infested soil were severely stunted and yielded less than those in control plots. (Fig. 1). Damage to sugar cane by P. christiei has subsequently been reviewed by Prasad (65).

Other species of Trichodorus have also been recorded as pathogens. T. viruliferus caused severe necrosis, arrested growth and caused swelling of extending roots of apple when observed through the window of an underground laboratory (62). P. lobatus caused stubby root damage to citrus seedlings in Australia (3) and in

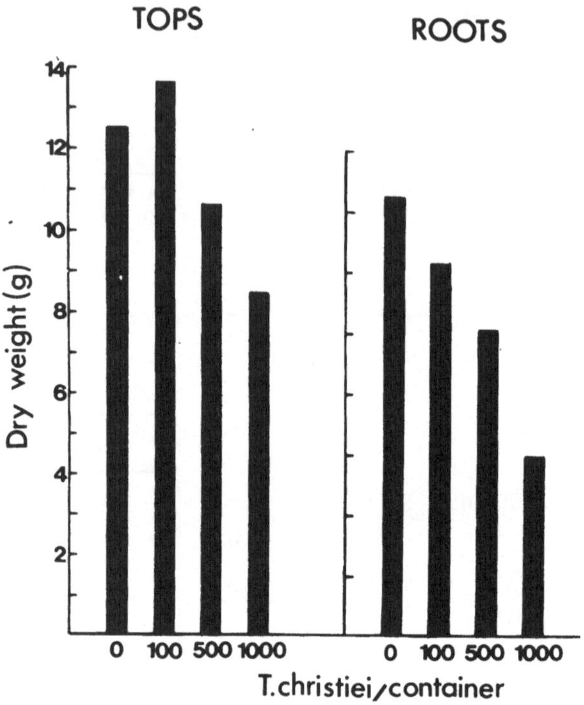

Fig. 1. Effect of P. christiei on dry weight of tops and roots of sugar cane seedlings (After Apt and Koike, (4)).

glasshouse tests P. porosus damaged maize and sorghum (Sorghum vulgare) (18). T. proximus caused brown lesions and stopped root growth on the roots of St. Augustine Grass (Stenotaphrum secundatum) growing in water agar and in glasshouse tests lessened growth and damaged roots (66); in Japan P. mirzai is considered to be the cause of 'yellow patch' in golf-link-green grass (Zoysia tenuifolia) (103).

Stunting of sugar beet in a new polder soil in the Netherlands was caused by P. teres (= T. flevensis) (53). In Britain Trichodorus spp. (and/or Longidorus spp.) cause the primary damage to sugar beet which results in a condition known as Docking disorder (26, 97, 100) named after the village of Docking, Norfolk where it was first studied. Seedlings attacked by Trichodorus nematodes may show typical damage symptoms with stubby lateral roots which turn grey-brown and later black as they die and decay. New roots that grow are in turn attacked. Often the tap root stops growing or is killed and lateral roots near the surface thicken and replace it resulting in a poorly yielding misshapen (fangy) root at harvest. Brown and Sykes (17) found significant negative regressions between numbers of Trichodorus spp at harvest and root numbers in one experi -ment, suggesting a loss of 12% of roots for each increment of 500 Trichodorus. Because of the damage to the root system plants are unable to absorb enough nutrients, and leaves may show symptoms of nitrogen or magnesium deficiency. Although a shallow root system is a common result of injury by Trichodorus effects differ because of the influence of secondary pathogens or soil conditions. Patches of affected plants occur in the sandier areas of infested fields; some plants in the patches are less damaged and may recover resulting in a 'hen and chick' effect in the field.

INTERACTION WITH OTHER PATHOGENS

Sol and Seinhorst (84) first successfully showed that P. pachydermus was a vector of tobacco rattle virus (TRV) and Taylor and Robertson (91) demonstrated the location of the virus particles in the oesophagus of that nematode. Since the first record of transmission, very many plants have been shown to be hosts of the Tobra viruses (viruses with tubular particles, exemplified by TRV), several species of Trichodorus have been shown to transmit them and surveys have been carried out in some European countries (24, 25, 28, 31, 32, 41, 42, 61, 88, 89, 93, 94). It is noticeable that although Trichodorus spp. attack a wide range of plants in most continents of the world, the Tobra viruses have been recorded almost exclusively in Europe, including the British Isles. This may signify merely a lack of survey work in other parts of the world but it might also mean that the Tobra viruses are at present mainly a European problem (see also Taylor and Robertson in these Proceedings). Recent evidence about the non-coincident nature of Tobra viruses and their vectors in Norfolk East England was collected by Seddon (78) in his yet unpublished work on pea early browning virus (PEBV). He surveyed 60 fields totalling 394 ha, about 3% of the total pea (Pisum sativum) crop in Norfolk where 17% of the pea crop in England and Wales is grown. Twenty-nine fields showed neither PEBV nor Trichodorus but both were found in 15 fields; the remaining 16 fields harboured the vectors but not the virus and in no case was the virus found without Trichodorus also being found albeit in small numbers in some instances. It seems probable that it is only a matter of time before the virus and vectors coincide on all the area of peas grown on soils suitable for Trichodorus. Soil sampling might help to locate virus susceptible crops on fields where there are no vectors, although few Trichodorus are needed to transmit virus and occasionally numbers may be too small for current sampling methods to detect (85). It is of the utmost importance therefore to obtain more evidence on soil texture and mechanical analysis so that these may be used to predict the likelihood of Trichodorus spp. being present. The recorded presence of potential vectors in most countries indicates the important part that international plant quarantine measures might play in delaying or preventing the spread of Tobra virus disease to hitherto uninfected areas.

Hollis and Martin (45) found no evidence of a disease complex involving P. christiei and other micro-organisms although maize showed typical stubby-root symptoms and the nematode reduced growth and yield. Similarly experiments testing the relationship between Trichodorus spp., other plant parasitic nematodes and several soil-borne fungi found associates with cotton stunt, were inconclusive (10). However, there is some evidence (54, 55) of a positive interaction between Fusarium monoliforme and P. christiei on root growth (but not top growth) of sugar cane.

Brown and Sykes (17) found evidence of competition between Longidorus elongatus and Trichodorus spp. in sugar beet fields.

EFFECT OF NON-BIOLOGICAL FACTORS ON ACTIVITY AND SURVIVAL

Soil type, texture and particle size

The particle size and crumb structure of soils are of fundamental importance to soil inhabiting nematodes. Pore space and moisture holding capacity are probably the most important features of a soil that determine whether or not it is a suitable habitat for nematodes. For the purpose of this paper we have adopted a modified size-grade system for naming soil particles (Table 2).

Table 2. Names and sizes of soil particles and the international size-grade system

Description	Particle size (µm)	
	This paper	International system
Coarse sand	200–2000	60
Fine sand	20–200	
Silt	2–20	2–60
Clay	< 2	2

Clay soils generally contain more than 20% of clay particles and have a texture that tends to prevent free movement of air and water. Loams contain around 8-15% clay particles and although they aerate and drain well, the clay fraction ensures that the soil can withstand short periods of drought. The pore space in loam soils expressed as a percentage of the total volume is about 35 compared with about 50 in a clay soil, although the individual pores are very small in both kinds of soil. Sandy soils contain about 5% or less clay and are therefore composed of relatively large particles. This gives such soils an open texture so that they are free-draining and subject to rapid drying out during periods of drought (90).

The mechanical analysis of ten soils in eastern England in which *Trichodorus* nematodes were abundant was as follows: coarse sand 32-60%, fine sand 22-42%, silt 6-12%, clay 7-12%. Organic matter was low and free calcium carbonate content varied between 1.8 and 4.6%. *Trichodorus* were not found in fine textured soils that contained much silt or clay (52).

In a survey of 143 sugar beet fields prone to Docking disorder *Trichodorus* nematodes were usually found in soils with 80% or more sand particles (Fig. 2), they were not found in 12 'control' fields which were not prone to the disorder and which contained on average only 56% sand particles (22).

The ability of a nematode to move through a soil is limited both by the size of the pore necks between particles or aggregates and by the mean free path (i.e. the mean distance that can be travelled before encountering blind ends or impassable necks). The length of the mean free path is probably more important for long nematodes (e.g. *Xiphinema*, *Longidorus*) than for shorter ones like *Trichodorus* (52). Adult *Trichodorus* vary in length approximately from 420 μ (female *T. renifer*) to 1,500 μ (female *T. proximus*) and the larvae from about 289-530 μ. Diameters of adults vary between 16-32 μ. It is probable that individual *Trichodorus* may squeeze through necks with diameters somewhat smaller than their own but their observed occurrence in coarser texture soils agrees well with the relationship between their adult diameter and the proportion of larger soil particles. Coarse sandy soils are inherently unstable and tend to compact easily; however even when compacted they provide sufficient space for *Trichodorus* to move.

Moisture

All soil nematodes are dependent on water for most of their activities and prolonged drought results at best in an inactive resistant stage (e.g. *Ditylenchus dipsaci*), at worst in death. Of all the plant parasitic nematodes that have been studied with respect to their resistance to desiccation *Trichodorus* is the most susceptible. For example, Simons (83) showed that only 1% of *P. pachydermus* survived being left on a membrane filter for 2 days in

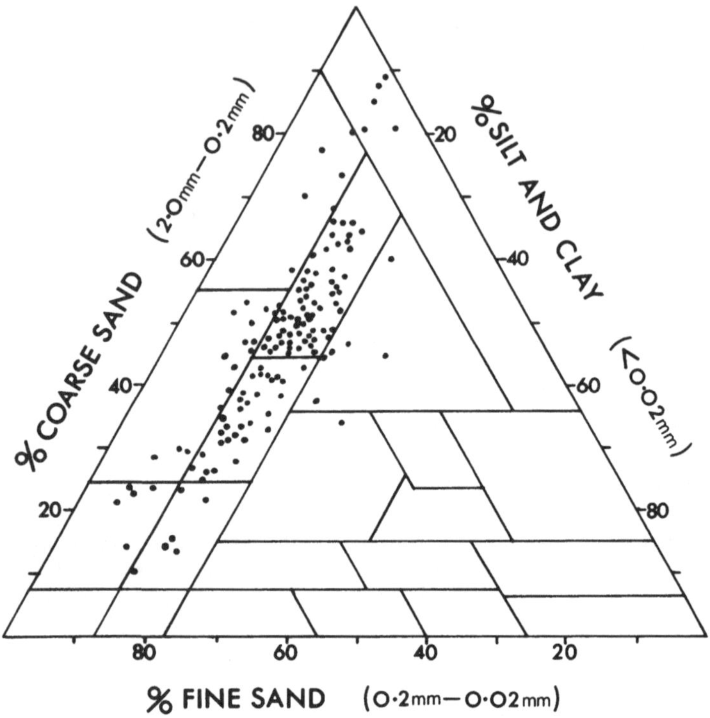

Fig. 2. Relationship between occurrence of Trichodorus spp. and soil texture. Schematic soil chart taken from Salter and Williams (76).

a relative humidity of 100%; the survival rate for five other genera of ectoparasitic nematodes varied between 91.6% and 100% (Table 3).

The fact that Trichodorus are almost entirely restricted to free draining sandy soils means that they are particularly likely to encounter very dry soil conditions so it is not surprising that numbers in the soil are often correlated with rainfall. Cooke and Draycott (33) found that the difference between the number of several nematode genera at two sampling dates was correlated with the rainfall before the second sampling date, and that the change in numbers was greater for Trichodorus than for other nematodes (Fig. 3.)

Rössner (73) found that Trichodorus were more susceptible to desiccation than Rotylenchus or Pratylenchus. Wyss (101) also studied the tolerance of migratory root nematodes towards increasing desiccation and high osmotic pressures and concluded that Tylenchida were less susceptible to both conditions than Dorylaimida. He found

Table 3. Percentage survival of different genera of ecto-parasitic nematodes after 2 days exposure to three relative humidities and water (After Simons (83)).

	Relative humidity			
	100	97.7	96.0	H_2O
Paratylenchus	98.5	98.1	99.6	97.9
Tylenchorhynchus	91.6	90.4	64.6*	96.4
Rotylenchus	96.7	94.4	93.9*	97.4
Helicotylenchus	97.0	93.4	86.4	98.3
Criconemoides	100.0	98.2	97.0	92.9
Trichodorus	1.0*	0.0*	0.0*	86.4

*significantly smaller than H_2O

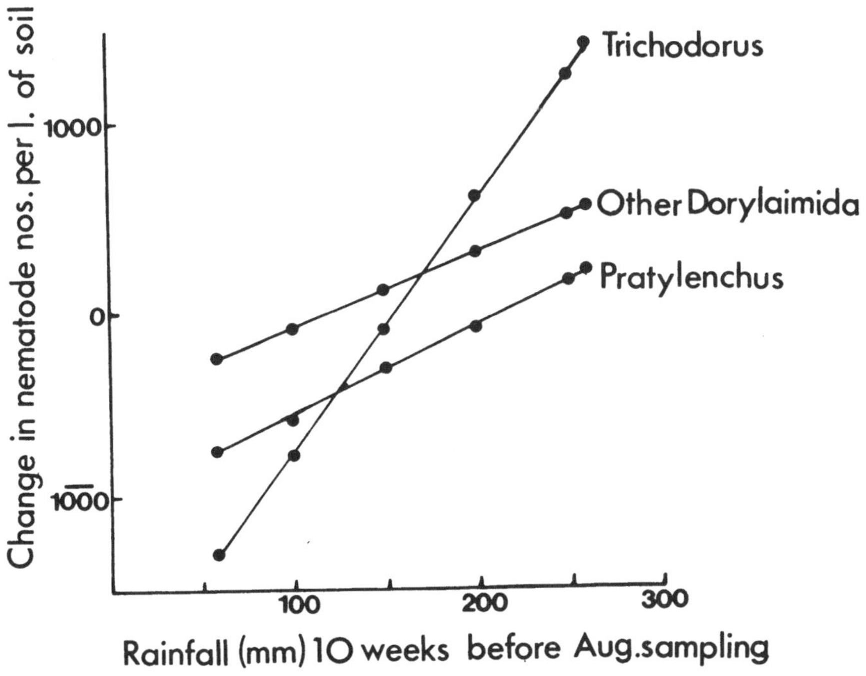

Fig. 3. Influence of accumulated rainfall ten weeks before Autumn sampling on the change in nematode numbers between Spring (n_1) and Autumn (n_2) (after Cooke and Draycott, (23)).

the highest tolerance shown by Tylenchorhynchus dubius; Trichodorus spp. were the most susceptible.

Evidence on the ability of Trichodorus to move in different moisture regimes in four different grades of sand was obtained by G.W. Storey (87). As these data have not previously been published they are given in full here. Soil moisture characteristics for the four sand particle sizes (obtained by sieving) were determined by the method described by Wallace (95). Movement of Trichodorus sp. was investigated in the four sand fractions at 13 suction pressures in an apparatus similar to that devised by Wallace (96) and modified by Evans (27). This consisted of a glass tube 8.5 cm long attached to one end of a tube containing a sintered glass plate. Inside the glass tube was placed a series of seven pieces of polythene tubing, each 1 cm in length. The tube containing the sintered glass plate was attached to a 100 cm long polythene tube with its free end immersed in water. The whole apparatus was filled with water and the sand fraction was poured into the polythene tubing and tapped vigorously to ensure that it was evenly distributed. The apparatus was then fixed horizontally and the appropriate suction exerted. After leaving for an hour to stabilize, a drop of water containing about 50 Trichodorus sp. was introduced at the open end after which the apparatus was corked. After 48 hours' storage at 23.3°C, the tube was uncorked and the individual polythene rings containing the soil were removed. The nematodes distributed along the rings from the point source were separated from the soil and were counted. The number found in each ring was expressed as a percentage of the total number caught; these percentages were multiplied by 1 in the case of the second ring, by 2 in the case of ring 3 and so on until ring number 7 for which the percentage catch was multiplied by 6; the sum of these products was called the 'movement score'. The relationships between movement scores and soil moisture (expressed as ml of water/100 ml of apparent soil volume) for the four sand fractions, together with their moisture characteristics are shown in Fig. 4.

Movement was greatest in the 200-400 μ fraction and slightly less in the 100-200 μ and 400-800 μ fractions; movement was least in the 800-1,400 μ fraction. In all four soil fractions, movement was greatest when the soil pores were half-full of water and least in waterlogged and dry soil. In practice the free draining sandy soils in which Trichodorus are usually found rarely remain waterlogged for more than a very short period of time.

The increased survival and activity of Trichodorus in conditions of high water content are reflected in the close correlation between the area of sugar beet in England affected by Docking disorder and total rainfall in May (the month in which most damage to the seedling sugar beet takes place). Fig. 5 illustrates the variation in area affected between the years 1963 and 1972.

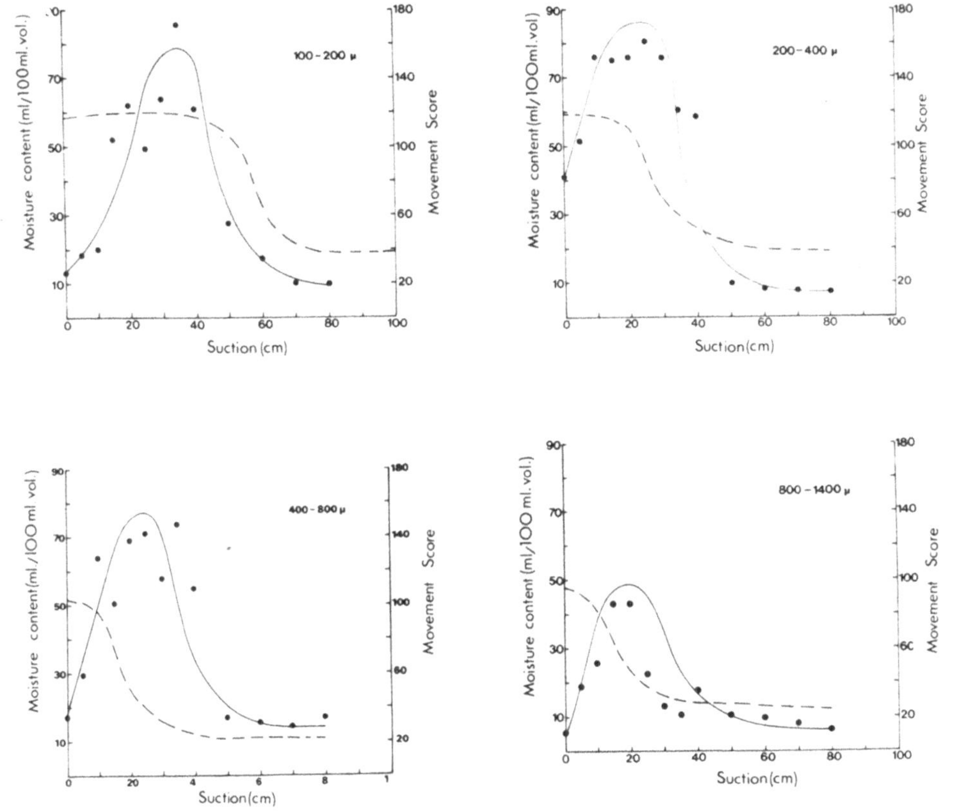

Fig. 4. Movement of Trichodorus sp. through four fractions of sand (after Storey (87)).

(Smallest area affected 174 ha in 1963; greatest area affected 7819 ha in 1969) and the correlation between log. area affected and May rainfall expressed as a percentage of the long term average.

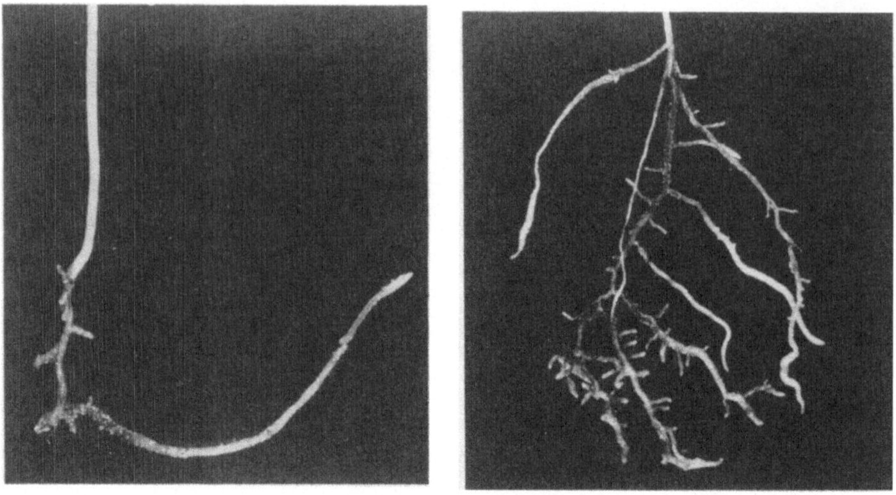

PLATE. Damage to roots of *Picea sitchensis* by *Trichodorus*

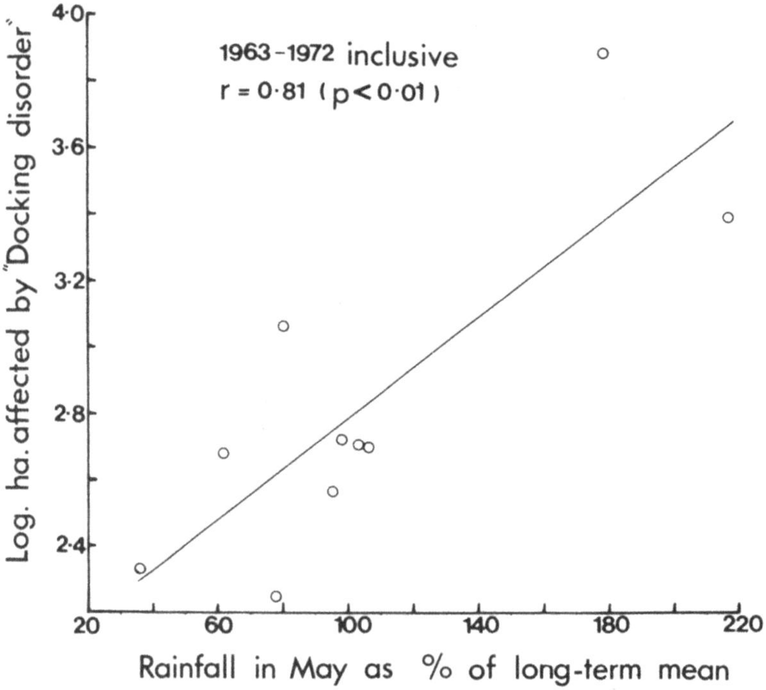

Fig. 5. Relationship between rainfall in May and the area of sugar beet crops affected by Docking disorder (after Cooke, (22)).

Mechanical injury

Bor and Kuiper (11) showed that Paratrichodorus teres and P. pachydermus are susceptible to mechanical injury by rough handling in transport of the soil samples. Soil carried carefully by car yielded 2,240 P. teres/l, whereas soil sent by post in a cardboard box yielded 628/l, and soil sent by post in a paper parcel yielded only 240/l.

The sampling corer (borer) also affected subsequent catch of P. teres. When a 10 cm diameter cylinder was used to take the soil sample in the field, 2,540/l were recovered, whereas corers of 2 cm and 1 cm diameter yielded only 580 or 390/l respectively.

An experiment was conducted in which polythene bags containing nematode-infested soil were dropped from various altitudes onto a hard floor. The results are given in Table 4.

Table 4. Percentage of dead nematodes when soil samples containing them were dropped. (after Bor and Kuiper, (11)).

	P. teres	P. pachydermus
Control - no dropping	3	· 0
Fall from 25 cm altitude	19	22
" " 100 cm "	76	82
" " 350 cm "	98	98
Fall - water added to bag	31	-
Mixing the soil	5	-

Temperature

Malek, Jenkins and Powers (57) found that the optimum temperature for growth and reproduction of P. christiei was around 25°C and experiments by Ayala, Allen and Noffsinger (5) testing potential hosts of P. allius suggested that reproduction and survival were far greater at 24° ± 2°C than at ambient temperature varying between 18 and 35°C under a wide variety of plant species. However, they found that P. porosus and P. christiei could reproduce on maize at all temperatures tested between 12-29°C and at greenhouse ambient temperatures (18-35°C), although the optimum was 24°C. Rohde and Jenkins (71) found development of Trichodorus sp. was inhibited at 35°C.

Soil samples containing nematodes are usually stored in a refrigerator at 2-5°C, and although van Gundy and Thomason (29) found that P. christiei in tap water survived better at 15°C than 3° or 27°C most nematodes in soil, water or plant tissues are best stored at low temperatures. Barker, Nusbaum and Nelson (6) showed that Tylenchorhynchus claytoni was greatly affected by high storage temperatures; soil stored for 1-16 wk at 36°C yielded fewer nematodes than soil stored at lower temperatures and storage at -15°C actually seemed to increase the catch. In a sampling exercise carried out in 1970 by nine members of the Agricultural Development and Advisory Service (ADAS) the collaborators extracted Trichodorus nematodes from subsamples of a single well mixed soil sample, stored further subsamples under conditions normal for the centre (some in refrigerators but some outdoors in shaded positions) for 2-3 months (May-July) then made a second set of extractions. Counts were always smaller than on the first occasion and catches varied from 10-86% of the previous largest count. Those centres that stored samples outdoors obtained the poorest results, and we attributed this to nematodes being killed when the soil heated up in the polythene bags, perhaps due to great diurnal fluctuations in temperature. All collaborators denied that their samples had dried out.

Chemicals

Van Gundy and Thomason (29) found that Cu^{++} ions in concentrations as low as 0.5 ppm killed P. christiei in 24 h whereas Hemicycliophora arenaria could tolerate concentrations of 4 ppm for 48 h without apparent injury. Cooper (24) tested the effect of exposing P. pachydermus, T. cylindricus, T. primitivus and other soil nematodes to three concentrations (1,4 and 16 ppm) of $CuSO_4$ or $MnSO_4$ for 36 h at 22°C. All Trichodorus spp. were immobilised by 16 ppm $CuSO_4$; 83% of P. pachydermus, 53% of T. cylindricus but no T. primitivus were immobilised by 4 ppm $CuSO_4$ and 65% P. pachydermus 31% of T. cylindricus but no T. primitivus were immobilised by 1 ppm $CuSO_4$. None of the Trichodorus spp. seemed adversely affected by exposure for 36 h to $MnSO_4$ solutions. No effect of $CuSO_4$ or $MnSO_4$ solutions was noted either on other stylet bearing or on saprozoic nematodes. These results showed that Trichodorus spp. are more sensitive to copper salts than many other soil-inhabiting nematodes and that of the three Trichodorus species tested P. pachydermus was the most and T. primitivus the least sensitive.

Hafkenscheid (30) also demonstrated the toxic effect of

increasing concentration of $CuSO_4$ solutions on P. pachydermus; 10^{-6}M $CuSO_4$ killed over 90% after 60 hours whereas less than 10% were killed in control solutions. He showed that the source of water used in the final separation of P. pachydermus may have a dramatic effect by plotting their survival rate in different types of water (Fig. 6).

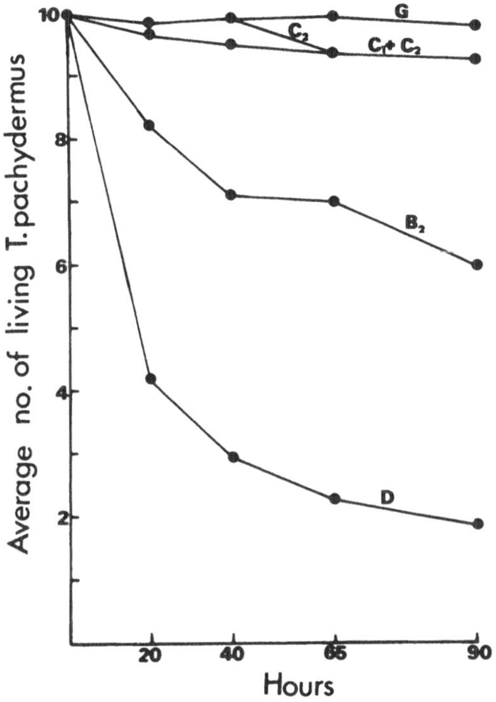

Fig. 6. Effect of water source on survival of P. pachydermus. C_1 water from a tap at Hengelo, Overijssl; C_2 Wageningen tap water; B_2 demineralised water; and D Wageningen tap water from an infrequently used tap (after Hafkenscheid, (30)).

Evidently from Hafkenscheid's evidence (30, fig.6) Trichodorus do not survive well in water in which there are high concentrations of copper and we lightly suggest that there might be a fairly brisk trade among nematologists for Hengelo tap water.

EFFECTS OF NON-BIOLOGICAL FACTORS ON DISTRIBUTION

Geographical distribution

Many references demonstrate that most Trichodorus species predominantly inhabit coarser textured soils. For example T. viruliferus was recorded from light sandy soils (47), T.cylindricus and P. teres from light sandy soil (46) and T. velatus and T. variopapillatus from light sand to sandy loam and wet sandy soil (48) respectively. T. hooperi was recorded from a sandy loam (56) and Seinhorst (80) first found T. similis in peaty soil containing much sand. Several surveys confirm most frequent occurrence in a sandy habitat. Persson (61) and Sturhan (88) found Trichodorus to be numerous only in light soils in Sweden and Germany respectively. Cooper (24) found nematodes of this genus in more than 75% of sandy potato fields in Scotland and Whitehead and Hooper (98) recorded Trichodorus in nearly all of the 64 fields in a survey of those prone to Docking disorder of sugar beet. Cooke (22) found Trichodorus spp. in 85% of sandy fields that suffered from Docking disorder and Rössner (72) recorded the nematodes in light, sandy forest-tree nurseries in Hesse, West Germany, but not in clay soils. However, Yeates (102) did not find Trichodorus in dune sand in New Zealand which suggests that the relative inactivity of this genus in very coarse sands may inhibit its distribution.

The terms 'clay' soil, 'light' soil and 'heavy' soil tend to be misleading because they might refer only to whether or not farm implements may pass easily and have little or no relationship to texture or type. Although it is fairly clear what is meant by 'light sandy soil' this term has neither national nor international recognition and we hope that in future authors will accompany their data with accurate mechanical soil analyses; this applied especially to descriptions of locations where new species are found.

The exception to this distribution pattern seems to be T. primitivus which Seinhorst (80) found in a clay soil. He claimed that this species seemed to occur on a wider range of soils than P. pachydermus and this was confirmed by Cooper (24).

Jones, Larbey and Parrot (52) suggested that silt and clay particles tend to clog the main soil pores and that this is the chief reason why Trichodorus are usually found in coarse sandy soils. However, it is questionable that this is the only reason and copper deficiency might also be important (24); many sandy soils are deficient not only in copper but in other heavy metals also. Cooper speculated that the wider distribution of T. primitivus might reflect this species' greater tolerance to difference in copper content of soils; pot tests however were inconclusive, adding copper salts to various soils did not result in either an increase or a decrease in the numbers of Trichodorus during the following 16 months.

Vertical distribution

Hijink and Kuiper (33) state that Trichodorus spp. like all other plant parasitic nematodes may be distributed irregularly in the soil both horizontally and vertically. Because of the extreme susceptibility of Trichodorus spp. to dry conditions and the fact that the sandy soils in which they are found are usually free draining and subject to rapid drying periods of drought Trichodorus are rarely numerous in the surface layer of soil except after periods of prolonged rainfall. Cooke and Draycott (23) found that in 14 experiments Trichodorus were usually fewest at 0-5 cm depth; however, in 1968 with prolonged rainfall before August sampling, Trichodorus were relatively more numerous in this surface soil layer, whereas in 1967 when there was less summer rainfall, numbers were smaller in August (Table 5).

Table 5. Percentage Trichodorus in four soil depth fractions 1967 and 1968 (after Cooke and Draycott, (23)).

Soil depth fraction (cm)	1967 April/May	1967 August	1968 April/May	1968 August
0-5	7.0	5.7	4.5	22.2
5-10	20.8	23.4	25.6	30.4
10-15	36.4	35.5	34.1	24.6
15-20	35.8	35.5	35.8	22.8
Avge. Trichodorus/ l soil	1369	608	642	1431
Rainfall (cm) 10 wk before sampling	10.9	10.5	7.4	22.0

Kuiper and Loof (53) found most Trichodorus teres (= T. flevensis) 5-10 cm below the soil surface in a field where they were pathogenic on sugar beet; this was where the vertical growth of infested beet roots stopped. There were fewer Trichodorus 15-30 cm deep, where there was little root growth, but numbers increased again below 30 cm (Table 6).

Richter (69) said that Trichodorus spp. inhabited deeper soil layers than other nematode genera. He found that males of P. pachydermus and T. viruliferus differed in their vertical distribution and judging by the proportion of males P. pachydermus occurs deeper in the soil than T. viruliferus.

Table 6. Depth distribution of T. teres below unhealthy sugar beet plants (after Kuiper and Loof, (53)).

Soil depth fraction (cm)	Number of T. teres per litre of soil
0–5	750
5–10	3520
10–15	620
15–20	270
20–25	70
25–30	150
30–35	280
35–40	270
40–45	580
45–50	760

Whitehead and Hooper (98) found no clear difference in the depth at which T. cylindricus or P. anemones occurred in the top soil in heavily infested sugar beet fields except in one field where there were few T. cylindricus in the surface 5 cm.

Hijink and Kuiper (33) sampled 16 sites and found that in 13 of them the greatest numbers of Trichodorus were below the tilth (20 cm deep) where they are less subject to mechanical injury as a result of soil cultivations and where soil is unlikely to dry out.

It seems that numbers of Trichodorus spp. down the profile of the cultivated soil layer (surface 20–30 cm) can vary enormously. When the soil remains at or above field capacity for long periods of time large populations can be maintained especially where host roots are abundant. However, in conditions of drought numbers can decrease rapidly. Below the tilth where nematodes are not subject to moisture or mechanical stress a relatively large and more stable population is maintained.

EXTRACTION FROM SOIL

Because of the susceptibility of Trichodorus spp. to mechanical damage any violent method of extraction is likely to reduce the final catch. Similarly because of the sluggish way in which

Trichodorus spp. move recovery is likely to be small if the extraction method is one which relies on the mobility of the nematode through soil (e.g. Baermann funnel or modifications). Nevertheless final stages of most extraction methods depend on the nematodes wriggling through a mesh and any method of diminishing damage for example by submerging sieves under a constant head of water onto which nematode suspensions are poured (98) may improve the catch; it has also been recommended that the final separation sieve should be free of metallic components (63). The skill and patience of different operators will also vary. An unpublished collaborative exercise carried out in England in 1970 highlighted the difficulties (see above). Nine workers, each stationed at one of the regional laboratories of ADAS extracted nematodes by the method used at their own centre and by the Seinhorst two-flask method (79). Each operator made four extractions for each of six kinds of nematode from well mixed bulk soil samples that had previously been divided into 9 equal parts. Centres varied greatly in their estimates of nematode numbers but the greatest discrepancies occurred in the counts made of *Pratylenchus neglectus*, *Trichodorus* spp. and *Hemicycliophora* sp. Average counts for pairs of soil samples varied from 780 - 6,153/1 *Trichodorus* but differences between extraction techniques within each centre were not great enough to be significant.

In April of this year one of us (A.L.W.) compared the Seinhorst two-flask method with the sugar centrifugation method given by Southey (85). The soil contained 17% coarse sand, 68% fine sand, 5% silt, 6% clay and 4% of other inorganic and organic matter. It was thought to contain between 2-3,000 *Trichodorus*/1; five samples of 200 ml of soil were processed by each of the two methods. A mean of 2790 *Trichodorus* (range 2545-3080) were extracted by the two-flask method compared with 1940 (range 1660-2660) by the centrifuge method. Counts from the centrifuge method were smaller and more variable than those from the two-flask method although the former method was quicker and gave nematodes in very good condition. It seemed that centrifuging lost many of the juveniles.

LIFE CYCLE, POPULATION DYNAMICS AND HOST-PARASITE RELATIONSHIPS

The embryogenesis of *P. christiei* was studied by Bird, Goodman and Mai (8). Eggs were laid in the single cell stage and the first two cleavages were transverse to the longitudinal axis. First stage juveniles were recognisable after 96 hours and emerged from the eggs 100-120 hours after they were laid.

Ayala, Allen and Noffsinger (5) found that the optimum temperature for reproduction of *P. christiei* was 16-24°C and *P. porosus* 24°C. Rohde and Jenkins (71) found that *P. christiei* completed its life cycle in 21-27 days at 22°C and 16-17 days at 30°C; Morton and Perry (59) state that it took 17-18 days at 27°C and at 20°C P. allius took 22 days to develop from second stage larva to fourth stage larva or adult but the total time for the complete life cycle was not ascertained (5). In all of these species (and in some others) males are rare and reproduction must be parthenogenetic.

The life cycle of T. viruliferus takes about 45 days egg to adult at 15-20°C (62, 64). Breeding continued throughout the year (64) but the proportion of females with developed oocytes was consistently greater near the root tips (80% of the mean) and it was concluded that root-tip colonies constituted the main breeding sites for T. viruliferus on mature apple trees. Pitcher and McNamara's laboratory studies suggested that the increase in numbers in root-tip colonies must be a form of aggregation, the mechanism of which remains unknown, rather than development of larvae from new eggs laid there. In root observation chambers several Trichodorus species aggregated around the roots of a variety of plants (98).

Several species of Trichodorus may occur together; in eastern England and eastern Scotland, P. pachydermus and T. primitivus are often present together (24, 98), whereas in West Germany P. pachydermus and T. similis are most common (88, 101). Whitehead, Fraser and Greet (99) noted a change in the relative abundance of different species in mixed populations of Trichodorus spp. In a field in Norfolk, eastern England, in November 1965 before the soil was ploughed nearly all active Trichodorus in the top 20 cm were T. cylindricus; in March after ploughing there were few Trichodorus with T. cylindricus and P. teres equally common.

The rate of recovery of populations of Trichodorus after fumigation varies considerably. Perry (60) reported that P. christiei re-established itself after fumigation much more quickly than any of the numerous other plant parasitic nematodes that occurred in his experiments. Similarly Rhoades (67) found that P. christiei attained higher population in cabbage plots four months after fumigation with DD or EDB than in unfumigated plots although numbers following fumigation with DBCP (which had a far more effective residual action) remained low. In another series of experiments (68) populations of P. christiei following fumigation with DD increased to several times those in unfumigated plots by harvest about 3 months later although some of the experimental nematicides tested gave a better control.

These results contrast with experience in England with other species of Trichodorus. In a trial in Yorkshire, northern England, where 93% of P. anemones were killed by fumigation in winter 1965-66 there were slightly fewer in fumigated than in unfumigated plots by October 1967. In a similar experiment made concurrently in Norfolk in eastern England there was virtually no recovery of populations of P. teres and T. cylindricus during 1966 and 1967 following a 99% kill (99). Cooke and Draycott (23) found very little multiplication of Trichodorus spp. by about 8 months after fumigation in any of fourteen trials, where the initial kill averaged 92%. Cooke (unpublished) found that the rate of re-establishment of a mixture of T. cylindricus and P. pachydermus following a 98% kill in February 1965 was very slow (Fig. 7).

The fact that P. christiei is parthenogenetic is a possible reason why populations recover more quickly than the European species where males are common, although other factors (e.g. duration of life cycle, efficiency of initial kill, soil temperature, host crop) may also be important.

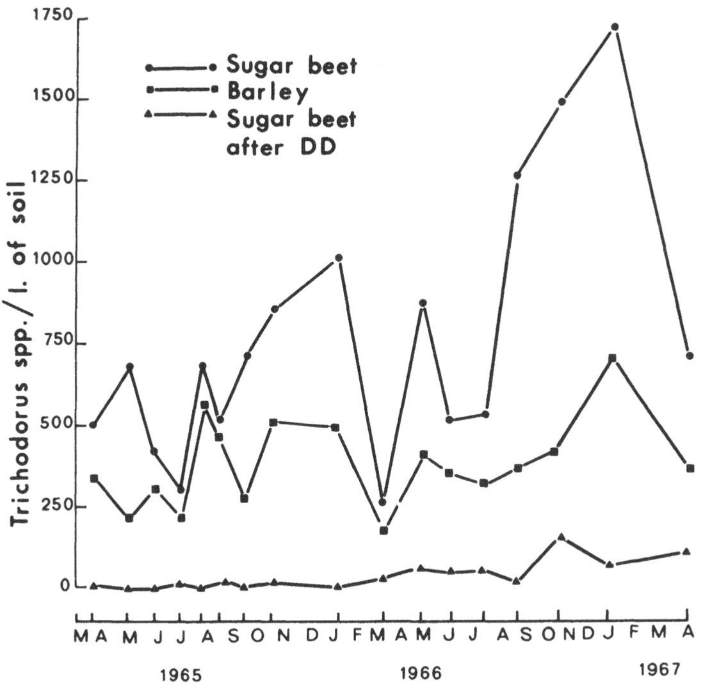

Fig.7. The effect of host plant and fumigation on the numbers of Trichodorus spp. found in soil. Sugar beet or barley was grown for two consecutive years; DD was applied before sugar beet in February 1965

Bird and Mai (9) found that populations of P. christiei followed a typical sigmoid growth pattern; after an initial inoculation of 25 nematodes to 7.5 cm pots containing lettuce (Lactuca sativa) seedlings populations had risen to 6,107/l after 45 days and 12,533/l after 90 days.

Although all Trichodorus species appear to have a wide host range populations are considerably influenced by cropping. Rohde and Jenkins (70) studied the effect of 42 different plant species, representing 14 families, on P. christiei. Fourteen of these were 'excellent' hosts (tenfold or greater increase over the inoculation level after 60 days) seventeen were 'good' hosts (an increase in number but less than tenfold) fourteen were poor hosts (a decrease in numbers but larvae were found), and only four were non-hosts (none survived).

Ayala, Allen and Noffsinger (5) made host range studies for P. christiei, P. porosus and P. allius testing 50, 25 and 38 plant species as potential hosts for the respective nematode species. Again, the great majority of plant species tested were excellent or good hosts.

Bird and Mai (9) reported differences in numbers of P. christiei caused not only by host species but also by host variety (three varieties of maize were tested), host age (populations could not be maintained on soil containing 0-15 day old tomato seedlings whereas they could in soil containing 10-25 or 20-35 day old seedlings), crop rotation, and mowing (densities increased to a greater extent on mown than on unmown clover (Trifolium pratense).

Brodie, Good and Jaworski (14, 15) and Brodie, Good and Marchant (16) made a series of studies of the population dynamics of T. christiei and other nematodes in cultivated soil. Their results can be summarised as follows:

(a) In newly cleared land relatively free of plant parasitic nematodes there was no substantial increase in numbers of P. christiei until the third year after clearance. Sudan grass (Sorghum sudanense) and cotton were the best hosts, increasing numbers to over 500/l of soil in the third year; hairy indigo (Indigofera hirsuta), millet (Panicum ramosum) and beggarweed (Desmodium tortuosum) gave moderate increases (to about 200/l) and marigold (Tagetes minuta) and Crotalaria (Crotalaria spectablis) appeared to suppress populations.

(b) On older agricultural land the effect of different cover crops grown between tomato crops was studied. Trends in nematode population differences were established after the first year's growth of cover crops which were maintained throughout the experiment. Millet was most favourable for the increase of P. christiei, bermuda grass (Cynodon dactylon) and hairy indigo gave slight increases. There was no increase on beggarweed or marigolds.

(c) In a series of rotations testing the effects of bermuda grass or bahia (Paspalum notatum) P. christiei was slightly favoured by a continuous row-crop (cotton-maize-peanut) rotation where cotton and maize (Zea mayr) increased numbers but peanut (Arachis hypogea) decreased numbers. The sequence of cropping after a grass ley that produced the smallest final densities of T. christiei was when cotton and maize did not follow each other.

Johnson and Burton (51) studied millet and sudangrass both of which are extensively-grown crops in south-eastern USA and support large populations of P. christiei. Poor yields of maize were said to follow millet very frequently, although Johnson and Burton's data suggest that the nematodes Pratylenchus brachyurus, P. zeae and Belonolaimus longicaudatus were responsible for the yield reduction and not P. christiei.

Whitehead, Fraser and Greet (99) found that Trichodorus spp. multiplied more under sugar beet than under barley (Hordeum sativum) an observation confirmed by one of us (D.A.C.) who also showed that numbers were greatest during the winter (Fig. 7).

The number of Trichodorus in the soil prior to planting a crop may be considered as the 'residue' from a previous crop or sequence of crops. A thorough knowledge of the species of Trichodorus present and host-status of other crops in the rotation may then be helpful in maximising the yields of susceptible crops. However, the extreme variability in distribution caused by the aggregation of the nematodes around host roots (33, 64) complicates the interpretation of soil sample results; further complications arise from the extreme susceptibility of Trichodorus to adverse physical conditions which may cause very large and rapid changes in population levels. Predictions of damage based solely on soil sampling may therefore be grossly misleading if these factors are not taken into consideration.

REFERENCES

1. Alhassan, S.A. & Hollis, J.P. (1966). Parasitism of Trichodorus christiei on cotton seedlings. Phytopathology 56, 573-574

2. Allen, M.W. (1957). A review of the nematode genus Trichodorus with descriptions of 10 new species. Nematologica 2, 32-62.

3. Anon (1970). New South Wales Department of Agriculture. Ann. Rep. 1969-70. Sydney, Australia.

4. Apt, W.J. & Koike, H. (1962). Influence of the stubby-root nematode on growth of sugar cane in Hawaii. Phytopathology 52, 963-964.

5. Ayala, A., Allen, M.W. & Noffsinger, E.M. (1970). Host range, biology, and factors affecting survival and reproduction of the stubby root nematode. J. Agr. Univ. P. R. 54, 341-369.

6. Barker, K.R., Nusbaum, C.J. & Nelson, L.A. (1969). Effects of storage temperature and extraction procedure on recovery of plant parasitic nematodes from field soils. J. Nematol. 1, 240-247.

7. Bird, G.W. (1969). Somatic musculature of Trichodorus porosus. [Abs.] J. Nematol. 1, 281.

8. Bird, G.W., Goodman, R.M. & Mai, W.F. (1968). Observations on the embryogenesis of Trichodorus christiei. (Nematodea: Diptherophoroidea). Can. J. Zool. 46, 292-293.

9. Bird, G.W. & Mai, W.F. (1967). Factors influencing population densities of Trichodorus christiei. Phytopathology 57, 1368-1371.

10. Bird, G.W., McCarter, S.M. & Roncadori, R.W. (1971). Role of nematodes and soil-borne fungi in cotton stunt. J. Nematol. 3, 17-22.

11. Bor, N.A. & Kuiper, K. (1966). Gevoeligheid van Trichodorus teres en T. pachydermus voor uitwendige invloeden. Meded. Rijks. Landbouw-wetensch. Gent 31, 609-616.

12. Bridge, J. (1970). A comparison of the feeding habits of Tylenchorhynchus species on ryegrass roots. Proc. 10th Int. Nematol Symp., Pescara, p.59.

13. Brodie, B.B. & Cooper, W.E. (1964). Pathogenicity of certain parasitic nematodes on Upland cotton seedlings. Phytopathology 54, 1019-1022.

14. Brodie, B.B., Good, J.M. & Jaworski, C.A. (1970a). Population dynamics of plant nematodes in cultivated soil; effect of summer cover crops in old agricultural land. J. Nematol. 2, 147-151.

15. Brodie, B.B., Good, J.M. & Jaworski, C.A. (1970b). Population dynamics of plant nematodes in cultivated soil; effect of summer cover crops in newly cleared land. J. Nematol. 2, 217-222.

16. Brodie, B.B., Good, J.M. & Marchant, W.H. (1970). Population dynamics of plant nematodes in cultivated soil; effect of sod-based rotations in Tifton sandy loam. J.Nematol. 2, 135-138.

17. Brown, E.B. & Sykes, G.B. (1971). Studies on the relation between density of Longidorus elongatus and growth of sugar beet, with supplementary observations on Trichodorus spp. Ann. of app. Biol. 68, 291-298.

18. Chevres-Roman, R., Gross, H.D. & Sasser, J.N. (1971). The influence of selected nematode species and number of consecutive plantings of corn and sorghum on forage production, chemical composition of plant and soil, and water use efficiency /Abs./. Nematropica 1, 40-41, 46.

19. Christie, J.R. (1959). "Plant nematodes. Their bionomics and control". Agric. Exp.Stat. Univ. Fla, Gainsville, Florida. 256 pp.

20. Christie, J.R. & Perry, V.G. (1951). A root disease of plants caused by a nematode of the genus Trichodorus. Science 113, 491-493.

21. Cobb, N.A. (1913). New nematode genera found inhabiting fresh water and non-brackish soils. J. Wash. Acad. Sci. 3, 432-444.

22. Cooke, D.A. (1973). The effect of plant parasitic nematodes, rainfall and other factors on Docking disorder of sugar beet. Pl. Path. 22, 161-170.

23. Cooke, D.A. & Draycott, A.P. (1971). The effects of soil fumigation and nitrogen fertilizers on nematodes and sugar beet in sandy soils. Ann. appl. Biol, 69, 253-264.

24. Cooper, J.I. (1971). The distribution in Scotland of tobacco rattle virus and its nematode vectors in relation to soil type. Pl.Path. 20, 52-58.

25. Dalmasso, A. (1967). Connaissances actuelles sur les nematodes phytophages et leurs relations avec les maladies a virus. Ann. Epiphyties 18, 249-272.

26. Dunning, R.A. & Cooke, D.A. (1967). Docking disorder. Brit. Sugar Beet Rev. 36, 23-29.

27. Evans, K. (1969). Apparatus for measuring nematode movement. Nematologica 15, 433-435.

28. Gibbs, A.J. & Harrison, B.D. (1964). Nematode-transmitted viruses in sugar beet in East Anglia. Pl. Path. 13, 144-150.

29. Van Gundy, S.D. & Thomason, I.J. (1962). Factors influencing storage life of Hemicycliophora arenaria, Pratylenchus scribneri and Trichodorus christiei. Phytopathology 52, 366-367.

30. Hafkenscheid, H.H.M. (1971). Influence of Cu^{++} ions on Trichodorus pachydermus and an extraction method to obtain active specimens. Nematologica 17, 535-541.

31. Heathcote, G.D. (1965). Nematode-transmitted viruses of sugar beet in East Anglia, 1963 & 1964. Pl. Path. 14, 154-157.

32. Heathcote, G.D. (1973). Nematode-transmitted viruses of sugar beet in England 1965-72. Pl. Path. 22, 161-170.

33. Hijink, M.J. & Kuiper, K. (1966). Waarnemingen over de verdeling van aaltjes in de grond. Meded. Rijks. Landbouwwetensch., Gent. 31, 558-571.

34. Hirumi, H., Raski, D.J. & Jones, N.O. (1969). Ultrastructure of somatic muscles of Trichodorus christiei and Longidorus elongatus : Comparative cytology of two plant parasitic nematodes. J. Nematol. 1, 291 (abstr.)

35. Hirumi, H., Raski, D.J. & Jones, N.O. (1971). Primitive muscle cells of nematodes : morphological aspects of platymyarian and shallow coelomyarian muscles in two plant parasitic nematodes, Trichodorus christiei and Longidorus elongatus. J. Ultrastruct. Res. 34, 517-543.

36. Hoff, J.K. (1964). Studies on the pathogenicity of Trichodorus christiei with special references to the distribution of mitotic divisions and cellular differentiation in the Allium cepa root tip. Diss. Abstr. 25, 17.

37. Hoff, J.K. & Mai, W.F. (1962). Pathogenicity of the stubby-root nematode to onion. Pl. Dis. Rptr. 46, 24-25.

38. Hőgger, C.H. (1972). Effect of Trichodorus christiei inoculum density and growing temperature on growth of tomato roots. J. Nematol. 4, 66-67.

39. Hőgger, C.H. (1973). Preferred feeding site of Trichodorus christiei on tomato roots. J. Nematol. 5, 228-229.

40. Van Hoof, H.A. (1962). Trichodorus pachydermus and T. teres, vectors of the early browning virus of peas. Tijdschr. Plantenziekten 68, 391.

41. Van Hoof, H.A. (1964). Trichodorus teres, a vector of rattle virus. Neth. J. Pl. Path. 70, 187.

42. Van Hoof, H.A. (1968). Transmission of tobacco rattle virus by Trichodorus species. Nematologica 14, 20-24.

43. Van Hoof, H.A., Maat, D.Z. & Seinhorst, J.W. (1966). Viruses of the tobacco rattle virus group in Northern Italy: Their vectors and serological relationships. Neth. J. Pl. Path. 72, 253-258.

44. Van Hoof, H.A., Maat, D.Z. & Seinhorst, J.W. (1967). Some data on the presence of tobacco rattle virus and its vectors in France. Meded. Rijks. Landbouw-wetensch. Gent 32, 939-947.

45. Hollis, J.P. & Martin, W.J. (1960). Greenhouse pathogenicity trials with nematode infested soil. Phytopathology 50, 639-640.

46. Hooper, D.J. (1962). Three new species of Trichodorus (Nematoda: Dorylaimoidea) and observations on T. minor Colbran, 1956. Nematologica 7, 273-280.

47. Hooper, D.J. (1963). Trichodorus viruliferus n.sp. (Nematoda: Dorylaimida). Nematologica 9, 200-204.

48. Hooper, D.J. (1972). Two new species of Trichodorus (Nematoda: Dorylaimida) from England. Nematologica 18, 59-65.

49. Hooper, D.J. (1973). Nematodes. In A.J. Gibbs (ed.) Viruses and Invertebrates. North-Holland Publishing Co. 673 pp.

50. Jensen, H.J., Martin, J.P., Wismer, C.A. & Koike, H. (1959). Nematodes associated with varietal yield decline of sugar cane in Hawaii. Pl. Dis. Reptr. 43, 253-260.

51. Johnson, A.W. & Burton, G.W. (1973). Comparison of Millet and Sorghum-Sudangrass hybrids grown in untreated soil and soil treated with two nematicides. J. Nematol. 5, 54-59.

52. Jones, F.G.W., Larbey, D.W. & Parrot, D.M. (1969). The influence of soil structure and moisture on nematodes, especially Xiphinema, Longidorus, Trichodorus and Heterodera spp. Soil Biol. Biochem. 1, 153-165.

53. Kuiper, K. & Loof, P.A.A. (1962). Trichodorus flevensis n.sp. (Nematoda: Enoplida). A plant nematode from new polder soil. Versl. PlZiekt. Dienst. Wageningen 136, 193-200.

54. Liu, L.J. & Ayala, A. (1970). Pathogenicity of Fusarium moniliforme and F. roseum and their interaction with Trichodorus christiei on sugar-cane in Puerto Rico. Phytopathology, 60, 1540. (abstr.)

55. Liu, L.J. & Ayala, A. (1971). Pathogenicity of Fusarium spp. and relation with Trichodorus christiei on sugar-cane in Puerto Rico. Nematropica 1, 7. (abstr.)

56. Loof, P.A.A. (1973). Taxonomy of the Trichodorus aequalis complex (Diphtherophorina). Nematologica 19, 49-61.

57. Malek, W.R., Jenkins, W.R. & Powers, E.M. (1965). Effect of temperature on growth and reproduction of Criconemoides curvatum and Trichodorus christiei. Nematologica 11, 41 (abstr.)

58. Milne, D.L. (1972). Nematodes of Tobacco. In J.M. Webster (ed.) Economic Nematology, Acad. Press. 563 pp.

59. Morton, H.V. & Perry, V.G. (1968). Life cycle and reproductive potential of Trichodorus christiei [Abstr.]. Nematologica 14, 11.

60. Perry, V.G. (1953). Return of nematodes following fumigation of Florida soils. Proc. Fla. State Hort. Soc. 66, 112-114.

61. Persson, S. (1968). Nematoder av släktet Trichodorus i sydsvenska åkerjordar och deras förmåga alt överföra rattelvirus. Meddn. St. Växtsk Anst 14, 123, 167-199.

62. Pitcher, R.S. (1967). The host-parasite relations and ecology of Trichodorus viruliferus on apple roots, as observed from an underground laboratory. Nematologica 13, 547-557.

63. Pitcher, R.S. & Flegg, J.J.M. (1968). An improved final separation sieve for the extraction of plant parasitic nematodes from soil debris. Nematologica 14, 123-127.

64. Pitcher, R.S. & McNamara, D.G. (1970). The effect of nutrition and season of year on the reproduction of *Trichodorus viruliferus*. Nematologica 16, 99-106.

65. Prasad, S.F. (1972). Nematode diseases of sugar-cane. In J.M. Webster (ed.). Economic nematology. Acad. Press. 563 pp.

66. Rhoades, H.L. (1965). Parasitism and pathogenicity of *Trichodorus proximus* to St. Augustine grass. Pl. Dis. Reptr. 49, 259-262.

67. Rhoades, H.L. (1968). Re-establishment of *Trichodorus christiei* subsequent to soil fumigation in central Florida. Pl. Dis. Reptr. 52, 573-575.

68. Rhoades, H.L. (1969). Nematicide efficacy in controlling sting and stubby-root nematodes attacking onions in Central Florida. Pl.Dis. Reptr. 53, 728-730.

69. Richter, E. (1969). Zur vertikalen Verteilung von Nematoden in einem Sandboden. Nematologica 15, 44-54.

70. Rohde, R.A. & Jenkins, W.R. (1957a). Host range of a species of *Trichodorus* and its host-parasite relationships on tomato. Phytopathology 47, 295-298.

71. Rohde, R.A. & Jenkins, W.R. (1957b). Effect of temperature on the life cycle of stubby-root nematodes. Phytopathology 47, 29 (abstr.)

72. Rössner, J. (1969). Phytoparasitäre Nematoden in Forstpflanzgärten. Z. Angew. Zool. 56, 1-64.

73. Rössner, J. (1971). Einfluss der Austrocknung des bodens auf wandernde Würzelnematoden. Nematologica 17, 127-144.

74. Ruehle, J.L. (1969). Influence of stubby-root nematode on growth of southern pine seedlings. Forest Sc. 15, 130-134.

75. Russell, C.C. & Perry, V.G. (1966). Parasitic habit of *Trichodorus christiei* on wheat. Phytopathology 56, 357-358.

76. Salter, P.J. & Williams, J.B. (1967). The influence of texture on the moisture characteristics of soil. IV. A method of estimating the available water capacities of profiles in the field. J. Soil Sc. 18, 174-181.

77. Sasser, J.N., Nelson, L.A. & Barker, K.R. (1972). Effects of Trichodorus christiei and Belonolaimus longicaudatus on growth and yield of soybean following chemical soil treatment. /Abs./ J. Nematol. 4, 233-234.

78. Seddon, J.C. (1970). Pea Early Browning Virus. Open Conference of Advisory Entomologists (ADAS England and Wales) Appendix N (ENT/OM/58) to Minutes of Meeting held at High Leigh Conference Centre, Hoddesdon Jan. 1970.

79. Seinhorst, J.W. (1955). Een eenvoudige methode voor het afscheiden van aaltjes uit grond. Tijdschr. Plantenziekten, 61, 188-190.

80. Seinhorst, J.W. (1963). A redescription of the male of Trichodorus primitivus (de Man), and the description of a new species T. similis. Nematologica 9, 125-130.

81. Seinhorst, J.W. (1967). The Relationships between Population increase and population density in plant parasitic nematodes. III Definition of the terms, host, host status and resistance. IV The influence of external conditions on the regulation of the population density. Nematologica 13, 429-442.

82. Siddiqi, M.R. (1973). Systematics of the genus Trichodorus Cobb, 1913 (Nematoda: Dorylaimida), with descriptions of three new species. Nematologica 19, 259-278.

83. Simons, W.R. (1973). Nematode survival in relation to soil moisture. Meded. Landbouwhogesch., Wageningen 73-3.

84. Sol, H.H. and Seinhorst, J.W. (1961). The transmission of rattle virus by T. pachydermus. Tijdschr. Plantenziekten 67, 307-311.

85. Southey, J.F. (ed.) (1970). Technical Bulletin No.2, Laboratory methods for work with plant and soil nematodes. HMSO London 1970, 148 pp.

86. Standifer, M.S. & Perry, V.G. (1960). Some effects of sting and stubby root nematodes on grapefruit roots. Phytopathology 50, 152-156.

87. Storey, G.W. (1967). The effect of moisture and soil particle size on activity of Trichodorus spp. Unpub. report Broom's Barn Exp. Sta.

88. Sturhan, D. (1967). Vorkommen von Trichodorus - Arten in Westdeutschland. Mitt. biol. Bund Anst. Ld u. Fort. 121 146-150.

89. Taconis, P.J. & Kuiper, K. (1963). Overdracht van het Nicotiana virus 5 door aaltjes ven het geslacht Trichodorus in Laailingen van 5 gewassen. Versl. Meded. Plziektenk. Dienst. Wageningen 141, 177-178.

90. Taylor, C.E. (1971). Nematodes as vectors of plant virus. In B.M. Zuckerman, W.F. Mai and R.A. Rohde (eds.) Plant Parasitic Nematodes. Volume II. Acad. Press. 347 pp.

91. Taylor, C.E. & Robertson, W.M. (1970). Location of tobacco rattle virus in the nematode vector Trichodorus pachydermus Seinhorst. J. gen. Virol. 6, 179-182.

92. Thomason, I.J. & Sher, S.A. (1957). Influence of the stubby root nematode on growth of alfalfa. Phytopathology 47, 159-161.

93. Todd, J.M. (1967). Soil-borne virus diseases of potato tobacco rattle virus. NAAS Quart. Rev. 77, 21-29.

94. Torrealba, P.A. (1969). Survey of plant-parasitic and free-living nematode genera from Venezuela. In J.E. Peachey (ed.) Nematodes of Tropical Crops. Comm. Agric. Bureaux. Farnham Royal, England. 355 pp.

95. Wallace, H.R. (1954). Hydrostatic pressure deficiency and the emergence of larvae from cysts of the beet eelworm. Nature (London) 173, 502-503.

96. Wallace, H.R. (1958). Movement of eelworms T. The influence of pore size and moisture content of the soil on the migration of larvae of the beet eelworm Heterodera schachtii Schmidt. Ann. appl.Biol. 46, 74-85.

97. Whitehead, A.G. (1965). Nematodes associated with Docking disorders of sugar beet. Brit. Sugar Beet Rev. 34, 77-78, 83-84, 92.

98. Whitehead, A.G. & Hooper, D. (1970). Needle nematodes (Longidorus spp) and stubby-root nematodes (Trichodorus spp.) harmful to sugar beet and other field crops in England. Ann. appl. Biol. 65, 339-350.

99. Whitehead, A.G., Fraser, J.E. & Greet, D.N. (1970). The effect of DD, chloropicrin and previous crops on numbers of migratory root-parasitic nematodes and on the growth of sugar beet and barley. Ann. appl. Biol. 65, 351-359.

100. Whitehead, A.G., Dunning, R.A. & Cooke, D.A. (1971). Docking disorder and root ectoparasitic nematodes of sugar beet. Rpt. Rothamsted Exp. Sta. 1970 (2), 219-236.

101. Wyss, U. (1970). Zur Toleranz wandernder Wurzelnematoden gegenüber zunehmender Austrocknung des Bodens und hohen osmotischen Drüken. Nematologica 16, 63-73.

102. Yeates, G.W. (1967). Studies on nematodes from dune sands 9. Quantitative comparison of the nematode faunas of six localities. New Zealand J. Sc. 10, 927-948.

103. Yokoo, T. (1964). On the stubby root nematodes from the Western Japan. Agric. Bull. Saga Univ. 20, 57-62.

104. Zuckerman, B.M. (1961). Parasitism and pathogenesis of the cultivated cranberry by some nematodes. Nematologica 6, 135-143.

105. Zuckerman, B.M. (1962). Parasitism and pathogenesis of the cultivated highbush blueberry by the stubby-root nematode. Phytopathology 52, 1017-1019.

DISCUSSION

Cohn asked whether any explanation could be offered on his observations in pot experiments that 8,000 Paratrichodorus christiei per litre of soil failed to produce stubby root symptoms on cotton and wheat host plants. Winfield suggested that the watering regime used may have affected the symptomology and Thomason added that P. christiei had a very wide host range and that its symptomology varied considerably.

In reply to questions asked about sampling procedures used for trichodorids, Winfield described his soil sampling procedure as being random within a grid-system. He said that the majority of trichodorids in the United Kingdom were found in sandy soils and requested that more detailed soil analysis should be given in descriptions of type habitats. **Loof** remarked that in the Netherlands trichodorids were also found in heavy soils, and this brought forth the remark that 'heavy' soils of the Netherlands were not clay and would not be classified as 'heavy' in the United Kingdom. It was generally agreed that the terms 'light' and 'heavy' used in describing soil texture were subjective, and there was a need for a more objective classification and identification of soil types when describing habitats.

Short Report

A PRELIMINARY NOTE ON THE GEOGRAPHICAL DISTRIBUTION OF TRICHODORIDS IN THE BRITISH ISLES

T.J.W. Alphey and B. Boag (Scottish Horticultural Research Institute, Invergowrie, Dundee, Scotland)

A survey conducted at the Scottish Horticultural Research Institute to determine the distribution of trichodorid species in the British Isles was initiated as a study of the association of plant parasitic nematodes with forest trees but was later expanded to include analysis of samples collected from other vegetation areas in a survey of *Longidorus* and *Xiphinema* (see Brown in these Proceedings).

Soil sampling was based on the 10km grid system of the British Ordnance Survey maps. For each soil sample the dominant vegetation type and altitude were recorded at the site. About 1kg of soil was collected and transported to the laboratory in polythene bags. Nematodes were extracted from sub-samples of soil using a modified wet-sieving technique (2); a further soil sample was processed for analysis of physical characteristics (1). Following extraction nematodes were heat killed in water and fixed in TAF fixative. Each sample was scanned under a dissecting microscope (mag X 40) recording the population size of trichodorids, if present. Permanent mounts of trichodorid specimens were made by a rapid glycerol method and identified under oil-immersion to species level whenever possible.

Over 2900 samples were examined for trichodorids. Only 191 positive samples were recorded. This figure is possibly an underestimate of the frequency of trichodorids in the British Isles as low density populations may have been missed in the samples which were primarily sieved for *Longidorus* and *Xiphinema* nematodes.

Figure 1 shows the distribution of trichodorids in the British Isles, each dot representing the presence of one or more trichodorid species. *Trichodorus primitivus* was present in 115 (60%) of the positive samples and appears to have a cosmopolitan distribution in the British Isles. In contrast *T. hooperi* was recorded in only 4 (2%) of the positive samples and was restricted to the south-west tip of England. Nine of the 11 species found in the British Isles were recorded in this survey (Table 1).

Table 1. Trichodorids recorded in the British Isles

Species	Country	Number of positive samples
Trichodorus primitivus	E. S. W. Ci.	115
T. velatus	E. S. W.	9
T. similis	E. S. Ci.	7
T. cylindricus	E. S.	7
T. hooperi	E.	4
T. viruliferus	E. S.	3
T. variopapillatus	-	0
Paratrichodorus pachydermus	E. S. W. Ci.	43
P. anemones	E. S.	3
P. teres	S.	1
P. nanus	-	0

E = England S = Scotland W = Wales Ci. = Channel Islands

These are the first records of T. viruliferus, T. similis, T. velatus, P. anemones and P. teres in Scotland.

Further analysis of the geographical distribution is being undertaken using the "Statistical Package for the Social Sciences" programme on an IBM 370/158 computer in conjunction with the Biological Records Centre, Monks Wood, England.

REFERENCES

1. BOAG, B. and ALPHEY, T.J.W. (1974). A Preliminary Study of the Factors Influencing the Distribution of Trichodorid Species in the British Isles. NATO Advanced Study Institute on Nematode Vectors of Plant Viruses.

2. COBB, N.A. (1918). Estimating the nematode population of the soil. Agric. Tech. Circ. I. Bur. of Plant Industry, U.S. Dept. Agric. 48 pp.

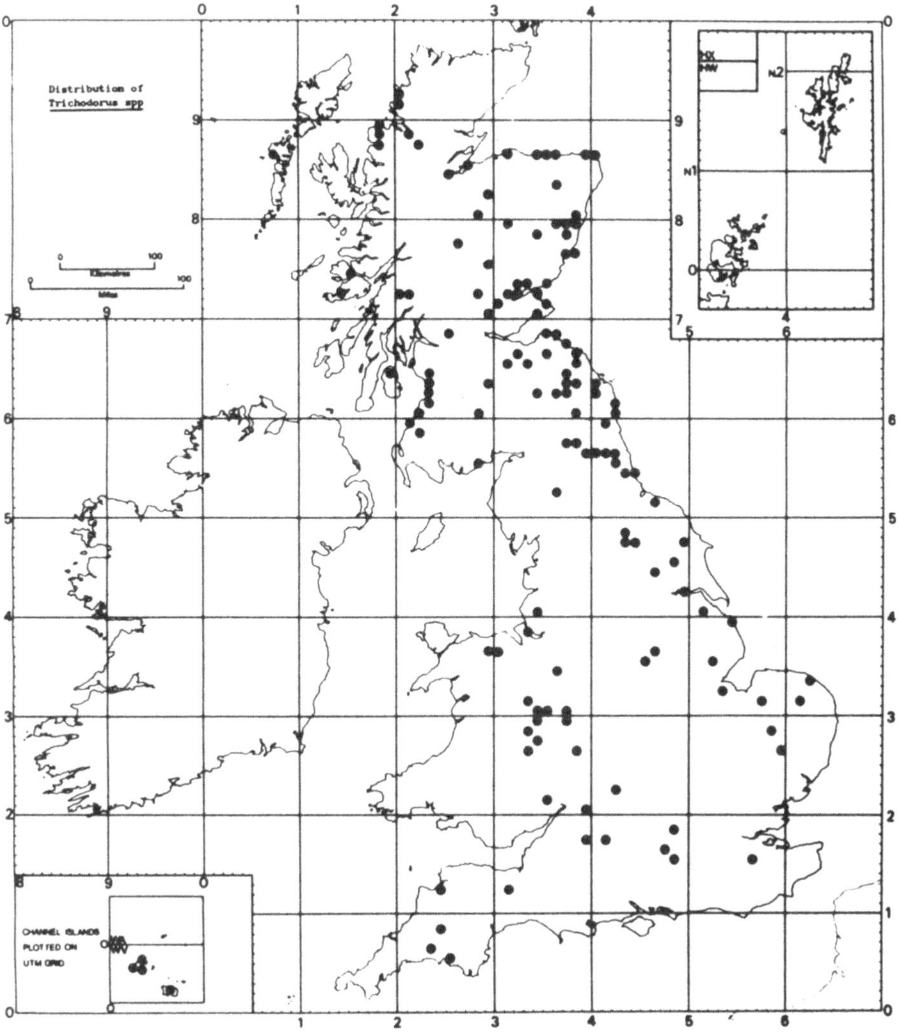

Fig. 1. Distribution of sites with trichodorid nematodes in Great Britain

Short Report

A PRELIMINARY STUDY OF THE FACTORS INFLUENCING THE DISTRIBUTION OF TRICHODORID SPECIES IN THE BRITISH ISLES

B. Boag and T.J.W. Alphey (Scottish Horticultural Research Institute, Invergowrie, Dundee, Scotland.)

The description of how soil samples were collected, the extraction technique and the identification of trichodorid species has been described (Alphey and Boag in these Proceedings). Most samples were also examined for moisture content, pH, percentage sand silt and clay. Statistical analysis of the results have not been completed and consequently only general observations are given here.

The vegetation from which samples were taken was initially split into 5 categories (Table 1). These results would suggest that deciduous woodland, or the vegetation associated with it, provides the environment where trichodorid species were most often recovered, followed by arable farmland, permanent grassland, coniferous woodland and finally scrubland. The distribution of the trichodorid species within the deciduous woodland category would indicate that nematodes are more often detected around sycamore, oak and elm trees than around beech.

The presence of trichodorid species would seem to be related to the percentage sand fraction in the soil. The majority of samples found to contain nematodes belonging to the genera *Trichodorus* and *Paratrichodorus* were from soils which contained over 80% sand. Only two samples out of 105 with percentage sand fractions below 40% were found to contain trichodorid species.

Allied to the percentage soil sand fraction was the frequency distribution of the nematodes with respect to the altitude at which sampling took place. The majority of samples containing trichodorid nematodes (58%) were taken below 200 ft above sea level (65 m). However, it is worth noting they were recovered up to 1,400 ft (500 m).

Another abiotic factor measured was pH. The results suggest that trichodorid species are rarely found in samples with pH below 4 but readily recovered up to pH 8, but too few samples were collected from soils above pH 8 to indicate whether or not it was the upper limit.

The results obtained from measuring the moisture content of the soil were difficult to interpret since they vary depending upon rainfall, season, soil type, etc. The limited amount of data examined indicated that trichodorid nematodes were more often recovered from the drier soils with moisture contents between 10 and 20%.

The collection of data for any one of the factors mentioned is of limited value by itself since they are often interrelated. However, the measurement and recording of many factors each time a sample is taken allows a composite picture to be built up of the environment in which the nematode lives. It is then possible to analyse the data statistically and suggest which are the most important factors.

All nematodes from samples collected during a survey should also be killed and fixed then stored for possible examination at a later date. This procedure has meant that nematodes in the samples collected for _Xiphinema_ and _Longidorus_ by Taylor and Brown (see Brown in these Proceedings) were available for examination for the trichodorid survey described here and since then for another subsequent survey of the criconematids of the British Isles. This approach enables the associations which exist between nematodes in the field to be studied.

The cost of surveys are considerable and it is only logical that as much information as possible is retrieved from them. It is hoped that this report has indicated that surveys can not only produce data on the geographical distribution of nematodes but also give some insight into their biology and the ecological niche they inhabit.

Table 1

Vegetation	No. of samples	No. of samples +ve for _Trichodorus_	% of _Trichodorus_ samples
Deciduous wood	752	83	43
Coniferous wood	647	26	14
Scrubland	175	5	3
Permanent grassland	683	38	20
Arable farmland	663	39	20
Total	2,920	191	100

Short Report

GEOGRAPHICAL DISTRIBUTION OF XIPHINEMA AND LONGIDORUS IN THE BRITISH ISLES

D.J.F. Brown (Scottish Horticultural Research Institute, Invergowrie, Dundee, Scotland.)

A survey of the distribution of Xiphinema and Longidorus in the British Isles began in 1970 as part of a co-operative project, financed by a NATO Scientific Research grant, between Professor F. Lamberti, Italy and Dr. C.E. Taylor, Scotland (2). At that time much was known about the ability of Xiphinema and Longidorus species to transmit viruses but there was little information about their geographical distribution and particularly in relation to soil, host plant and climatic factors, all of which have a bearing on their distribution. The survey therefore aimed at getting soil samples throughout the British Isles from five broad vegetation types - arable crops, permanent grassland, deciduous woodland, coniferous wood and scrub or moorland - and at the same time to test each soil sample for virus infectivity to establish the relationship between vector populations and the occurrence of virus in the field. The survey was undertaken with the help of the various agricultural advisory services in the British Isles and in all over 2,000 soil samples have been collected and processed. Together with about 2,000 records obtained from previous local samples by the advisory services and from a survey of the literature, and with records from Dr. B. Boag's investigation on nematodes associated with forest and woodland trees (1), there are about 5,000 records relating to the distribution of Xiphinema and Longidorus in the British Isles.

There are some 13 representatives of the Longidoridae family consisting of 8 Longidorus species, 1 Paralongidorus species and 4 Xiphinema species in Britain. The results of the survey show that the genus Longidorus tends to have a more northerly distribution than does Xiphinema although both genera seem to prefer the drier eastern side of the country. Numerous differences were observed in the distribution patterns of the nematode species, Longidorus attenuatus being found mainly in East Anglia whilst L. macrosoma and L. profundorum have so far only been detected in southern England. Longidorus caespiticola was mostly found in the west, especially Wales, where it frequently occurred in association with Xiphinema diversicaudatum. Xiphinema diversicaudatum is widespread in England, particularly in the south, and it has also been frequently identified from Scotland with the most northerly record being Dundee. Longidorus leptocephalus was widely distributed though showing a preference for the drier eastern side of the country, as did L. elongatus, which appeared to have its distribution centre in north east Scotland and this species has been recorded from the Shetland Isles in the north to the Channel Islands in the south. Populations of Longidorus vineacola, Paralongidorus maximum, Xiphinema coxi, X. mediterraneum and X. vuittenezi have also been infrequently identified and it is believed that these species could be recent introductions to the British Isles.

By efficient use of computers and modern data processing techniques and equipment, distribution maps have been produced for the first time on standard grid reference formats to provide quantitative information on the relative incidence of the nematode species being investigated. Comprehensive site data together with virus association, pH and soil texture data have also been collected during the course of this survey and will be analysed to establish which factors are most important in relation to the distribution and establishment of the different species of Longidorus, Paralongidorus and Xiphinema.

REFERENCES

1. Boag, B. (1974). Nematodes associated with forest and woodland trees in Scotland. Ann. appl.Biol. 77, 41-50.

2. Taylor, C.E. and Brown, D.J.F. (1973). NATO-financed survey (Distribution of Longidorus and Xiphinema). Rpt. Scott. Hort. Res. Inst. 1972, pp. 74-75.

Short Report

KNOWN DISTRIBUTION OF SPECIES OF *LONGIDORUS* AND *XIPHINEMA* IN NORWAY

Magne Støen (Norwegian Plant Protection Institute, 1432 Ås-NLH, Norway.)

In the years 1966-73 2,300 soil samples were examined for *Longidorus* and *Xiphinema*. The bulk of the samples came from fields planned for clonal stocks of potato, strawberry and raspberry. This means that lighter soils were selected, and there were few samples with heavy loam soil. Nematodes were extracted from 250 g samples in a Seinhorst elutriator.

Longidorus spp. were found in 16% of the samples, but this figure cannot be considered representative for the cultivated soils of the country.

Longidorus elongatus is widely distributed, especially in the coastal area, from Oslofjord in the east and along the south, west and northern coasts. The northernmost locality so far is at 67°N. *L. elongatus* often occurred in high densities, several localities having more than one specimen per gram of soil.

Longidorus leptocephalus was found inland and especially around the lake Mjosa in samples collected in September, October and November. This is one of our main areas for potatoes, where they are grown in rotation with grass and cereals, mainly barley. *L. leptocephalus* occurred in low densities. Half the localities had only 1 or 2 specimen per 250 g of soil and therefore this species might be distributed much more widely than was found in this survey. All the females and first stage larvae measured fit into the "large form" of *L. leptocephalus*; body length of females in the range of 4.5-5.7 mm, odontostyle 71-78 μm and that of first stage larva 43-48 μm.

Xiphinema diversicaudatum was found in the county Østfold, some 60 km south of Oslo. It seems to be well established in an area of several square kilometres, where it was found in association with agricultural and nursery crops, along roadsides and occasionally also in woodland in sandy loam soils.
X. diversicaudatum occurred in low densities, and was often mixed with *L. elongatus*. Outside this restricted area the species was found only in a single locality in the south in association with roses. No *Xiphinema* species were found in a survey of roses grown in glasshouses. It is concluded that *Xiphinema* is scarce in Norway.

Short Report

LONGIDORUS AND *XIPHINEMA* IN SWEDEN

K.B. Eriksson (Agricultural College, Uppsala, Sweden.)

Soil samples have so far been taken from about 200 localities, 80% of which are south of the 60° latitude i.e. "through" Uppsala. The samples were taken at random from various soil types and crops, and cultivated and natural habitats. *Longidorus elongatus* was found in 11 localities and *L. leptocephalus* (large form) in one, most of these in light soils in southern Sweden. *Xiphinema diversicaudatum* was recorded in southwest Sweden from land in which potatoes were growing; this constitutes the first record of this genus in Sweden.

Short Report

NEMATODES OF THE FAMILY LONGIDORIDAE IN BULGARIA

Boriana Choleva (Institute of Plant Protection, Kostinbrod, Sofia, Bulgaria)

In Bulgaria there have been occasional surveys of longidorid nematodes on vine and the following species have been reported(1): X. bulgariensis, X. index, X. turcicum, L. closelongatus, L. goodeyi. Recently such surveys have been extended by investigating the geographical distribution of longidorids throughout the country and the host range of different species among cultivated and wild plants. About 360 soil samples were collected from several regions throughout the country and from twelve vegetation types. The following is data from the survey, which is being continued.

X. mediterraneum was found in 95% of the samples from vine, peach, apricot, almond, fig, plum, rose (Rosa damascena), danewort or dwarf elder (Sambucus ebulus) and nettle (Lamium sp.) in light to heavy textured soils and occurred throughout the country.

X. index was found in 23% of the samples from vineyards and fig gardens. It occurred in five different regions and in light and heavy textured soils.

X. italiae (syn. bulgariensis) was found in 18% of the samples from vineyards, apricot, plum and peach orchards in light-textured and fairly dry soils. The species was found in five different regions.

X. vuittenezi occurred in 14% of the samples from vine, plum, peach, danewort and nettle in moist locations in light and light to medium textured soils.

X. turcicum occurred in 9% of the samples from vine, almond, peach, wheat and danewort in sandy soils around the Black Sea.

X. brevicolle was found in 6% of the samples from vine and almond in three different regions in soils of light and medium light texture.

X. neovuittenezi occurred sporadically in 4% of the samples from vine and fig in light sandy soils around the Black Sea.

L. siddiqii (L. brevicaudatus) occurred in 16% of the samples in vineyards, and in peach and almond orchards, in light alluvial soils in the southern regions of the country.

L. attenuatus was found in 11% of the samples from vine, apricot, tomato and roses in medium light soils in two regions.

L. closelongatus was found in 61% of the samples on vine, danewort and nettle in alluvial and forest drift soils in two different regions.

L. profundorum was found as single specimens in 4% of the samples from apricot, peach and apple in sandy loam soils in the Black Sea region.

L. vineacola was found as single specimens in 3% of the samples on vine and peach in sandy soils in the southern part of the country.

Longidorus sp. (near to sylphus) was found in 12% of the samples from tomatoes growing in greenhouses, from apples, almonds, cabbages and vines and in light textured soils.

REFERENCE

Stoyanov, A. (1964). A contribution to the nematodofauna of the grape vine (in Russian). Rastin Zashita. Moscow 12, 16-24.

Short Report

GEOGRAPHICAL DISTRIBUTION OF *XIPHINEMA* IN SPAIN

Maria Arias (Instituto Espanol de Entomologia, Madrid, Spain)

Over a period of more than 10 yr about 6,000 soil samples have been taken throughout Spain from agricultural, forestal, pastural, and natural (uncultivated) habitats, involving in total about 200 plant species. About 30% of the samples contained *Xiphinema*, and 12 species are now known to occur in Spain. *X. brevicolle*, *X. index*, *X. italiae* and *X. mediterraneum* are the most widespread; *X. diversicaudatum*, *X. sahelense* and *X. ingens* are moderately widespread; and *X. turcicum*, *X. vuittenezi* and *X. neovuittenezi* fairly restricted in their distribution. *X. rivesi* and *X. pyrenaicum* have been found only once.

X. mediterraneum is the most widespread of the species, having been found throughout the Peninsula, but mainly in the Mediterranean region and in the central area of the country. It has been found in association with about 80 different plant species and in cultivated and uncultivated soils.

X. brevicolle and *X. italiae* are also widespread in the Mediterranean region, often occurring with *X. mediterraneum*. *X. brevicolle* was found first on *Citrus* but later in association with about 40 different species of cultivated or wild plants; it has been found in the central and northern regions. *X. italiae* was found in association with several different plant species in a few places in the central region and once in the northern region.

The distribution of *X. index* is correlated with that of vineyards. It frequently occurred in the south, east and central regions most usually in cultivated soils but sometimes in natural ones.

X. diversicaudatum has been found only in the north, mainly in uncultivated soils, in association with *Fagus*, *Pinus* or *Quercus* and occasionally with fruit trees. *X. sahelense* occurred in the central and south-west regions associated with several crops and wild plants. *X. ingens* has only occasionally been found, in the north-west and the south in association with grapevine and in the central region on the Levante in uncultivated soils.

X. turcicum was found mainly in the south and east where it is widespread on different crops; single records have been obtained from the central and west regions. *X. vuittenezi* occurs in the central, south and east regions in association with several crops and wild plants. X. neovuittenezi was found in two separate localities in the east region and it is believed it may be fairly widespread there.

There is a single record for X. rivesi from the central region, near Madrid, but deeper sampling may be required to locate this species. The only record for X. pyrenaicum in the south region may represent an introduction of the species.

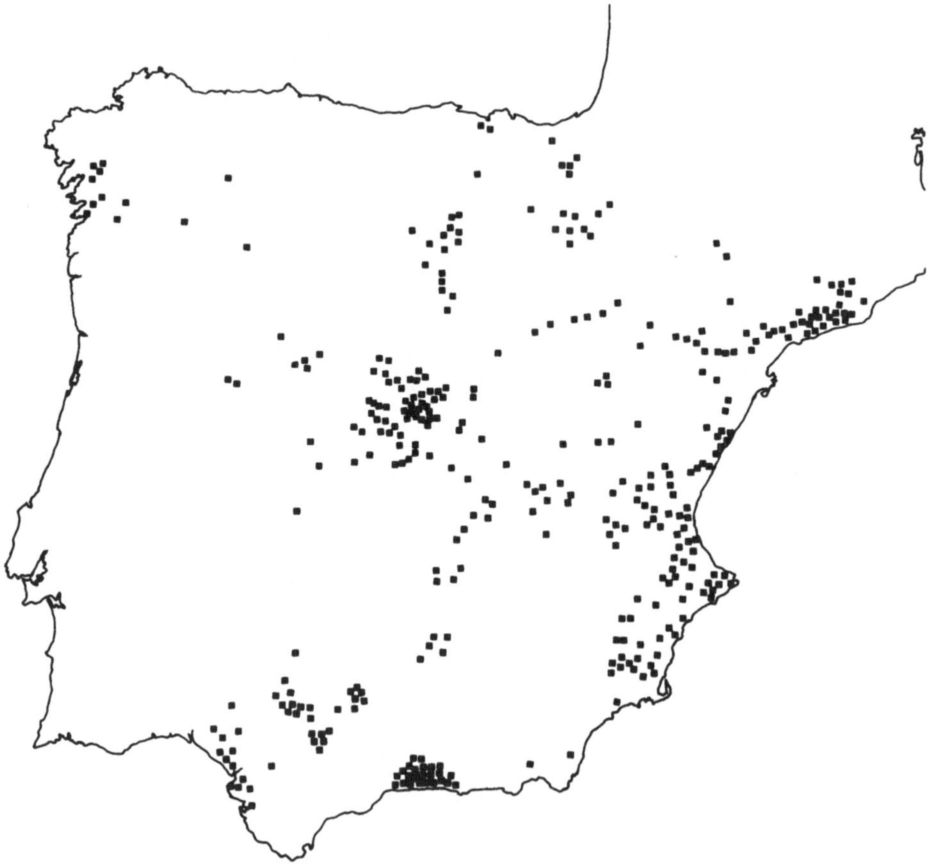

Fig. 1. Distribution of Xiphinema mediterraneum in Spain.

Short Report

THE DISTRIBUTION OF VIRUS-TRANSMITTING NEMATODES IN UNDISTURBED BIOTOPES OF LOWER SAXONY

Jürgen Rau (Institut für Pflanzenkrankheiten und Pflanzenschutz der Technischen Universität Hannover, BRD.)

Investigations on the occurrence of virus-transmitting nematodes in undisturbed biotopes in different areas of Lower Saxony (W. Germany) were made mainly in woods selected with a characteristic natural plant community considered to represent the climax vegetation, where stable biocoenoces with distinctive forms have developed. Assuming that defined plant communities could be used as a working hypothesis, conclusions can be drawn on the natural occurrence and on the ecology of these nematodes.

When deciduous forests of the Hanoverian hilly countryside were examined, it became evident that *Longidorus* and *Xiphinema* species occurred in definite vegetation zones. They were for instance never found in sites with the plant community termed Luzulo-Fagetum[1], but they were fairly frequent in the Melico-Fagetum. *L. macrosoma* was common in the Melico-Fagetum allietosum, whereas *X. diversicaudatum* occurred there only on carboniferous but not on Jurassic limestone. *L. elongatus* was never found in Fagetum types, but it occurred in the Querco-Carpinetum. The habitat of *X. vuittenezi* is apparently the Melico-Fagetum elymitosum which is typical of a more continental climate than that of the other areas examined.

Trichodorus spp. were common in the diluvial region of Lower Saxony but were never found in the hilly countryside. The only exception was *T. similis* which occurred in woodlands near rivers with the plant communities Fraxino-Ulmetum and Carici remotae-Fraxinetum.

---

[1]For the specific terms of the plant communities see: Ellenberg, H.: Vegetation Mitteleuropas mit den Alpen. *In* H. Walter (ed.), Einführung in die Phytologie. IV/2. Verlag E. Ulmer, Stuttgart 1963, 943 pp.

Short Report

LONGIDORUS AND XIPHINEMA IN THE SUDAN

A.M. Yassin (Gezira Agricultural Research Station, Wad Medani, Sudan.)

Longidorus africanus, *L. brevicaudatus* and *L. laevicapitatus* were found around the roots of various field crops throughout the Sudan, in soils ranging from light sandy to heavy clay (up to 55% clay content). The most important is *L. africanus* which causes stubby and swollen roots of cotton plants when tested in the glasshouse; there was a direct correlation between the degree of severity of symptoms and the number of nematodes per test. The life cycle of *L. africanus* is 6-8 months at 30°-35°C.

Among *Xiphinema* species, found mainly around the roots of trees and woody plants, were *X. basiri*, *X. simillimum* Loof and Yassin 1970 and possibly also *X. elongatum*. *Xiphinema* species were also of wide occurrence in light and heavy clay soils. The commonest was *X. basiri* which was often associated with root stubbiness, but sometimes discolored roots, of *Citrus* (bitter lemon), *Mangifera indica* (mango) and *Azadiracta indica* ("nim") trees. The life cycle extended up to about 8 months under the experimental conditions.

Short Report

DISTRIBUTION OF *XIPHINEMA* SPP. IN SOUTH AFRICAN VINEYARDS

P.C. Smith (Plant Protection Research Institute, Stellenbosch 7600, South Africa.)

Viticulture in South Africa is practised mainly in the Southwestern Cape Province, but smaller plantings also occur in various irrigation schemes, especially along the Olifants River and the lower Orange River in the northwestern and northern Cape. Recent new plantings in irrigation schemes in the Orange Free State are not yet of great importance.

Geographical distribution. The systematics and distribution of the South African Longidoridae are presently being studied by Professor Juan Heyns of the Rand Afrikaans University. Two species of the subgenus *Xiphinema* have been found. *X. brevicolle* seems to be the most common species in vineyards, although *X. americanum* also occurs. The subgenus *Elongiphinema* is represented by the species *X. italiae*, *X. elongatum* and a new *Xiphinema* species, of which only the last two have been found in vineyards (Heyns, in press).

During a survey of phytoparasitic nematodes in vineyards in the southwestern Cape, *Xiphinema* occurred in 70% of the samples examined and *Longidorus* spp. were found in 27% of the samples. Identification of this material was unfortunately only to generic level. However, it is known that at least the following *Xiphinema* species occur in our vineyards: *X. brevicolle*, *X. americanum*, *X. elongatum*, *X. hallei*, *X. variabile*, *X. zulu*, *X. meridianum* and *X. index*. So far *X. index* has been found only in 4 or 5 vineyards. Mixed populations of 2 or 3 *Xiphinema* species are common but in such cases one species is usually overwhelmingly predominant.

Vertical distribution. In one field experiment the vast majority of *X. elongatum* occurred between 8 and 16 cm depth in the soil and very few individuals were found below 20 cm, whereas *X. hallei* was found only below 40 cm. In another experiment with *X. meridianum*, over 40% of the population occurred between 20 and 40 cm. On a deep sandy soil a *Xiphinema* species was found as deep as 3.9 m.

Seasonal population fluctuations. From field experiments a little is known of the seasonal history of *Xiphinema* species. On a loamy sand soil a population of *X. elongatum* built up during winter to a peak in spring (September) when soil temperatures fluctuated between 8°–15°C and soil moisture was at field capacity.

This was followed by a sharp decline to a low level during the hot summer months. Very few adults (only 10% of the total population) were found, possibly because of the extraction techniques employed (Oostenbrink elutriator/cotton wool filter method). Gravid females were few, occurring mainly in spring.

With a population of *X. meridianum*, two annual peaks normally occurred, one in midwinter and a higher one in midsummer. The latter peak was absent during one very dry summer. About 80% of this population consisted of juveniles. Gravid females occurred mainly in spring and early summer. Juveniles seemed to be relatively insensitive to the soil temperature and/or moisture levels encountered as no correlation between their numbers and these factors could be determined. However, numbers of adult females were closely correlated with soil moisture and temperature, with moisture seeming to have the greater effect.

From the above information it seems fairly obvious that a great deal remains to be learned about the behaviour of *Xiphinema* species under South African conditions and the possible virus/nematode vector situation occurring in local vineyards.

REFERENCE

Heyns, J. (1974). The genus *Xiphinema* in South Africa I. Subgenus *Xiphinema* (Nematoda: Dorylaimida). II. Subgenus *Elongiphinema* (Nematoda: Dorylaimida). *Phytophylactica* (in press)

RELATIONS BETWEEN *XIPHINEMA* AND *LONGIDORUS* AND THEIR HOST PLANTS

ELI COHN

ARO, The Volcani Centre,

Bet Dagan, Israel.

INTRODUCTION

Two major difficulties have plagued the plant nematologist in his efforts to prove pathogenicity of nematodes. One is the problem of obtaining sufficient numbers of a defined nematode species for investigations in a controlled system; the other is the problem of isolating the nematode from the many organisms with which it interacts in its natural environment. Progress made in recent years in developing techniques for culturing plant parasitic nematodes has been largely limited to the endoparasitic forms. Once the initial difficulty of creating contact between the endoparasitic nematode and its host has been overcome, the parasite is able to thrive in its natural environment in the plant tissues, and the medium for further study exists. On the other hand, in working with the ectoparasite it is necessary to maintain a simulated environment through-out its entire life cycle, and this is certainly a more difficult task particularly if the life cycle happens to be a long one.

It is therefore not surprising that proof of endoparasitism among plant nematodes preceded the discovery of ectoparasitism by almost 200 years. The ability of ectoparasitic nematodes to attack plants was first demonstrated only as recently as 1951 in Florida (3). The discovery in 1958 by Hewitt *et al* (30) that an ectoparasitic nematode can transmit a virus disease in plants acted as a stimulant to more intensive research on this group of nematodes.

HOST PLANTS OF *XIPHINEMA* AND *LONGIDORUS*

A host plant of a nematode species is one from which the nematode can derive sufficient food to sustain growth and multiplication; on a non-host, a nematode cannot compete its life cycle (59). Hosts may be poor or good, depending on the rate of nematode multiplication and the size of nematode populations supported by them. Association of a nematode species with a particular plant variety or even the presence of symptoms are not valid evidence of host status. The literature abounds with reports of species of Xiphinema and Longidorus associated with certain plants. Such reports are definitely useful, especially if more precise information is not available, since they may indicate possible tendencies in host preference. An overall glance at the natural habitats of Xiphinema and Longidorus species may give some idea of the expected host preferences of the two genera. Text-fig. 1 shows a qualitative analysis of the type localities of as many species of Xiphinema and Longidorus that I was able to trace. It can be seen that the vast majority of Xiphinema species were described from around woody perennial plants, whereas most Longidorus species were found around herbaceous perennials or herbaceous annuals.

Evaluation of plants as hosts of Xiphinema and Longidorus has been done by determining - (a) rate of total nematode population increase in relation to time, or, (b) the time needed to complete the nematode life cycle. Work in this field has progressed relatively slowly, and the information at hand, is consequently scant. The two interrelated factors which are primarily responsible for this information lag are difficulties in culturing, and the apparently long life cycle of most Xiphinema and Longidorus species: these are briefly discussed below.

CULTURING AND REPRODUCTION RATE OF XIPHINEMA AND LONGIDORUS

Culturing of most Xiphinema and Longidorus species, even on whole plants growing in a soil medium, has proved extremely problematic. Among Xiphinema species, only X. index and X. brevicolle appear to be cultured relatively easily (10, 50, 51) and of the Longidorus species, only L. africanus and L. siddiqi (10,35). The reasons for the difficulties are not entirely clear, but better results have sometimes been obtained by improving culture techniques e.g. by maintaining more favourable moisture and temperature conditions (10) and especially minimizing moisture and temperature fluctuations (30). Nonetheless, inherent differences between nematode species in their ability to subsist in cultures obviously exist, and specific differences have been encountered also in their sensitivity to various handling procedures, such as extraction from soil, passage through filters, and hand-picking for inoculation. X. index and X. brevicolle for instance, show no apparent susceptibility to handling, but it is extremely difficult to obtain large numbers of viable specimens of X. mediterraneum or X. bakeri (43, 66) for inoculation purposes, even when extracting from a heavily infested soil.

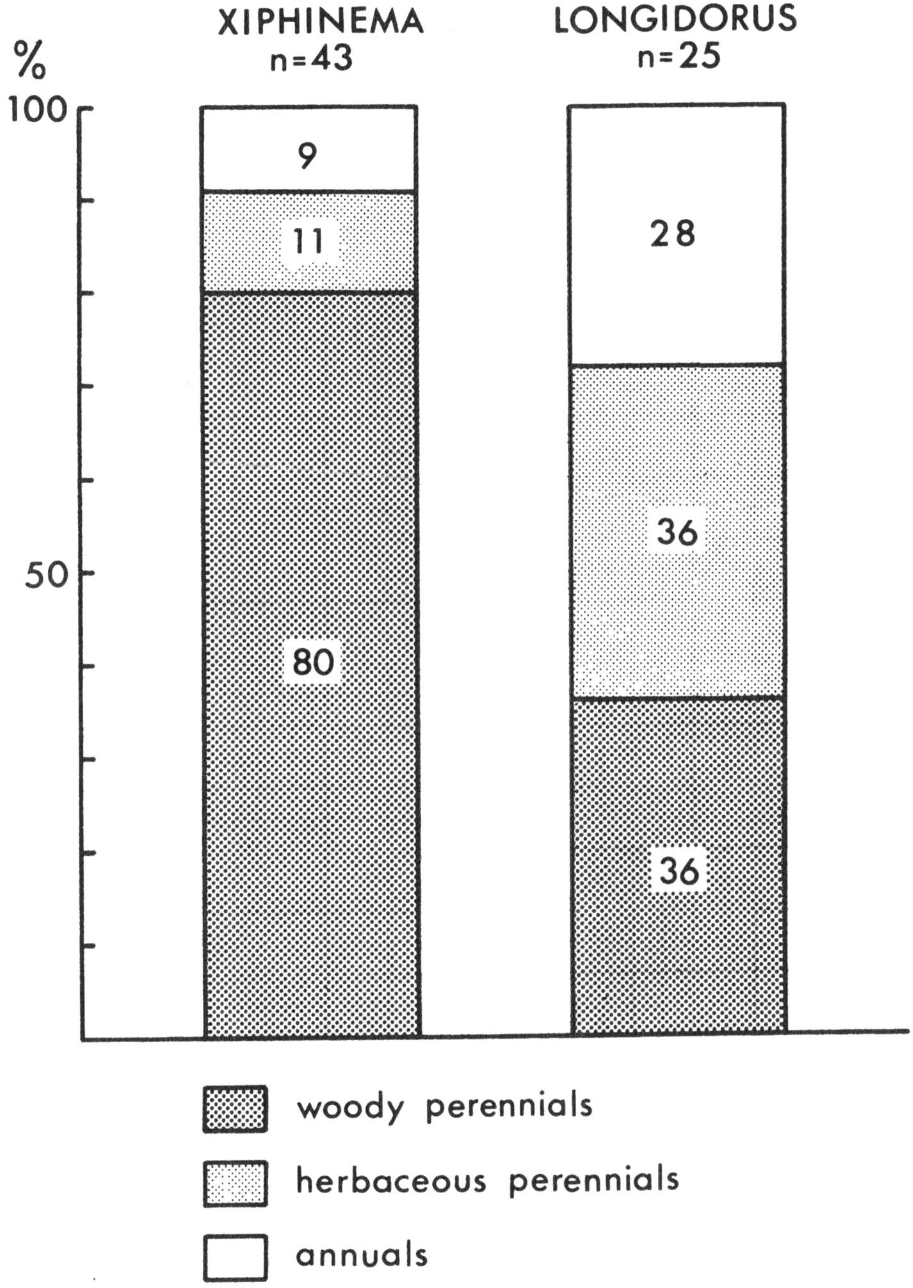

Fig. 1. Type habitats of Xiphinema and Longidorus species

Another limiting factor in determining accurately host plants of Xiphinema and Longidorus is the apparently long period that most species require to complete their life cycles. Although, possibly, this is partly due to imperfect culture techniques, evidently even under natural conditions many species of Xiphinema and Longidorus in fact do have unusually long life cycles. Among the Xiphinema species, only X. index completed its life cycle on grapevine under controlled conditions within 21 days (50); in laboratory cultures at 20-23° C, an Israeli population of X. index took 7-9 months to complete its life cycle on grapevine (9), while under Sardinian vineyard conditions, Prota and Garau (49) have estimated that 12-14 months elapse before the life cycle of X. index is completed. All other Xiphinema species on which there is information have life cycles ranging from several months for X. bakeri under greenhouse conditions (43) to an estimated 2 years for X. diversicaudatum and X. vuittenezi under orchard conditions (22). L. profundorum and L. macrosoma appeared to develop at a similar rate to that of X. diversicaudatum under field conditions (22). In greenhouse cultures L. africanus completed its life cycle within 3-4 months (9) and L. elongatus within 9 weeks (77). However, L. elongatus in the field apparently produces only one generation per year in a temperate climate (70).

How, then, are such high population levels of Xiphinema and Longidorus so often attained under specific conditions? Probably by a combination of a long life span - of all the individual life stages of the nematode (22) - and a high egg production capacity. At the present time, however, data on this subject are inadequate for critical assessment.

Host range

For the afore-mentioned reasons, most of our knowledge on the host range of Xiphinema and Longidorus is based on observed population increases in naturally-infested soils cropped with different plant species, rather than critical studies of population build-up from a known initial inoculum. For several nematode species, many plants have been assessed for host status and appropriate lists have been prepared. These include X. americanum (37), X. diversicaudatum (29, 71), X. index (51, 72), X. bakeri (43), L. elongatus (67, 69, 78), L. africanus (34) and L. maximum (= ? Paralongidorus) maximus (62, 63). Only a few plant species were tested for X. chambersi (56), X. brevicolle (9), X. italiae (9), L. attenuatus (27, 74) and L. cohni (11). These data again support the general conclusion that most Xiphinema species have a preference for woody plants, while such plants form a smaller part of the host spectrum of Longidorus species. Many of the above-mentioned species appear to be fairly polyphagous, particularly the three most ubiquitous ones - X. americanum, X. diversicaudatum and L. elongatus - but some degree of specificity is nevertheless evident in all of them, and poor and even non-hosts have been reported for them. For at least two species, X. index and L. cohni, more non-hosts than hosts were recorded among the tested plants.

PLANT - NEMATODE INTERACTIONS

Ability of the nematode to feed on a particular plant does not in itself indicate that the plant is a host. X. brevicolle, for instance, has been observed to feed readily on bur marigold (Bidens tripartita) in the laboratory, although this plant is a non-host of the nematode (9). Fritzsche and Hofferek (25, 26) have found it useful to differentiate between the "host-plant range" (Wirtspflanzenkreis) of nematode species and their "feeder-plant range" (Nährpflanzenkreis) and have reported lists of feeder plants for X. diversicaudatum and L. macrosoma which do not necessarily correspond with the known host ranges of these nematodes. This distinction is indeed practical, particularly in investigations on virus transmission, where the feeding ability of Xiphinema and Longidorus is perhaps of more significance than their reproductive capacity. Since the general subject of nematode feeding is treated separately by Wyss in these Proceedings, I shall restrict myself here to those aspects of feeding that are intimately connected with the plant root tissues and their response.

Nematode feeding

X. americanum was observed to feed on maple at root tips, at branching points and along the side of young roots (14). Sutherland (64) observed X. bakeri feeding on Douglas fir and Silka spruce seedlings apparently at random on various sites along the root; after one week most feeding was at root tips. Fisher and Raski (20) reported that larvae of X. index fed well back from the root tip of grapevine seedlings on the outer cortical cells of the piliferous region; pre-adults and adults fed mainly at tips. On dwarf nettle, an herbaceous plant, X. index and X. brevicolle fed mainly along the root and rarely near the tips (4). Feeding of X. diversicaudatum on petunia was observed on all parts of the roots, though tips were preferred (25). Depth of stylet penetration in relation to root tissues has been found in most instances to depend on the diameter of roots; in young thin roots stylets reach the vascular tissues, whereas in older, thicker roots the nematodes insert stylets into the cortical parenchyma layers (4, 20, 25, 48). Feeding times at one site for most Xiphinema species can be very long particularly when feeding at sites somewhat distant from the root tip, and feeding periods of some hours to several days have been reported for most of the species studied (4, 25, 57, 64).

Both L. africanus on seedlings of bur marigold (4) and L. elongatus on Fragaria vesca (76) were observed to feed exclusively

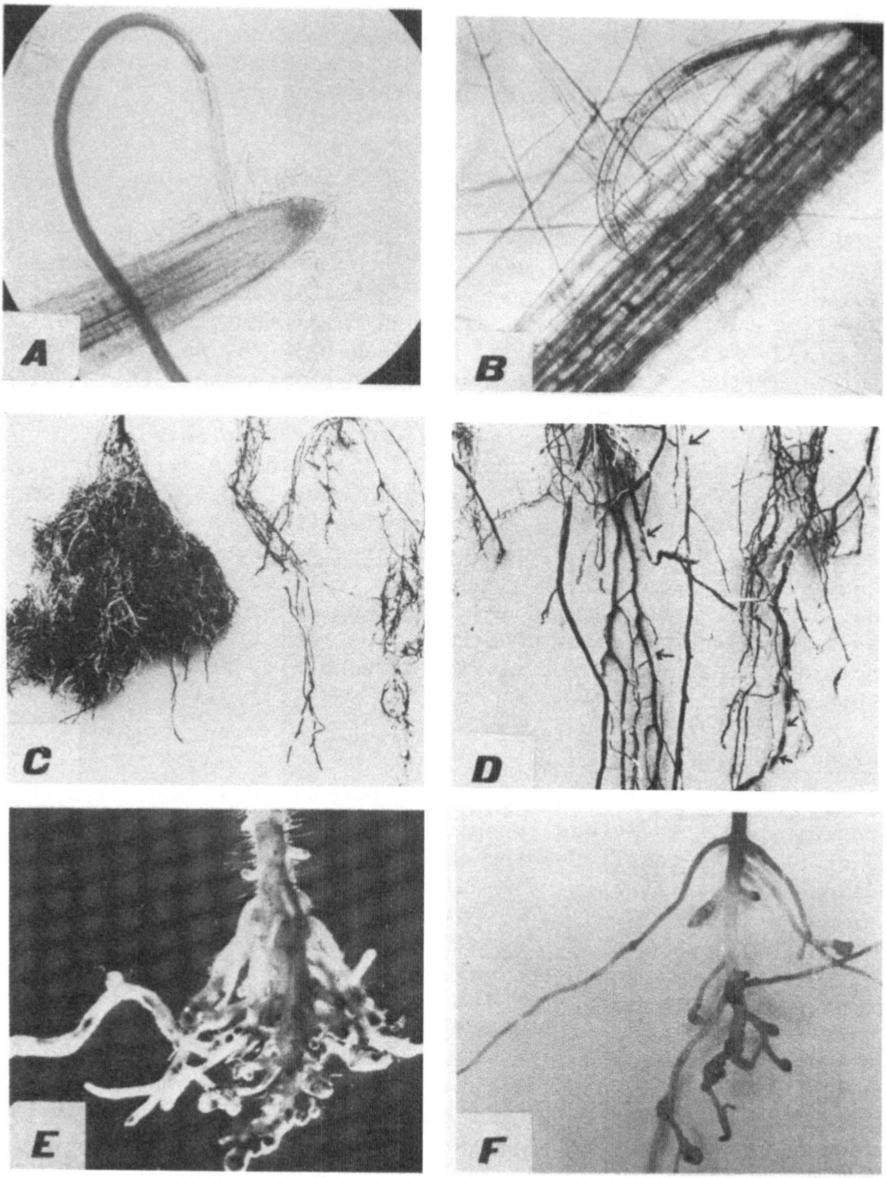

PLATE I: Feeding and symptoms of *Xiphinema* and *Longidorus*. A: *L. africanus* feeding at root tip of bur marigold. B: *X. brevicolle* feeding along root of *Urtica urens*. C: *X. americanum*-infested sugar maple roots (left), control (right). D: Cortical tissue breakdown (arrows) in infested sugar maple roots. E: Tomato roots infested by *X. bakeri*. F: Bur marigold roots infested by *L. africanus*. (C, D courtesy C.P. Di Sanzo & R.A. Rohde; E, courtesy F.D. McElroy).

at root tips, inserting their stylets deep into the tissue and feeding for relatively short periods. Maximum feeding time for L. africanus was 15 min; L. elongatus fed for as long as 70 min.

The main points which emerge from these observations may be summarized as follows:

(1) Both Xiphinema and Longidorus insert their stylets to a considerable depth into the plant tissue and are capable of penetrating the vascular elements, particularly in young root parts, at or near the tips.

(2) Xiphinema species differ in their selection of feeding sites on roots; some usually, but not exclusively, feed at root tips e.g. X. bakeri, X. diversicaudatum, X. index; others e.g. X. americanum, X. brevicolle rarely do so. On the other hand Longidorus invariably feeds at root tips.

(3) Feeding times at root tips are apparently short, while duration of feeding on older root parts can be very long. Hence, some Xiphinema species feed for long periods, while Longidorus apparently has shorter feeding times.

Plant reactions to nematode feeding

Plant response to the nematode's activity is closely related to the feeding site. Feeding on old and thicker parts of the root precludes the nematode from reaching the vascular tissues and reactions are restricted to the cortical tissue. These are not pronounced, however, and only slight disturbance of parenchyma cells, evidently due to physical rupture during stylet insertion, are visible (48). However, these sites under natural conditions may serve as loci for invading secondary organisms which can result in local disintegration of the cortex at a later stage. Thus, plant reactions to feeding of Xiphinema at points along roots, especially on older roots are typical of those to cortical parasites, i.e. initial lesion formation which spreads and intensifies with increased activity of secondary organisms. Feeding at root tips, however, and to some extent apparently also on thin roots where vascular penetration can be effected (25), results in various types of swellings. Fisher and Raski (20) reported that X. index formed galls on grapevine seedling roots only when the nematode fed at the root tips, and these galls were already visible 24 hours after inoculation. L. elongatus also formed root tip galls on Fragaria vesca one day after inoculation (76), while L. africanus formed sizable terminal swellings on bur marigold roots 20 hours after feeding was observed to end (4). Terminal root gall formation by these species was accompanied by a cessation of growth of the apical meristem.

Symptomatology (Plate I). The symptoms on plant roots caused by Xiphinema and Longidorus closely fit the pattern of parasitism

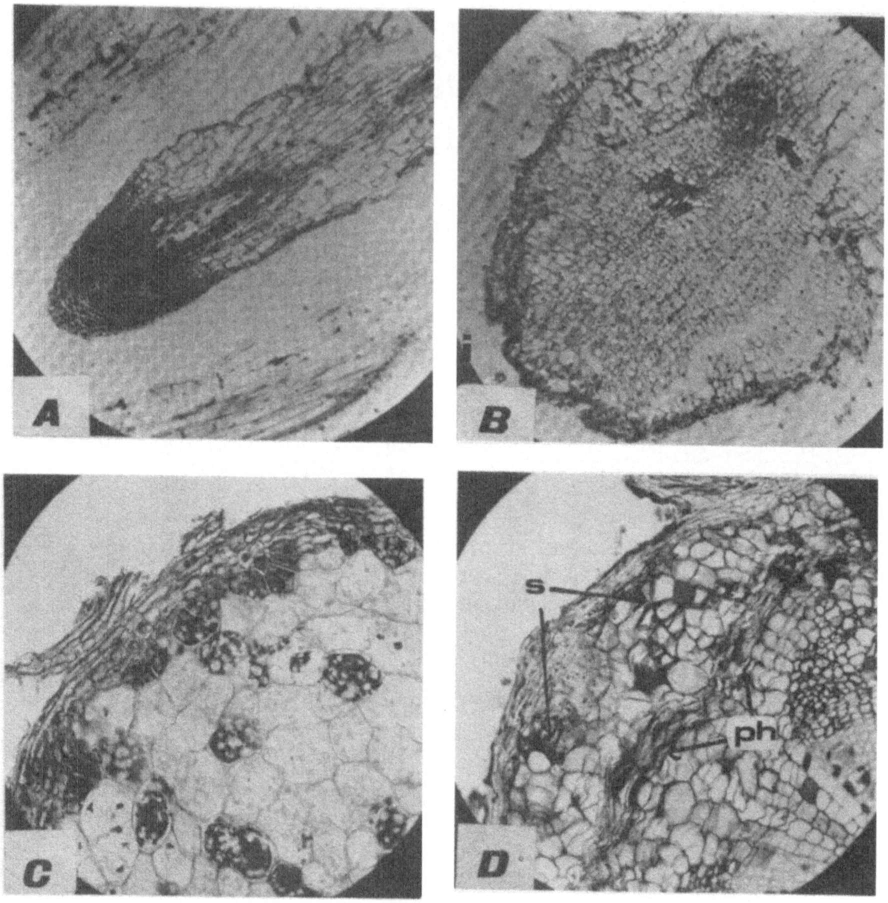

PLATE II: Histopathology of Xiphinema and Longidorus. A: Healthy growing root of bur marigold. B: Longitudinal section of terminal root gall on bur marigold caused by L. africanus, showing lack of apical meristem, hyperplasia of cortical tissue and laterial root initiation (arrow). C: Cross-section through healthy grapevine root. D: Cross-section through X. index - infected grapevine root showing groups of suberized cells in cortex (s) and phellogen layer (ph).

just described. Some Xiphinema species, such as X. americanum, X. chambersi and X. brevicolle, cause primarily a marked reduction in root systems due to the destruction of feeder roots. Remaining roots bear numerous lesions, sometimes leading to disintegration and decay of the cortical tissue at various points, and a distinctly discoloured and often blackened appearance of the roots. Some young roots may have slightly swollen and curved tips, but this is not a dominant feature. Such symptoms have been reported for X. americanum on many plants e.g. periwinkle (15), strawberry (47), alfalfa (45), maple (14) and pine (55); for X. chambersi on strawberry (47) and sweetgum (56); for X. brevicolle on grapevine, rose and citrus (4); and for X. italiae on grapevine (4). From personal observations in the field, it is probable that X. mediterraneum also belongs to this group of nematodes. Other Xiphinema species cause prominent terminal or subterminal swellings on roots which causes the tip to curl in addition to a reduction of lateral roots and extensive cortical necrosis. Such symptoms have sometimes been described as "curly tip" (57) and "fish-hook tip" (43). They form a characteristic feature of attack by X. diversicaudatum on rose (57), plum (60), celery, strawberry (29) and other herbaceous plants (71); by X. bakeri on forest tree seedlings (65) and several herbaceous plants (43); and by X. index on grapevine (54). The size of these swellings may differ according to age of the roots (60) or the plant species (29) and in some hosts may be lacking entirely (71).

Symptoms of attack by Longidorus species evidently are more uniform. Root systems are greatly reduced, lateral and sometimes tap roots are severely stunted, and small sometimes discoloured galls appear at the tip , primarily on the young roots. A necrotic spot, presumably at the point of stylet insertion, is often visible at the end of galls. On perennial plants the galls are often more elongated and tips may be slightly curved. Little or no necrosis is prominent along roots at sites other than root tips. Reports of root symptoms caused by different species of Longidorus are remarkably similar, as instanced by L. elongatus on strawberry (42, 76), sugar beet, carrot (75) and clover (69); L. attenuatus on strawberry (76) and sugar beet (73); L. africanus on lettuce (52); L. siddiqii, (4) on grapevine; L. cohni on oats (6); L. vineacola on onion and celery (5); and L. maximus on onion and other plants (63).

Histopathology (Plate II). The histology of roots infected by Xiphinema and Longidorus has not been studied to any great extent. Nevertheless, some common features of different nematode-plant combinations within each of the two genera have been described. X. diversicaudatum feeding on thick roots of rose resulted in the production of groups of dark suberized cells within the cortical parenchyma (13). This also occurred in grapevine infected by X. index (12) but an underlying phellogen layer also formed in the cortex. In rose parasitized by X. diversicaudatum (13), in Douglas fir roots infected by X. bakeri (2) and in grapevine infected by X. index (51) apical meristems were found reduced in size and activity and a marked hyperplastic response of cortical cells was evident, particularly on the inside of the root curve in rose. Hypertrophy of cortical cells and in some cases even a multinucleate condition of enlarged cortical cells in the feeding zone were observed (13, 51). These cells, however, were only slightly larger than normal ones and contained no more than 2-3 nuclei which exhibited slight hypertrophy, thereby differing markedly from the typical syncytia produced by heteroderid nematodes. Both in rose parasitized by X. diversicaudatum and Douglas fir roots parasitized by X. bakeri, some cortical cell walls were thickened and amorphous substances, possibly tannins, were present. In Douglas fir the enlarged cortex persisted and formation of secondary vascular tissues was retarded.

Galls caused by L. africanus on bur marigold (12) and L. elongatus on Fragaria vesca (76) appeared to be more compact than those caused by Xiphinema. Meristematic activity was drastically reduced and the apical meristem was almost wholly destroyed; at the same time an early maturation of the root tip region was evident and vascular tissue reached the apex. In bur marigold roots initial formation of lateral roots in the tip region was a prominent phenomenon. Hyperplasia of the cortical parenchyma and some hypertrophy (76) occurred and probably accounted for gall formation. The necrotic spot so often present on galls appeared to be the site of stylet insertion. Whitehead et al (74) observed a row of necrotic cortical cells extending into the root from this spot in galls on sugar beet roots infected with L. elongatus.

Gall formation. The galls induced by Longidorus and Xiphinema are strong evidence that the nematode discharges some substance into the plant during feeding that triggers off a sequence of events leading to gall formation. Although L. africanus fed for only 15 minutes on a bur marigold root tip, the swelling of tissue began 20 hours after feeding had ceased and continued for 3 days - long after the nematode had left the feeding site (4). A marked increase in amino acids (particularly proline), carbohydrates (including inulin, the storage product of bur marigold) and nucleic acids was noted in galls (16, 17, 18). This perhaps, is only to be expected if considerable cell hyperplasia and hypertrophy took place. Also there was an increase in phenols and peroxidase activity (16), which is not surprising considering that cells in the galls underwent accelerated divisions and tissues matured early. However, a rapid rate of increase in nucleic acids following nematode inoculation is particularly interesting and an increased absorption rate of the RNA precursor, uridine, was observed in the root tissue as early as 48 hours after inoculation (17). This leads to the suggestion that the rapid changes within the plant tissue leading to gall formation may be initiated by nucleic acids, primarily RNA, or their viral precursors injected by the nematode (19). The ability of *Longidorus* species to transmit viruses in plants would also add some weight to this hypothesis. Alternatively, the nematode might stimulate synthesis of plant RNA during feeding.

PLANT DAMAGE CAUSED BY *XIPHINEMA* AND *LONGIDORUS*

Plant resistance and tolerance

The degree of excellence as a host of a plant species to a particular nematode, does not give us an indication of its proneness to damage by that nematode, whether it be direct damage or indirect damage (by transmission of viruses). *X. americanum* readily acquires tobacco ringspot virus from cucumber, which is evidently a non-host of the nematode (37), and transmits it to tobacco which is a poor host (24); Taylor (67) has demonstrated that *L. elongatus* can transmit raspberry ringspot virus from *Rubus idaeus*, a non-host of the nematode, to *Stellaria media*, a good host, but not to ryegrass which is also a good host. This lack of correlation between host status and susceptibility to virus transmitted by a nematode is also found when the nematode affects plants directly. Thus, for instance, *L. elongatus* caused severe stunting of glue-cured tobacco and typical terminal root galls although this plant is a poor and possibly even a non-host for the nematode (40). On the other hand, growth of Pearson tomato, a good host of *L. africanus*, was unaffected by the nematode and the typical galls on root tips were not produced, despite rapid nematode reproduction (36). Instances of tolerance have also been reported for species of *Xiphinema*. Thus, although *X. americanum* parasitized both *Pinus edulis* and *Juniperus monosperma* seedlings, it reduced growth only of the latter (55); *X. bakeri* while reproducing well on 16 different plants, caused severe damage only on a few representatives of Solanaceae and Rosaceae (43).

In addition to resistant and tolerant plant species, there are some indications that "enemy plants" of *Xiphinema* and *Longidorus* also exist. McElroy (43) has reported a detrimental influence of some cruciferous and cucurbit plants on population levels of *X. bakeri*, and Thomas (71) has observed a reduction in populations of *X. diversicaudatum* by *Chrysanthemum coronarium*.

Population density and plant damage

Recurring reports, from different parts of the world, of certain plant-nematode associations related to visual effects on crops provide some indication of the probable economic significance of these animals. A summary of the damage caused to plants by species of the two genera is presented in Table 1; it is based primarily on proven cases of pathogenicity and sometimes on recurring reports of constant associations. Incidental cases of plant-nematode associations, transmission of virus diseases, or mere records of plants

Table 1 **SUMMARY OF PLANT DAMAGE CAUSED BY XIPHINEMA AND LONGIDORUS**

A. Xiphinema

Nematode species	Type of damage	Locality of damage	Selected references
X. americanum	premature decline in a large number of plant species, primarily a wide range of fruit and forest trees; also unthriftiness of perennial forage and ornamental crops, and strawberry.	North America	14, 28, 32, 39, 46, 47, 55
X. bakeri	decline in several forest trees, particularly in nurseries; reduces growth rate also in some herbaceous plants	Western Canada; northern California	2, 43, 65
X. brevicolle	growth retardation in citrus, grapevine and rose	Israel, possibly also in other Mediterranean countries	4, 12
X. chambersi	decline in strawberry and some varieties of forest trees	Eastern USA	47, 56
X. diversicaudatum	decline in rose, fig, strawberry and raspberry; probably retards growth of other woody and herbaceous crops and weeds	Europe and eastern USA	21, 23, 29, 57, 58
X. index	decline in grapevine and fig	California; Mediterranean and Near East regions	12, 33, 41, 49, 51, 54

Table 1. cont..

B. Longidorus

Nematode species	Type of damage	Locality of damage	Selected references
L. africanus	severe stunting and wilting of lettuce, sugar beet, sorghum and possibly other seasonal crops; mild decline in grapevine.	southern California and Israel	12, 35, 52
L. attenuatus	severe stunting of strawberry and sugar beet, and possibly other herbaceous crops.	central and northern Europe	73, 74, 76
L. cohni	severe stunting of oats; mild decline in Rhodes grass and alfalfa	coastal region of Israel	6, 11
L. elongatus	severe stunting of peppermint, strawberry, tobacco and sugar beet; probably also retards growth in other crops and weeds	Europe and north America	31, 40, 61, 69, 73, 75, 76
L(=? Paralongidorus) maximus	severe stunting of onion, lettuce, celery, gladiolus, beet, carrot cucumber and many other herbaceous crop plants	central Europe	62, 63
L. siddiqii	pre-emergence death of cotton; mild decline in grapevine.	Egypt and Israel	1, 4
L. vineacola	severe stunting of onion, celery and carrot.	Israel	5, 7, 8

as hosts were not considered. The information contained in the table again indicates a greater significance of Xiphinema species on woody crops, and of Longidorus species on herbaceous crops (see also Fig. 1).

Gross symptoms of above-ground damage to plants by Xiphinema and Longidorus are, as is the case with other nematodes, not generally distinguishable from those caused by many other soil pathogens, or by adverse conditions in the soil environment. The drastic effects of a damaged and reduced root system usually leads initially to an increased shoot/root ratio (60, 76) and eventually to an inhibition of plant growth and development. It is, of course, the severity of retardation and hence the magnitude of the damage that is of particular interest to the nematologist. The degree of damage largely is dependent upon the nature of the plant, the population density of the nematode, and the environmental conditions influencing these two factors.

The density of nematode populations feeding on plant roots obviously has a marked effect on the amount of damage caused to the plant. The curve for plant growth plotted against increasing nematode densities is sigmoidal in shape, but tolerances vary according to plant and nematode species and other environmental factors. Ideally, plant tolerances to nematode densities can be determined by inoculating with varying numbers of nematodes in the initial inoculum. This has been demonstrated clearly by Seinhorst (60) who showed that reduced top weight of Fragaria vesca can be expected at population levels exceeding 2 L. elongatus per 10 g soil. Similarly, Schindler (57) has shown that an inoculum of 500 X. diversicaudatum in the rhizosphere of rose caused a 50% reduction in plant weight after 109 days, whereas 10,000 nematodes in the inoculum arrested growth completely. Such effects of nematode densities on plant performance have also been reported for X. index on grapevine (12). Naturally, for different nematodes plant damage begins at different population levels. Thus, a population of 50 L. elongatus per 100 g soil caused root damage and severe stunting of strawberry (61) whereas inoculation with 1,000 X. diversicaudatum per litre had no apparent effect on growth of strawberry despite heavy galling and discolouration of roots (23). Moreover, the same nematode species has different effects on different plants. Thus, for instance, an initial population of 300 L. africanus per 500 ml soil was sufficient to kill bur marigold plants within 3 weeks, while a population of 1,000 nematodes caused a weight reduction in grapevine of only 18% over a 6-month period (12).

REFERENCES

1. Aboul Eid, H.Z. (1970). Pathogenicity of Longidorus siddiqi on Egyptian cotton (Gossipium barbadense) Pl.Dis.Reptr. 56, 699-700.

2. Bloomberg, W.J. & Sutherland, J.R. (1971). Phenology and fungus-nematode relations of corky root disease of Douglas fir. Ann.appl.Biol. 69, 265-276.

3. Christie, J.R. & Perry, V.G. (1951). A root disease of plants caused by a nematode of the genus Trichodorus. Science 113, 491-493.

4. Cohn, E. (1970). Observations on the feeding and symptomatology of Xiphinema and Longidorus on selected host roots. J.Nematol. 2, 167-173.

5. Cohn, E. & Ausher, R. (1971). Seasonal occurrence of Longidorus vineacola on celery in Israel and its control. Israel J. agr.Res. 21, 23-25.

6. Cohn, E. & Ausher, R. (1973). Longidorus cohni and Heterodera latipons economic nematode pests of oats in Israel. Pl.Dis.Reptr. 57, 53-54.

7. Cohn E. & Krikun, J. (1966). A disease of onion associated with an ectoparasitic nematode of the genus Longidorus. Pl.Dis.Reptr. 50, 711-712.

8. Cohn, E., Krikun, J. & Yisraeli, U. (1968). Further observations on the effect of Longidorus vineacola on onion. Pl.Dis.Reptr. 52, 525-527.

9. Cohn, E. & Mordechai, M. (1969). Investigations on the life cycles and host preference of some species of Xiphinema and Longidorus under controlled conditions. Nematologica 15, 295-302.

10. Cohn, E. & Mordechai, M. (1970). The influence of some environmental and cultural conditions on rearing populations of Xiphinema and Longidorus. Nematologica 16, 85-93.

11. Cohn, E. & Mordechai, M. (1974). Specificity of Longidorus cohni as a pathogen of oats. Phytoparasitica 2, 45-46.

12. Cohn, E. & Orion, D. (1970). The pathological effect of representative Xiphinema and Longidorus species on selected host plants. Nematologica 16, 423-428.

13. Davis, R.A. & Jenkins, W.R. (1960). Nematodes associated with roses and the root injury caused by Meloidogyne hapla Chitwood, 1949, Xiphinema diversicaudatum (Micoletzky, 1927) Thorne, 1939, and Helicotylenchus nanus Steiner, 1945. Bull.Md.Agr.Exptl.Sta. A 106, 16 pp.

14. Di Sanzo, C.P. & Rohde, R.A. (1969). Xiphinema americanum associated with maple decline in Massachusetts. Phytopathology 59, 279–284.

15. Epstein, A.H. & Barker, K.R. (1966). Pathogenicity of Xiphinema americanum on Vinca minor. Pl.Dis.Reptr. 50, 420–422.

16. Epstein, E. (1972). Biochemical changes in terminal root galls caused by an ectoparasitic nematode, Longidorus africanus: phenols, carbohydrates and cytokinins. J.Nematol. 4, 246–250.

17. Epstein, E. (1974). Biochemical changes in terminal root galls caused by an ectoparasitic nematode, Longidorus africanus: nucleic acids. J.Nematol. 6, 48–52.

18. Epstein, E. & Cohn, E. (1971). Biochemical changes in terminal root galls caused by an ectoparasitic nematode, Longidorus africanus: amino acids. J.Nematol. 3, 334–340.

19. Epstein, E. & Cohn, E. (1973). Biochemical changes in roots infected by the nematode, Longidorus africanus. Phytoparasitica 1, 58–59.

20. Fisher, J.M. & Raski, D.J. (1967). Feeding of Xiphinema index and X. diversicaudatum. Proc.Helm.Soc.Wash. 34, 68–72.

21. Flegg, J.J.M. (1968a). The occurrence and depth distribution of Xiphinema and Longidorus species in south eastern England. Nematologica 14, 189–196.

22. Flegg, J.J.M. (1968b). Life-cycle studies on some Xiphinema and Longidorus species in south England. Nematologica 14, 197–210.

23. Flegg, J.J.M., Baxendale, M. & Popham, A.M. (1970). The reproductive potential of Xiphinema diversicaudatum on strawberry. Nematologica 16, 398–402.

24. Flores, H. & Chapman, R.A. (1968). Population development of Xiphinema americanum in relation to its role as a vector of tobacco ringspot virus. Phytopathology 58, 814–817.

25. Fritzsche, R. & Hofferek, H. (1969a). Beitrage zum Saugverhalten und Nährpflanzenkreis von Xiphinema diversicaudatum (Mikoletzky) Thorne. Arch.Pflanzenschutz 5, 111–118.

26. Fritzsche, R. & Hofferek, H. (1969b). Nahrungsaufnahme und Nährpflanzenkreis von Longidorus macrosoma Hooper. Arch.Pflanzenschutz 5, 423–429.

27. Green, C.D. (1967). Docking disorder of sugar beet. Depth distribution of the needle nematode in soil. Rep.Rothamsted exp.Stn., 1966, 148.

28. Griffin, G.D. & Epstein, A.H. (1964). Association of dagger nematode, Xiphinema americanum, with stunting and winter kill of ornamental spruce. Phytopathology 54, 177-180.

29. Harrison, B.D. & Winslow, R.D. (1961). Laboratory and field studies on the relation of arabis mosaic virus to its nematode vector Xiphinema diversicaudatum (Micoletzky). Ann.appl.Biol. 49, 621-633.

30. Hewitt, W.B., Raski, D.J. & Goheen, A.C. (1958). Nematode vector of soil-borne fanleaf virus of grapevines. Phytopathology 48, 586-595.

31. Horner, C.E. & Jensen, H.J. (1954). Nematodes associated with mints in Oregon. Pl.Dis.Reptr. 38, 39-41.

32. Johnson, D.E., Lear, B., Miyagawa, S.T. & Sciaroni, R.H. (1969). Multiple applications of 1,2-dibromo-3-chloropropane for control of nematodes in established rose plantings. Pl.Dis.Reptr. 53, 34-37.

33. Katcho, Z.A. & Allow, J.M. (1968). Dagger nematodes, Xiphinema index, on grapevines in Iraq. Pl.Dis.Reptr. 52, 626-627.

34. Lamberti, F. (1968). The effect of cropping on the population levels of Longidorus africanus. Pl.Dis.Reptr. 52, 748-750.

35. Lamberti, F. (1969a). Pathogenicity of Longidorus africanus on selected field crops. Pl.Dis.Reptr. 53, 421-424.

36. Lamberti, F. (1969b). Effect of temperatures on the reproduction rate of Longidorus africanus. Pl.Dis.Reptr. 53, 531.

37. Lownsbery, B.F. (1964). Effects of cropping on population levels of Xiphinema americanum and Criconemoides xenoplax. Pl.Dis.Reptr. 48, 218-221.

38. Lownsbery, B.F. & Mitchell, J.T. (1965). Some effects of chemical amendments and cultural conditions on population levels of Xiphinema americanum. Pl.Dis.Reptr. 49, 994-998.

39. Malek, R.B. (1968). The dagger nematode, Xiphinema americanum, associated with decline of shelterbelt trees in South Dakota. Pl.Dis.Reptr. 52, 795-798.

40. Marks, C.F. & Elliot, J.M. (1973). Damage to flue-cured tobacco by the needle nematode, Longidorus elongatus. Can.J.Plant Sci. 53, 689-692.

41. Martelli, G.P. & Raski, D.J. (1963). Osservazioni su Xiphinema index Thorne et Allen, Fico e degenerazione infettiva della Vite. Inform.Fitopat 13, 416-420.

42. McElroy, F.D. (1971). Longidorus elongatus damaging strawberry in British Columbia. Pl.Dis.Reptr. 55, 266-267.

43. McElroy, F.D. (1972). Studies on the host range of Xiphinema bakeri and its pathogenicity to raspberry. J.Nematol. 4, 16-22.

44. Mountain, W.B. (1960). Theoretical considerations of plant-nematode relationshipd. In "Nematology, Fundamentals and Recent Advances" (Sasser, J.N. & Jenkins, W.R., Eds.) Univ. N.Carol.Press, Chapel Hill. pp.419-421.

45. Norton, D.C. (1963). Population fluctuations of Xiphinema americanum in Iowa. Phytopathology 53, 66-68.

46. Norton, D.C. (1967). Xiphinema americanum as a factor in unthriftiness of red clover. Phytopathology 57, 1390-1391.

47. Perry, V.G. (1958). Parasitism of two species of dagger nematodes (Xiphinema americanum and X. chambersi) to strawberry. Phytopathology 48, 420-423.

48. Pitcher, R.S. & Posnette, A.F. (1963). Vascular feeding by Xiphinema diversicaudatum (Micol.) Nematologica 9, 301-302.

49. Prota, U. & Garau, R. (1973). Indagini sulla Biologia di Xiphinema index Thorne et Allen in vigneti Sardi. Nematol. Medit. 1, 36-54.

50. Radewald, J.D. & Raski, D.J. (1962a). A study of the life cycle of Xiphinema index. Phytopathology 52, 748.

51. Radewald, J.D. & Raski, D.J. (1962b). Studies on the host range and pathogenicity of Xiphinema index. Phytopathology 52, 748-749.

52. Radewald, J.D., Osgood, J.W., Mayberry, K.S., Paulus, A.O. & Shibuya, F. (1969a). Longidorus africanus, a pathogen of head lettuce in the Imperial Valley of Southern California. Pl.Dis.Reptr. 53, 381-384.

53. Radewald, J.D., Osgood, J.W., Mayberry, K.S., Paulus, A.O., Otto, H.W. & Shibuya, F. (1969b). Results of a preplant fumigation trial for the control of Longidorus africanus on lettuce. Pl.Dis.Reptr. 53, 519-523.

54. Raski, D.J. & Radewald, J.D. (1958). Reproduction and symptomatology of certain ectoparasitic nematodes on roots of Thompson seedless grapes. Pl.Dis.Reptr. 52, 941-943.

55. Riffle, J.W. (1972). Effect of certain nematodes on the growth of Pinus edulis and Juniperus monosperma seedlings. J. Nematol. 4, 91-94.

56. Ruehle, J.L. (1972). Pathogenicity of Xiphinema chambersi on sweetgum. Phytopathology 62, 333-336.

57. Schindler, A.F. (1957). Parasitism and pathogenicity of Xiphinema diversicaudatum, an ectoparasitic nematode. Nematologica 2, 25-31.

58. Schindler, A.F. & Braun, A.J. (1957). Pathogenicity of an ectoparasitic nematode Xiphinema diversicaudatum on strawberries. Nematologica 2, 91-93.

59. Seinhorst, J.W. (1961). Plant-nematode inter-relationships. Ann.Rev.Microbiol. 15, 177-196.

60. Seinhorst, J.W. (1966). Longidorus elongatus on Fragaria vesca. Nematologica 12, 275-279.

61. Sharma, R.D. (1965). Direct damage to strawberry by Longidorus elongatus. Meded.Rijkslandb.Hogesch.Gent. 30, 1437-1443.

62. Sprau, F. (1960). Über ein vermutlich Pflanzenschadigendes Auftreten eines freilebenden Nematoden, Longidorus maximus (Butschli) an einer Reiheyvon Kulturpflanzen. Nematologica (Suppl.2), 49-55.

63. Sturham, D. (1963). Der pflanzenparasitischer Nematode Longidorus maximus, seine Biologie und Ökologie, mit Untersuchungen an L. elongatus und Xiphinema diversicaudatum, Z. angew. Zool. 50, 129-193.

64. Sutherland, J.R. (1969). Feeding of Xiphinema bakeri. Phytopathology 59, 1963-1965.

65. Sutherland, J.R. (1970). Some forest nursery seedling hosts of the nematode, Xiphinema bakeri. Can.J.Plant Sci. 50, 588-590.

66. Sutherland, J.R. & Ross, D.A. (1971). Temperature effects on survival of Xiphinema bakeri in fallow soil. J.nematol. 3, 276–279.

67. Taylor, C.E. (1967). The multiplication of Longidorus elongatus (de Man) on different host plants with reference to virus transmission. Ann.appl.Biol. 59, 275–281.

68. Taylor, C.E., Thomas, P.R. & Converse, R.H. (1966). An outbreak of arabis mosaic virus and Xiphinema diversicaudatum (Micoletzky) in Scotland. Pl.Path. 15, 170–174.

69. Thomas, P.R. (1969a). Crop and weed plants compared as hosts of viruliferous Longidorus elongatus (de Man). Pl.Path. 18, 23–28.

70. Thomas, P.R. (1969b). Population development of Longidorus elongatus on strawberry in Scotland, with observations on Xiphinema diversicaudatum on raspberry. Nematologica 15, 582–590.

71. Thomas, P.R. (1970). Host status of some plants for Xiphinema diversicaudatum (Micol.) and their susceptibility to viruses transmitted by this species. Ann.appl.Biol. 65, 169–178.

72. Weiner, A. & Raski, D.J. (1966). New host records for Xiphinema index Thorne & Allen, 1959. Pl.Dis.Reptr.50,27–28.

73. Whitehead, A.G. & Hooper, D.J. (1970). Needle nematodes (Longidorus spp.) and stubby-root nematodes (Trichodorus spp.) harmful to sugar beet and other field crops in England. Ann.appl.Biol. 65, 339–350.

74. Whitehead, A.G., Fraser, J.E. & Greet, D.N. (1970). The effect of D-D, chloropicrin and previous crops on numbers of migratory root-parasitic nematodes and on the growth of sugar beet and barley. Ann.appl.Biol. 65, 351–359.

75. Whitehead, A.G., Dunning, R.A. & Cooke, D.A. (1971). Docking disorder and root ectoparasitic nematodes of sugar beet. Rep.Rothamsted exp.Stn. 1970, 219–236.

76. Wyss, U. (1970a). Parasitierungsvorgang und Pathogenität wandernder Wurzelnematoden an Fragaria vesca var. semperflorens. Nematologica 16, 55–62.

77. Wyss, U. (1970b). Untersuchungen zur Populations dynamik von Longidorus elongatus. Nematologica 16, 74–84.

78. Yassin, A.M. (1969). Glasshouse and laboratory studies on the biology of the needle nematode, Longidorus elongatus. Nematologica 15, 169-178.

DISCUSSION

Several questions concerning the sites and the efficiency of feeding of Longidorids were raised. Jatala asked for further information concerning the specificity of feeding sites of Longidorus elongatus and remarked that during field trials he had found L. elongatus around the base of mint (Mentha) stem-cuttings; the nematodes were feeding and producing symptoms similar to root-pruning. Cohn reiterated the fact that L. elongatus is primarily a root-tip feeder and this was confirmed by Weischer. It was pointed out that in field trials secondary pathogens present in the soil may confuse the symptomatology of diseases.

Following questions on the life-cycle of Xiphinema index from different locations Cohn suggested that biotypes could be used to explain the differences in duration of the life-cycle recorded in different localities. He thought that biotypes probably played as large a role as climate or any other factor. These statements were supported by Thomason who added that host biotype, as well as nematode biotype, would affect the life-cycle. The presence or absence of galling on root-stocks was given as evidence of different biotype interaction.

The production of root galls was discussed in detail. Cohn stated that X. americanum did not produce galls as it was a small nematode with a small stylet; the stylet caused only cortical tissue damage whereas damage to the vascular tissue was necessary to produce a galling response. It was then suggested that host reaction and selection of feeding sites was an advanced, evolved state. Galling is more evolved than simple necrosis and therefore X. diversicaudatum and X. index could be considered more evolved than the X. americanum group. Wyss described experiments in which swellings were produced in celery root tips by indole acetic acid (6ppm) or by crushed L. elongatus nematodes; these swellings were similar to those produced at the feeding sites of L. elongatus. It was postulated that the swellings at feeding sites could have been induced by the injection of RNA by nematodes, as high concentrations of RNA had been found at those sites. However, Taylor stated that RNA production was stimulated in plants subjected to all mechanical injury.

The histopathology of galls produced by longidorids and Heterodera spp. was compared. It was concluded that the giant cells in both types of gall were similar but could be distinguished on two characteristics. Hypertrophied cells caused by longidorid feeding were smaller than those caused by Heterodera and they contained a maximum of three nuclei whereas Heterodera giant-cells always contained more than three nuclei.

Short Report

THE EFFECT OF LONGIDORUS ELONGATUS ON SEEDLING FRUIT TREE ROOTSTOCKS IN POLAND

A. Szczygiel (Research Institute of Pomology, Experimental Station, Brzezna, Nowy Saçz, Poland.)

Longidorus elongatus is the most commonly found longidorid species associated with fruit crops in Poland. Several investigations have shown that L. elongatus is harmful to strawberry but prior to the investigations described here there was no information on its damaging effects on fruit trees commonly grown in Poland. A study therefore was undertaken on the pathogenicity of L. elongatus to five seedling rootstocks commonly used in Poland: apple (Pyrus malus cv. Antonowka), pear (Pyrus communis cv. caucasiae), cherry (Prunus avium and P. mahaleb) and plum (Prunus insititia cv. olivaricata). Seedlings grown in sterilized soil were transplanted at the 5-6 leaf stage into pots containing 400-600 g soil containing different population densities of L. elongatus, each one replicated 5 or 6 times. The nematodes had been cultured on strawberry or celery plants growing in heat-sterilized soil, and the different populations densities were obtained by mixing this soil with the soil in which the host plants were raised. The experiment was conducted in the laboratory under artificial illumination and lasted 4-5 months. Nematode populations were determined in each pot and measurements made of plant growth.

All the rootstocks showed typical symptoms of Longidorus feeding on the roots i.e. swelling and curving of root tips, necrosis and dwarfing of small feeder roots. The severity of symptoms was in most instances correlated with the initial nematode density, the highest nematode populations added to the pots causing the greatest damage. However, there was no observable correlation between severity of symptoms and the nematode populations in the pots at the end of the experiment.

Differences between rootstocks were noted in the effect of nematodes on plant growth. Pear and cherry suffered the most pronounced growth reduction, and P. avium was affected more severely than P. mahaleb. There were larger differences in weight reduction of the aerial parts than of the roots. Apple and plum apparently were unaffected by the nematode infestations and no significant differences were observed in the growth of the plants. The experiments demonstrate that pear and cherry rootstocks are susceptible to L. elongatus but that apple and plum rootstocks are tolerant.

FACTORS INFLUENCING THE MOVEMENT OF NEMATODES IN SOIL

R.S. PITCHER

East Malling Research Station,

Maidstone, England.

INTRODUCTION

The soil constitutes a very complex environment. Variations in its many constituents could produce readily detectable soil differences within areas of a few m^2 or even less, while on a world-wide scale the scope for variety is infinite. The constituent parts of a soil are, in turn, subject to a wide range of factors, both internal and external, which exert major influences on the soil as an environment. All these factors are of the greatest importance to soil-inhabiting nematodes, controlling many aspects of their behaviour as well as largely determining the ability of any given species to survive or multiply.

A few of the factors, such as the basic texture and structure, are comparatively stable, as they depend on the proportions of the major components sand, silt and clay, together with other soil minerals. These constituents influence the flora and so are indirectly responsible for the presence of varying amounts of dead organic matter, the fourth major component of most soils. The pH, also depending mainly on the major soil components and the flora, changes only slowly in nature, but it can be modified more rapidly e.g. by the addition of alkaline dressings. Under natural conditions the two most important factors liable to considerable change are temperature, which is subject to diurnal and seasonal rhythms, and moisture, subject to seasonal rhythms and to more erratic, short-term, weather-induced changes. Aeration, although basically structure-dependent, is governed in the short term by soil moisture content. Chemicals such as plant nutrients and salts, although largely structure-dependent in nature, are also influenced by the biological components of the soil. The importance of living elements in the soil cannot be overstressed; soil should never be regarded as inert. Soil flora, ranging from bacteria to higher plants, and the micro-fauna largely determine the amount of carbon dioxide (CO_2) in the soil atmosphere and also produce many more complex metabolites, including antibiotics and biotoxins. The osmotic potential of soil, although largely dependent on soluble salts, is also influenced by more complex organic compounds.

The above outline, although inadequate in many respects, is intended to stress that none of the factors mentioned operates in isolation in the soil environment. Similarly, their influences on nematode movement are closely inter-dependent, as Wallace (21) has illustrated in the schema reproduced in Fig. 1.

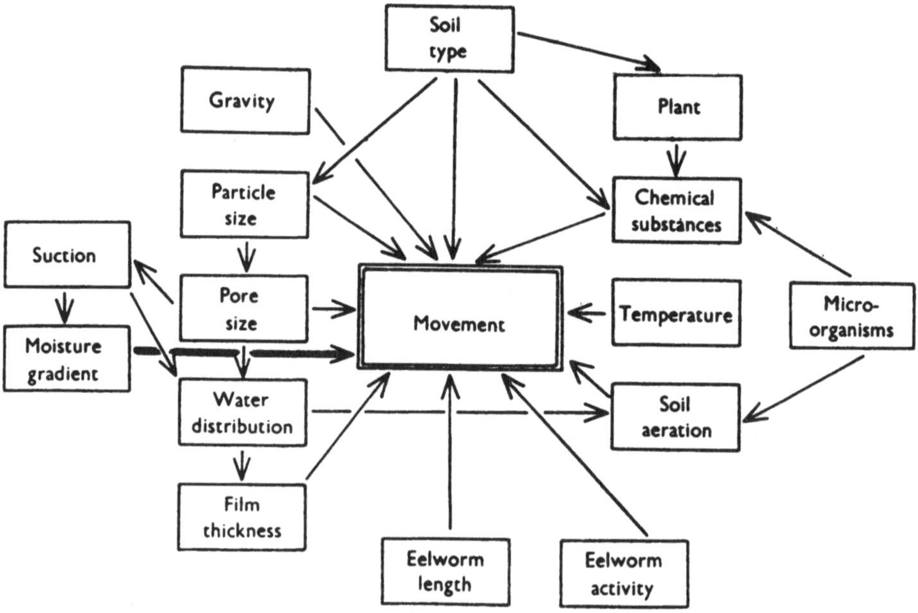

Fig. 1. Diagrammatic representation of the inter-relationship of factors which influence nematode movement in the soil. (From Ref. 21).

In these circumstances it is difficult to distinguish the effect of any one factor, and an experimenter attempting to do so must recognise that many factors will be affected by his manipulation of the one he has chosen to study. Nevertheless, it is hard to suggest any more practical approach than that of designing experiments in which the effects of variations of a single factor can be observed. Side-effects can often be reduced by using highly

Fig. 2. Diagrammatic representations of (a) soil particles enclosing a pore space, either water-filled (left), partly drained (middle) or fully drained (right), (b) soil aggregates, or crumbs; small circles indicate each crumb is composed of numerous particles (several thousands or millions).

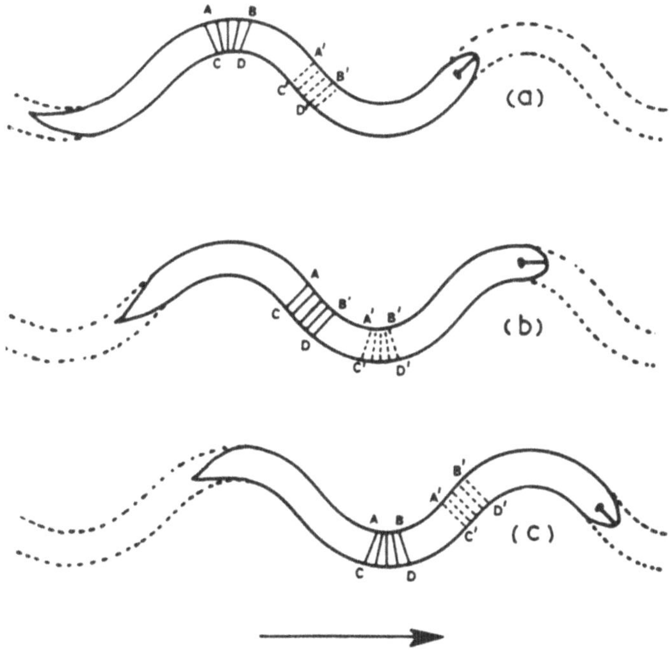

Fig. 3. The progressive change in form and curvature of segment ABCD as the nematode moves to the right. (a) dorsal muscles relax, ventral muscles contract; (b) dorsal muscles begin to contract, ventral muscles to relax; (c) dorsal muscles are now contracted and ventral muscles relaxed. (From Ref. 25).

artificial media, such as glass beads of a uniform size, and the information gained applied to the design of more complex experiments. All those interested in the behaviour of nematodes in the soil owe a large debt of gratitude to H.R. Wallace, whose classic series of experiments has done much to illuminate this formerly little-understood phenomenon. Many of his conclusions are summarised and most of his papers listed in his valuable review article (25). The following is intended as a brief and less formal review of the salient points of present knowledge, with notes on their relevance to the three known virus vector genera: Xiphinema, Longidorus and Trichodorus. Before proceeding further, it is desirable to consider the basic structure of an 'average' soil.

TYPICAL SOIL STRUCTURE

Individual particles, which rarely have plane surfaces, mostly lie in contact with their neighbours at a minimum of six points, ensuring their spatial stability. In most soils a labyrinth of spaces, or pores, exists between the particles, the size of the pores being dependent on that of the particles. Pores contain either gas (mainly air) or liquid (mainly water) or, more often, a mixture of both. Neighbouring pores are connected by narrow gaps, known as pore necks, whose size governs the movement of living and inanimate objects within the soil. In all but the lighter, very sandy soils, soil particles are usually grouped in aggregates, or crumbs, a few millimetres or a few centimetres across, each composed of many thousands, or millions of individual particles, bound together by organic matter (Fig. 2).

ACTIVE MOVEMENT

Unlike larger soil animals, such as earthworms, nematodes move between particles or crumbs, rarely displacing them. Movement, of the smaller nematodes at least, is dependent on the presence of water films on the surface of particles or crumbs. Nematodes normally progress by wave motion, identical in principal to that of snakes and most fish, especially the elongate eels, which they so closely resemble in shape. In order to move, the nematode positions its body in a series of waves, arranged alternately to left and right of its long axis. Each individual wave passes down the body from head to tail, propelling the nematode forward along a sinuous track (Fig. 3).

Unlike the majority of vertebrates which propel themselves by wave motion, a nematode moves on its side. To allow of this the main muscle bands lie dorsally and ventrally, instead of laterally, as in vertebrates using wave propulsion (Fig. 4).

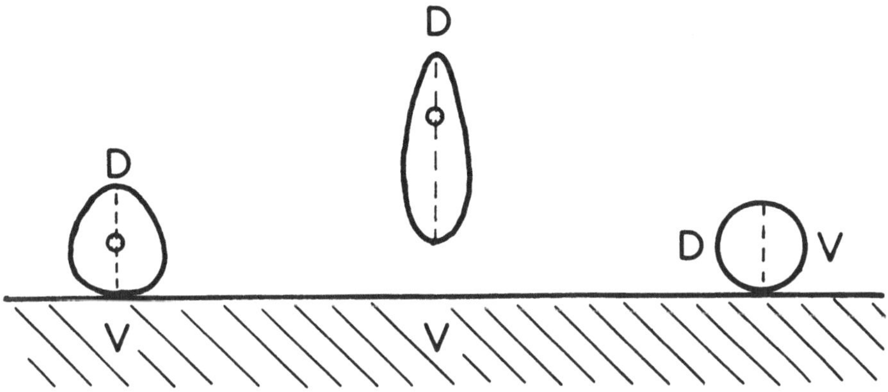

Fig. 4. Diagrammatic representation of transverse sections of a moving snake (left), fish (middle) and nematode (right), indicating the orientation of dorsal (D) and ventral (V) surfaces relative to the substratum. Dotted lines indicate the division between the major banks of body muscles and shaded circles the vertebral column.

Wave motion depends for success on the presence of a rigid longitudinal 'skeleton' against which the muscles can act. Nematodes possess neither a vertebral column nor a hard exoskeleton but have developed a high internal hydrostatic pressure, which holds their body rigid as does the air in an inflated tubular-shaped balloon. Although their bodies are very flexible laterally, the overall dimensions of nematodes are kept virtually constant by a mesh of over-lapping, diagonally-arranged, inextensible, cuticular fibres, running from head to tail. With this equipment a nematode can move forwards or backwards with equal facility.

Forward motion is achieved by pulses of muscular contraction passing from head to tail, alternately on the dorsal and ventral sides of the body ('left' and 'right', or vice versa, in relation to the direction of motion - Fig. 5).

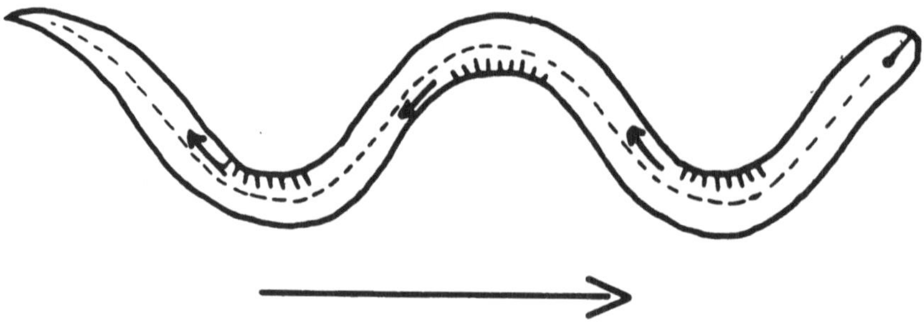

Fig. 5. A nematode moving to the right along a sinusoidal track. Zones of muscle contraction are 'hatched', the arrows show the direction of movement of the waves of contraction (from head to tail).

The mathematics of wave propulsion in snakes have been elegantly analysed by Gray (6) and the same principles apply to nematodes (24). It is possible, however, to explain wave motion in non-mathematical terms. Pressure is exerted on the substratum at right angles to the long axis, principally by the straighter portions of the body, lying midway between consecutive waves (Fig. 6a).

If one considers the tail of a nematode, pressing against a solid particle, two consequences are possible. Either the particle is pushed aside (Fig. 6b), in which case the nematode's tail moves to its 'right', but no forward progress results. If, however, the particle remains in place (e.g. because of the presence of neighbouring particles, as in soil) the pressure exerted by the nematode will generate a component acting in the general direction of its head, causing it to move forward along a sinuous path whose mean long axis coincides with that of the nematode (Fig. 6c). Comparable forces act at all four pressure points shown in Fig. 6a. Under ideal conditions the nematode will glide forward at a steady speed without sideways slip, as the lateral forces produced by the alternate waves are in balance and the total forward thrust components equal the total backward components of the drag exterted by friction. A broadly analagous energy balance exists in a car travelling at a steady speed, when the propulsive forces exerted via the wheels equal the frictional drag of the tyres on the road. For acceleration to take place, either in car or nematode, the propulsive force must, of course, exceed frictional drag. Most nematodes cannot swim in water because lateral resistance is low, allowing excessive slip (a situation comparable to that in Fig. 6b).

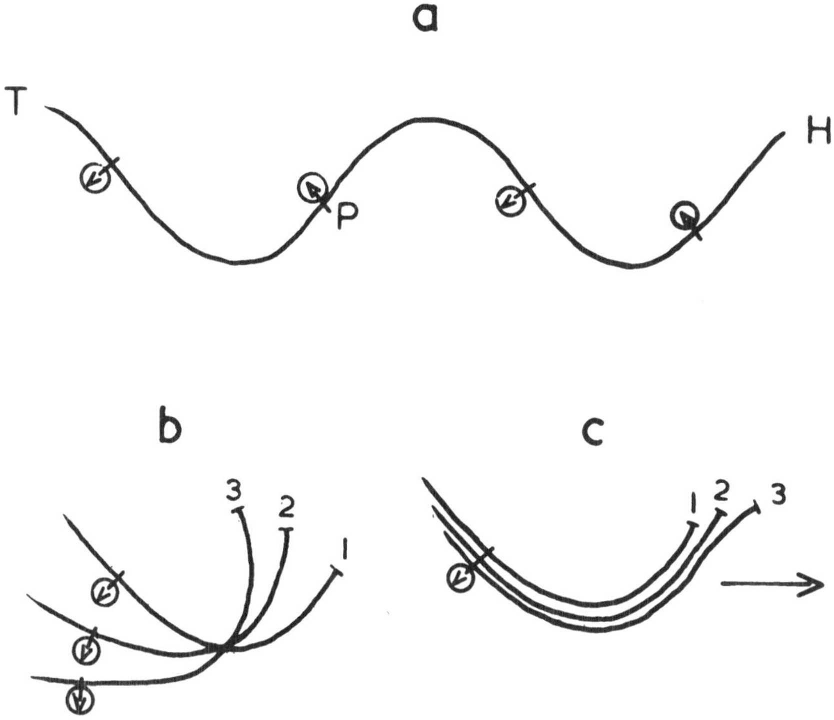

Fig. 6. Diagrammatic representation of (a) a nematode (H=head, T=tail) moving to the right along a sinusoidal path amongst soil particles, some of which are shown as circles. Arrows indicate the pressures exerted by the nematode on individual particles (b) the situation if the nematode's tail pushes the soil aside (nematode does not move forward) (c) the situation if the soil resists the pressure of the nematode's tail (nematode moves forward). 1, 2 & 3 indicate successive positions of the point P shown on the nematode body in (a).

However, nematodes, like eels, can swim in water if they can propagate waves of sufficient power and speed (Turbatrix, a good swimmer propagates 250 waves per minute, compared with Heterodera, which achieves only 8).

The principles applying to nematode movement in a particular medium such as soil apply equally to their movement on a plane substratum, provided it has either a soft but viscous surface or is covered by a fairly thin water film. The nematode obtains the necessary lateral resistance either from the viscous medium itself or from the surface tension of a water film of requisite thickness. A thicker water film provides only near-swimming conditions, while the surface tension forces of a thinner film hold the nematode so firmly against the substratum that excessive frictional forces are produced (Fig. 7).

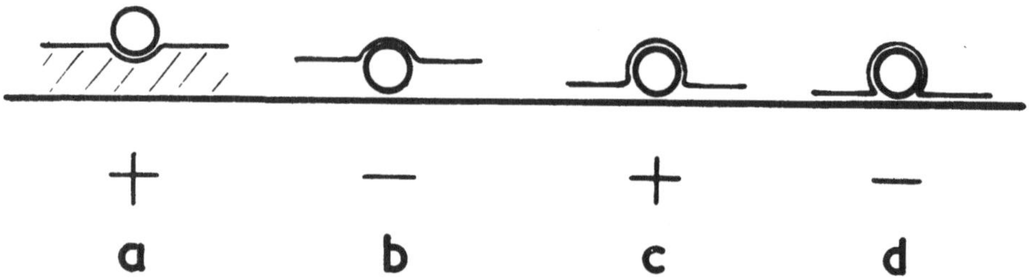

Fig. 7. Transverse sections of a nematode attempting to move by wave motion (a) on a viscous medium (b) in a thick water film (c) in a thin water film (d) in a very thin water film (+ = efficient progress, − = inefficient progress).

Unfavourable conditions oblige the nematode to exert a greater lateral thrust, resulting in an increase in wave amplitude, body angle to direction of movement and number of waves per unit length and in a decreased wavelength. These limitations result in slower progress (Fig. 8).

Most of the above principles were derived from work on tylenchid nematodes, ca. 1 mm long, and therefore probably apply equally to Trichodorus spp. Adult longidorids and the larger larvae, with body lengths upwards of 2.5 mm, seem to be less dependent on moisture films and wave propagation. Their long bodies enable them to bridge soil cavities which smaller nematodes could not cross, by stretching forward while maintaining an anchorage with their tails (Flegg, pers. comm.).

PASSIVE MOVEMENT

Nematodes are sometimes conveyed passively in moving water. Maximum progress, in speed and distance, occurs in surface drain-

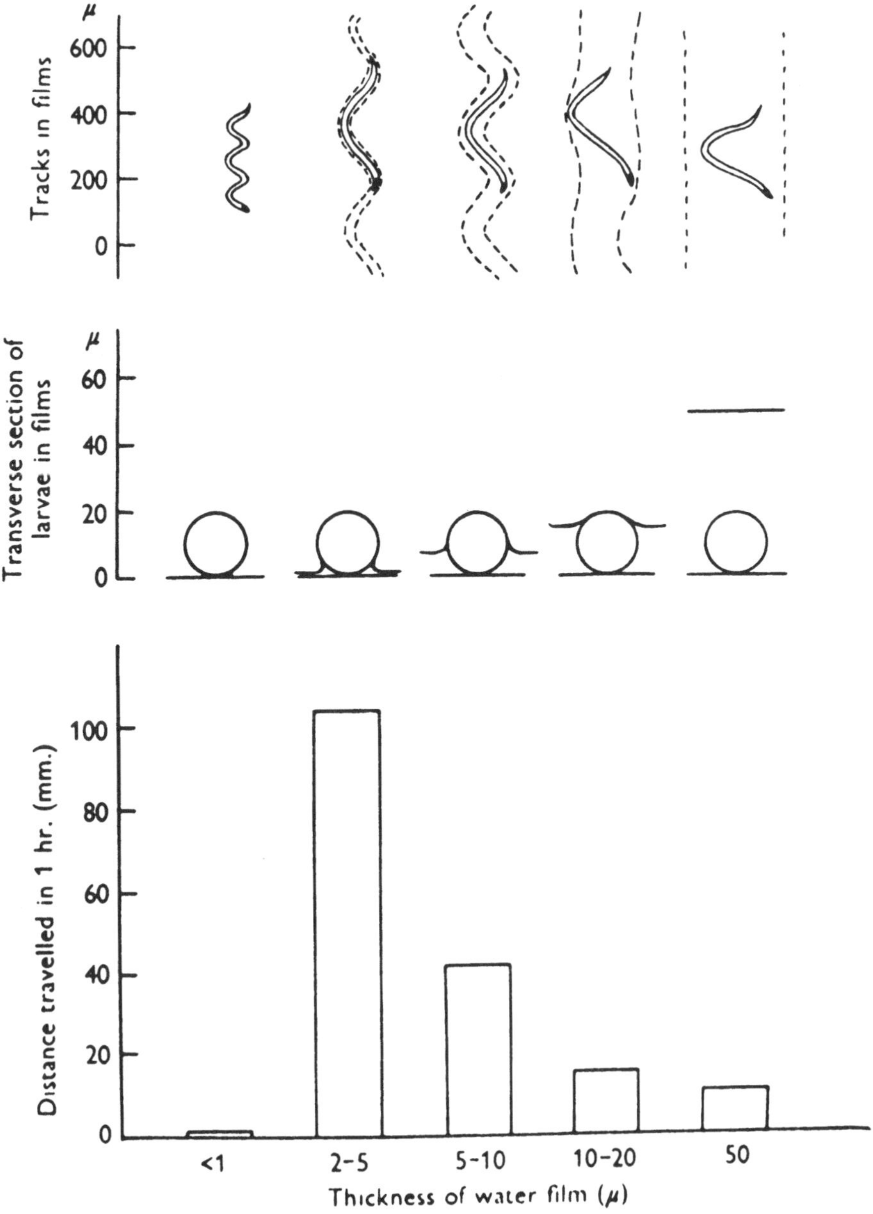

Fig. 8. The relationship between film thicknesses and larval speed, with diagrams showing the relationship of the eelworm to the film and the boundaries of tracks made in the films (From Ref.19).

age water, but this is of limited importance in practice. More frequent, but less effective, is downward passage in percolating water. Substantial progress is possible only in open-textured soils; small pore size restricts passive movement considerably. Active nematodes move further than dead or inactive specimens, which are frequently obstructed by the pore necks of even an open soil, whereas an active nematode will usually reorientate itself and pass through to an adjoining pore.

SOIL TEXTURE AND STRUCTURE

The mean size of particle is an important factor controlling nematode movement in soil. Wallace (20) has shown that maximum mobility is achieved when the mean particle size is about one third of the body length of the nematode. In medium to heavy-textured soils the progress of nematodes, especially adults and large larvae, is more dependent on the space between crumbs or aggregates than on that between particles. Jones *et al* (10) pointed out that aggregates are probably the most important micro-environment in the soil, being the site of ultimate contact between soil, water and micro-organisms. In a clay soil they found that, at best, only 10–20% of the total space was available to *Xiphinema diversicaudatum* and that this proportion declined to 5% at 30 cm below the surface. Movement of *Xiphinema* and *Longidorus* is considerably limited by their large body diameter (40–60 μ for adults) in relation to pore neck size and, to a lesser extent, by their body length in relation to blind-ended soil cavities. *Trichodorus* spp., with a body diameter of 30–35 μ are less constrained, but nematode movement is virtually impossible in densely-backed soils with a mean particle size below 50 μ. Sands provide the best conditions for movement with loams intermediate between them and clay soils.

MOISTURE

The moisture content of a soil is dependent upon the amount of tension, or suction, applied to it subsequent to its saturation. In nature the main sources of tension are drainage under the influence of gravity, and loss of moisture into the atmosphere, either directly from the soil surface or *via* plants. When a saturated soil has drained fully under the influence of gravity it is said to be 'at field capacity'. In nature this usually represents a tension of about 0.1 atmospheres, although the experimental reproduction of a comparable condition often requires 0.3 to 0.5 atmospheres. A soil is said to be 'at wilting point' when the moisture has fallen to a point at which plants begin to wilt because their transpiration losses exceed their intake *via* their roots. Atmospheric conditions clearly influence the onset of this

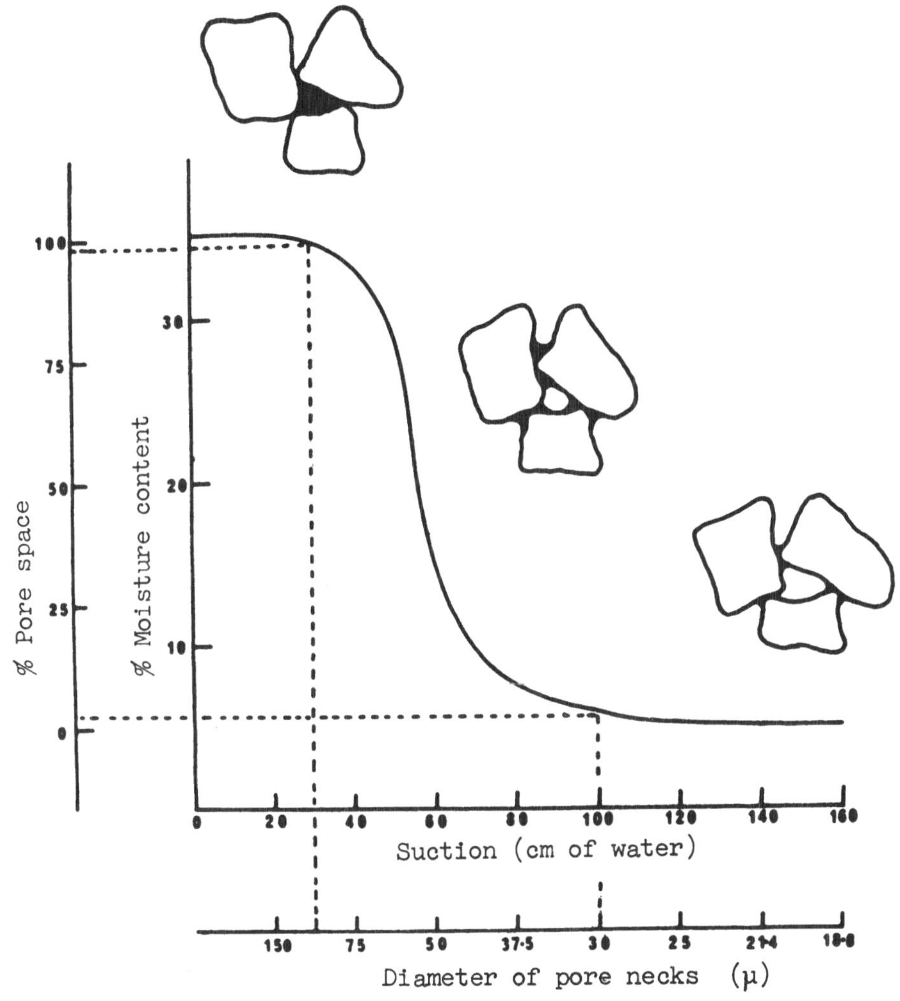

Fig. 9. The moisture characteristic of a hypothetical soil, showing decreasing moisture content as suction increases and drains the soil pores (From Ref. 25).

condition, but it usually involves tensions of the order of 15 atmospheres.

By gradually applying artificial tensions (e.g. by using a manometer) and plotting these against moisture content, a sigmoid curve is produced, characteristic of the soil. A typical moisture characteristic curve is shown in Fig. 9.

The term 'flex point' is used to define the tensions over which the moisture content falls sharply from a value close to 100%. The flex point represents the emptying of the medium-sized soil pores, which constitute the majority; this emptying process is also illustrated pictorially in Fig. 9. Individual pores change from the completely full to the well-drained state quite suddenly, when the applied suction overcomes the surface tension forces retaining water within the pore. Similar principles govern the sudden bursting of a soap bubble when the drainage of fluid to the base of the bubble has thinned its skin to a point at which surface tension is no longer capable of holding it together. In the soil, pore diameter and neck size are the major determining factors, so that the largest pores empty first and the smallest last. The flattening 'foot' of the curve represents both the emptying of the smaller pores and the removal of all except lingering traces of water in the pore necks of the medium and large-sized pores. There is a big variation between the moisture characteristics of extreme soil types, as shown in Fig. 10.

Maximum nematode activity in soils is closely associated with the flex point of the moisture curve ((19) and Fig.10). This appears to be due to a coincidence of two favourable factors, namely adequate aeration arising from the emptying of so many pores, coupled with water films of the medium thickness which allows maximum efficiency, long-wavelength motion. In peat, however, nematode activity continues to increase as tension rises beyond the flex point. The reasons for this are not fully understood, but may result from fundamental differences between frictional forces on organic and mineral particles, especially in very thin water films. As in the case of wave motion, the large longidorids behave rather differently from the small tylenchids which were used by Wallace, but *Trichodorus* spp. probably conforms to the rule relating maximum activity to the flex point. Nevertheless *Trichodorus* spp. and the much longer *Longidorus* spp. seem to be equally limited by dry spring conditions (e.g. damage to sugar beet seedlings is much more severe in wet springs). Moisture content may also have indirect effects on nematode movement. Flooding, in addition to producing sub-optimal aeration and water film conditions, may encourage the growth of anaerobic bacteria, such as *Clostridium* spp., which produce nematotoxic fatty acid metabolites (7), or other bacteria which produce hydrogen sulphide (18).

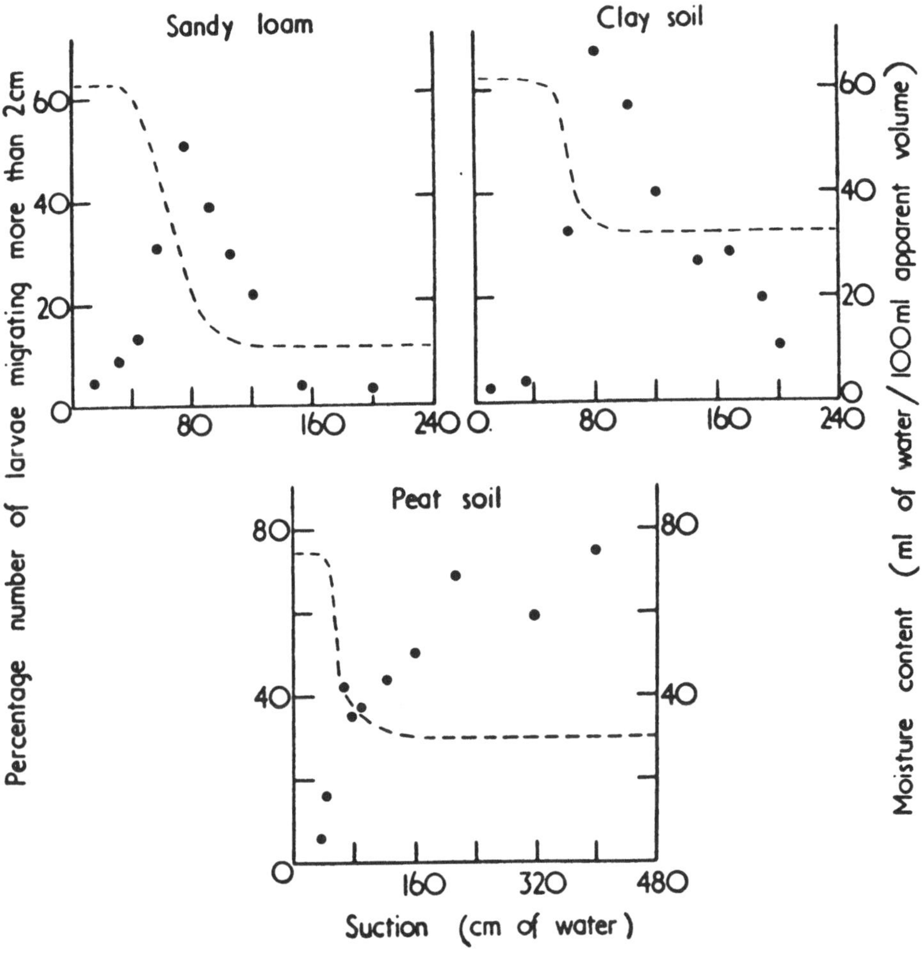

Fig. 10. Mobility of *Heterodera rostochiensis* in three soil types of crumb size 150 μ to 250 μ. The dotted line indicates the moisture characteristic. (From Ref. 24).

ORIENTATION

The complex and often controversial topic of the orientation of plant parasitic nematodes towards host roots has been well reviewed by Klingler (12). There are two major theories as to how nematodes reach plant roots.

(1) Undirected actions (kineses)

e.g. Orthokinesis - nematodes move at random in the soil, but, on encountering a root, further movement is inhibited, perhaps by the presence of food or of a short-range attractant substance or some factor which slows down their activity.

(2) Directed actions (taxes)

e.g. Klinotaxis - nematodes move actively up a concentration gradient, frequently turning from side to side to sample the environment and then moving in the direction which they judge to be more attractive. More than one attractant stimulus may be involved, such as long-range attraction by a readily-diffusable substance coupled with a stronger, more specific but shorter-range attractant which identifies the root as suitable for the nematode.

Attraction to a variety of stimuli has been demonstrated ranging from the obvious but ill-defined influences of growing plants (4, 21, 14), through known chemicals and electrical fields to unidentified organic metabolites. The best-documented chemical attractant is CO_2, which Klingler (11) showed to be effective from within a few mm to 2 cm from roots, provided a minimum concentration gradient of ca. 0.1% volume per cm is present. CO_2 is, of course, not a host-specific stimulant, but the active migratory stages (usually larvae) of many plant-parasitic nematodes are produced in fairly close proximity to a suitable host plant, which then provides the strongest source of CO_2 (e.g. eggs laid close to the host plant on which the parent female had been feeding). There is some evidence of attraction by amino-acids, which could well be more host-specific but their influence seems to be weak (1).

Electric potentials have been shown to influence many nematodes, either *in vivo* or *in vitro* e.g. *Ditylenchus dipsaci* moves towards the cathode (9). In the former case the reduced potential (redox) found in the zone of elongation of many plant roots has been shown to be especially attractive to larval heteroderids (1). However, further work (2) showed that redox potential is neither a universal stimulus nor necessarily the strongest. Little is known of the attractiveness of complex biological metabolites. The rhizosphere is normally a zone of much microbial activity, which probably enhances the CO_2-gradient as well as producing more complex substances. Microfloral biproducts are not the only nor probably the major source of attraction, however, as plant roots grown under aseptic conditions also have considerable attractant powers (16, 4).

Repulsive stimuli seem rare, although Bird (1) showed that both extremes of pH could repel. There are, however, many reports of ineffective stimuli e.g. oxygen gradients, both ascending and descending (12); nitrogen, temperature, except perhaps in relation to vertical movement (23); osmotic potential (3); pH within the ranges found in most soils (1); sugars; and light, gravity and water percolation (22), although the last two may affect passive movement.

SWARMING AND AGGREGATION

The terminology of these phenomena has sometimes been used rather confusingly in the literature and the distinctions made tend to be artificial. For the sake of clarity, an arbitrary distinction will be made here between 'swarming', implying the accumulation of nematodes (plant-parasitic or otherwise) in the absence of plant roots and 'aggregation', referring to accumulations on or near plant roots.

(1) Swarming

This can facilitate dispersal (e.g. of microphagous nematodes on compost) or drought resistance (e.g. 'wool' formed by *Ditylenchus dipsaci* on heavily-infected bulbs) or it can occur for unknown reasons (e.g. of *Tylenchorhynchus martini* in water). The swarming of *T. martini* has received much attention from Hollis and his colleagues (15, 18). It appears to be associated with high nutrition and cuticular abnormalities and may represent a diseased condition.

(2) Aggregation

The attractiveness of the 'elongating zone' of roots, ca. 1-3 mm behind the apical meristem, has long been recognised. Attraction seems to be at a maximum when conditions favour vigorous root growth and is best known in larval heteroderids (13, 27, 16). Motile fungal zoospores also accumulate in this zone (28). More recently many species of *Trichodorus* have been noted tó accumulate in the elongating zone of a variety of host plants (17, 26). Such behaviour is widespread in this genus and seems to play an important role in the development of the stubby root symptoms associated with most *Trichodorus* spp.

Data on the feeding habits of longidorids is mostly confined to observations of agar cultures. Cotten (5) and Wyss (unpubl.) reported root-tip feeding by *Xiphinema index* and *Longidorus* spp. respectively and small numbers of *X. index* were seen to aggregate on some roots. Other evidence, however, suggests that the feeding of some *Xiphinema* spp., especially those allied to *X. americanum*, may be dispersed more widely along the fine feeder roots and Valdez (unpubl.) could find no aggregation of *X. diversicaudatum* in the rhizosphere of hop (a poor host) grown in observation boxes filled with field soil containing a natural population of this nematode. There appear to be no reports of large aggregations of longidorids comparable to those commonly observed in Trichodorus spp.

The above brief review indicates that, while the broad scope of the subject has been recognised by many workers and some valuable groundwork has been done, much remains to be elucidated, especially in the difficult field of the interactions of many of the factors which have, as yet, been studied separately. While the movements of the virus vectors seem to be influenced by most of the factors which have been studied, the behaviour of the large longidorids often differs from that of the tylenchids, on which most ofthe available data is based.

ACKNOWLEDGEMENT

I am indebted to Professor H.R.Wallace for permission to reproduce Figs. 1, 3, 8, 9 & 10.

REFERENCES

1. Bird, A.F. (1959). The attractiveness of roots to the plant parasitic nematodes Meloidogyne javanica and M. hapla. Nematologica 4, 322-335.

2. Bird, A.F. (1960). Additional notes on the attractiveness of roots to plant parasitic nematodes. Nematologica. 5, 217.

3. Blake, C.D. (1961). Importance of osmotic potential as a component of the total potential of the soil water on the movement of nematodes. Nature (London) 192, 144-145.

4. Blake, C.D. (1962). Some observations on the orientation of Ditylenchus dipsaci and invasion of oat seedlings. Nematologica. 8, 177-192.

5. Cotten, J. (1973). Feeding behaviour and reproduction of Xiphinema index on some herbaceous test plants. Nematologica. 19, 516-520.

6. Gray, J. (1953). Undulatory propulsion. Quart. J. micr. Sci. 94, 551-578.

7. Hollis, J.P. & Rodriguez-Kabana, R. (1966). Rapid kill of nematodes in flooded soil. Phytopathology. 56, 1015-1019.

8. Ibrahim, I.K.A. & Hollis, J.P. (1973). Electron microscope studies on the cuticle of swarming and nonswarming Tylenchorhynchus martini. J. Nematol. 5, 275-281.

9. Jones, F.G.W. (1960). Some observations and reflections on host finding by plant nematodes. Med. Landb. Hoogesch. Gent. 25, 1009-1024.

10. Jones, F.G.W., Larbey, D.W. & Parrott, D.M. (1969). The influence of soil structure and moisture on nematodes, especially Xiphinema, Longidorus, Trichodorus and Heterodera spp. Soil Biol. Biochem. 1, 153-165.

11. Klingler, J. (1963). Die Orientierung von Ditylenchus dipsaci in gemessenen künstlichen und biologischen CO_2-Gradienten. Nematologica. 9, 185-199.

12. Klingler, J. (1965). On the orientation of plant nematodes and of some other soil animals. Nematologica. 11, 4-18.

13. Linford, M.B. (1939). Attractiveness of roots and excised shoot tissues to certain nematodes. Proc. helminth. Soc. Wash. 6, 11-18.

14. Lownsbery, B.F. & Viglierchio, D.R. (1961). Importance of response of Meloidogyne hapla to an agent from germinating tomato seeds. Phytopathology, 51, 219-221.

15. McBride, J.M. & Hollis, J.P. (1966). Phenomenon of swarming in nematodes. Nature (London). 211, 545-546.

16. Peacock, F.C. (1959). The development of a technique for studying the host-parasite relationship of the root-knot nematode, Meloidogyne incognita, under controlled conditions. Nematologica. 4, 43-55

17. Pitcher, R.S. (1967). The host-parasite relations and ecology of Trichodorus viruliferus on apple roots, as observed from an underground laboratory. Nematologica. 13, 547-557.

18. Rodriguez-Kabana, R., Jordan, J.W. & Hollis, J.P. (1965). Nematodes: biological control in rice fields: role of hydrogen sulphide. Science. 148, 524-526.

19. Wallace, H.R. (1958). Movement of eelworms. I. The influence of pore size and moisture content of the soil on the migration of larvae of the beet eelworm Heterodera schachtii Schmidt. Ann. appl. Biol. 46, 74-85.

20. Wallace, H.R. (1958). Movement of eelworms. III. The relationship between eelworm length, activity and mobility. Ann. appl. Biol. 46, 662-668.

21. Wallace, H.R. (1960). Movement of eelworms. VI. The influence of soil type, moisture gradients and host plant roots on the migration of the potato root eelworm Heterodera rostochiensis Wollenweber. Ann. appl. Biol. 48, 107-120.

22. Wallace, H.R. (1961). The bionomics of the free-living stages of zoo-parasitic and phyto-parasitic nematodes - a critical survey. Helminth. Abstr. 30, 1-22.

23. Wallace, H.R. (1962). Observations on the behaviour of Ditylenchus dipsaci in soil. Nematologica. 7, 91-101.

24. Wallace, H.R. (1963). The biology of plant parasitic nematodes. Edward Arnold, London. 280 pp.

25. Wallace, H.R. (1968). The dynamics of nematode movement. Annu. Rev. Phytopath. 6, 91-114.

26. Whitehead, A.G. (1967). In: Nematology Department. Rep. Rothamsted exp. Stn, 1966. 145-147.

27. Widdowson, E., Doncaster, C.C. & Fenwick, D.W. (1958). Observations on the development of Heterodera rostochiensis Woll. in sterile root cultures. Nematologica. 3, 308-314.

28. Zentmeyer, G.A. (1961). Chemotaxis of zoospores for root exudates. Science. 133, 1595-1596.

DISCUSSION

Hooper expressed concern about the emphasis Pitcher had put on the importance of soil particle and soil crumb size, almost to the exclusion of soil pore size and Weischer continued by asking whether the nematodes followed the trace of the host plant root through the soil. Pitcher replied by quoting the work of Jones et al (F.W.G. Jones, D.W. Larby and D.M. Parrott. Soil Biol. Biochem. 1, 153-165 (1969)) in which the importance of pore size was recognised. He continued by explaining that plant roots follow the line of least resistance through the soil and thus pore size is, indirectly, taken into consideration. He illustrated the point using Trichodorus viruliferus feeding on apple root; the nematodes initiating the infestation feed on the root elongation zone and follow it as it grows through the soil, but additional nematodes appear to migrate directly to the feeding zone from the rhizosphere, and not along the root.

Estey asked why it was thought that water was necessary for nematode movement and remarked that he had observed nematodes migrating through 'dry' soil. Furthermore if the nematodes did require water for movement then how could they recognise moisture gradients. In the discussion which followed, it was stated that nematodes can move under most conditions but were most efficient when held against soil particles by the surface tension of water films around the soil particles. It was postulated that nematodes could indirectly recognise gradients by detecting differences in oxygen tension in the soil water, amongst other possible factors.

DYNAMICS OF THE PLANT NEMATODE SYSTEM

J.W. SEINHORST

Institute of Phytopathological Research

Wageningen, The Netherlands

INTRODUCTION

Plant parasitic nematode species feed in different ways and different places on plants and cause different reactions of the plant tissue. However, at least the root infesting species have in common that they are continuously active and develop slowly but continuously at wide ranges of temperature and soil moisture content. Moreover, most species also multiply more or less continuously. As a result the dynamics of the plant-nematode system are very similar for the various species.

The plant nematode system has two main aspects:

1. What effect have the nematodes on the development and economic yield of the plant
2. How does the feeding and its effects on the plant affect growth and population increase of the nematode.

NEMATODES AND YIELD

Economic yield can be affected by nematode attack in two ways:

1. plants infested with nematodes are worthless, irrespective of the number of nematodes present in or on them e.g. onions and flower bulbs attacked by stem nematodes, propagation material with root knot, and, indirectly because of attack by a virus, plants attacked by virus transmitting nematodes.

2. nematode infestation reduces the rate of growth of the plant and thereby the economic product.

In both cases there is a strong relation between density of the nematode population at the time of sowing or planting of annual crops and the ultimate volume or weight of product, due to the continuous activity but slow multiplication of the nematodes (with the exception of Aphelenchoides species). In perennial crops this relation may, over a longer period, be obscured by a high rate of multiplication. Knowledge of this relation is essential for the solution of the problems of determining the degree of control resulting in the maximum economic return and for determining the efficacy of systemic nematicides or any other method of control that does not result in the death of nematodes before sowing or planting.

NEMATODE DENSITY AND PROPORTION OF PLANTS ATTACKED

According to Seinhorst (8) the relation between nematode density P and the proportion y of the plants that is not attacked by nematodes is described by the equation $y = z^P$ (eq. (1)) (Nicholson's competition hypothesis) if the nematodes are distributed randomly; z is a factor <1. If the nematode distribution is not random it can be thought to be built up of small areas with a random distribution of the nematodes. Fig. 1. gives the relation between nematode density and proportion of plants not attacked by nematodes according to eq. (1) and when the nematodes are distributed very irregularly. Field observations have shown that generally eq. (1) gives a good description of the relation between density of stem nematode populations before sowing and attack in onions in small plots, if differences in nematode density were obtained by differences in crops grown previously. However, values of z appeared to be closer to 1 at low nematode densities than at higher ones in fields with a natural patchiness in the distribution of the nematode (3). This apparent difference in the activity of the nematode was also expressed in the higher multiplication rate in the centre than at the edge of the patch.

Theoretically eq. (1) could also apply to virus transmission by nematodes, but no results of experiments are available yet to corroborate this. It has been found, however, that the same density of Trichodorus pachydermus may be associated with very different degrees of attack of plants by tobacco rattle virus in different places in a field. The cause is unknown but it is certainly not only differences in degree of infestation of the nematode population with the virus.

The proportion of diseased plants can increase during the season because of multiplication of the nematodes in or on the plants that were infested initially and subsequent spread of the nematode to neighbouring plants. This does not materially change the situation except for a decrease of the value of z. Eq. (1) gives a good

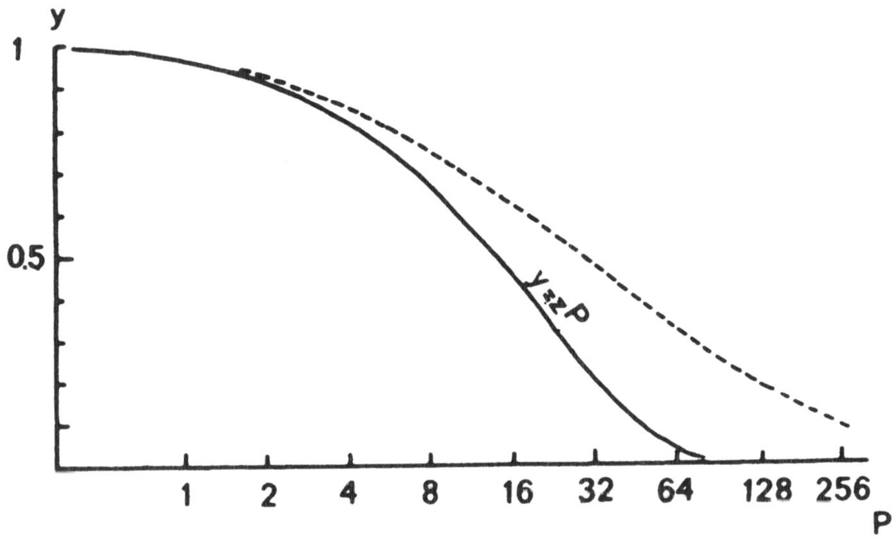

Fig. 1. Relations between density of stem nematode (Ditylenchus dipsaci) P and relative yield of healthy onions y. Solid line: according to $y=z^P$ (eq.(1)), dotted line: if P is the average density of a very irregularly distributed population (range of densities 1 to about 200, all densities with low frequency) (According to Seinhorst (8, 10))

description of the total area covered by varying numbers of randomly distributed areas of the same size and more or less the same shape. This remains true while these areas are growing.

NEMATODE DENSITY AND CROP WEIGHT

Although changes could occur during the growing season in the previous case, a statical probabalistic approach sufficed for a satisfactory theoretical treatment. Final weights of plants or parts of plants are the result of a certain period of development of the plant during which, even under constant environmental conditions, growth conditions change continuously because of increase in size and change in the nature of the plant and decrease of the space available for expansion. Very generally it can be said that the growth of a plant for various reasons is increasingly inhibited during its development because of its very growth. If nematodes are also interfering with this growth a complicated interaction between the various inhibitions is bound to arise. No doubt a good insight in the effect of nematode attack on the growth and yield of crops can only be obtained by unravelling these interactions. There-fore this should be one of the main aims of research in plant nematology if it is to provide a sound base for economic nematode control. Fortunately, however, a number of pot and field experiments with various plant and nematode species have shown (without revealing much about the dynamics of the interactions) that the relation between nematode density P at the time of sowing or planting and yield in many if not almost all cases can be described by the simple, arbitrary equation $y = m+(1-m)z^{P-T}$ for $P \geq T$ and $y = 1$ for $P \leq T$ (eq. (2)) in which y is the ratio between weight of the whole plant or parts of it at nematode density P and in the absence of nematodes, z is a constant <1, $z^{-T} = 1.05$ to 1.1 (mostly 1.05) and m the so called minimum yield i.e. the proportion of the yield in the absence of nematodes that is not affected even at very high initial nematode densities (Fig. 2). The constant z could be called the tolerance of the plant to the nematode species involved under the environmental conditions of the place and time of the observations. The less the damage and yield reduction at a given nematode density the

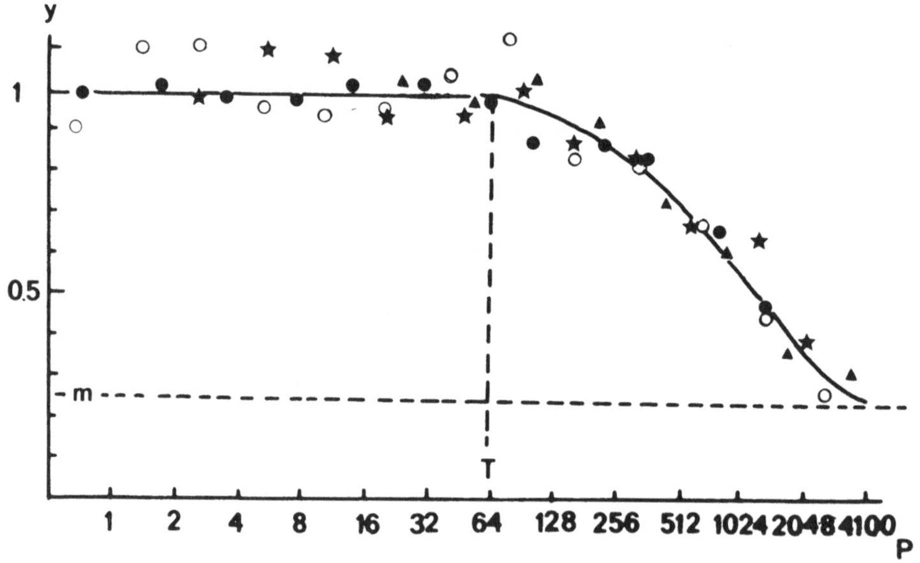

Fig.2. Line according to eq. (2) ($y = m+(1-m)z^{P-T}$) fitted to results of experiments on the relation between population density of Pratylenchus penetrans and the relative yield of various plants. T=64 m=0.25

closer to unity is the value of z. However, this value also depends on the unit used in expressing P and is therefore difficult to handle. A much better measure for the tolerance is T, which is not a number but a density. According to eq. (2) y = 1 up to nematode density T, which suggests that the nematodes do not cause yield reduction up to this density. It is, therefore, called the "tolerance limit". Indeed the results of pot experiments strongly suggest that, unlike according to eq. (1), nematode attack does not lead to measurable reduction of plant weight up to about a density T. The reason for this has not yet been clearly established. Therefore a discussion of whether such a tolerance limit exists in reality is irrelevant. It only has a distinct meaning in association with eq. (2) or a similar equation. A practical interpretation of the meaning of T is that up to that density no growth reduc-tion of the plant can be demonstrated. On the other hand the value of the tolerance limit cannot be derived with any degree of accuracy from the results of most experiments, without fitting a curve according to eq. (2) or a more sophisticated theory to these data, in this way taking yield measurements over a wide range of densities into account. The more irregular the distribution of the nematode the higher the apparent value of m when a curve according to eq. (2) is fitted to yield data at initial average densities not much higher than 100 T (13).

The results of pot experiments could not always be described by eq. (2). To explain those of an experiment with *Heterodera rostochiensis* on potato, Seinhorst and den Ouden (14) derived a relation between nematode density and plant weight at different times after planting, from a model of the growth of the potato plant at various nematode densities. This model was based on the assumption that growth was affected in two ways: by invasion of the root tips, and by giant cell formation. Only the first, which was the most important, is further considered here. A larva could only do damage in this way during a short time. The rate of growth of the plant (increase of plant weight per unit weight of plant and per unit duration of growth) as reduced by a nematode density P was assumed to be $r_P = yr_0$ (eq. (3)) in which r_0 is the rate of growth in the absence of nematodes and $y = m+(1-m)z^P$ (meanings of m and z as in eq. (2). Further it was assumed that plant growth was exponential as long as nematodes were available to penetrate into new root tips i.e. up to a size of plant at which roots had filled the pot sufficiently to make all eggs hatch. After that, weight increase of the plant was considered to be proportional to time and the same at all nematode densities (as far as invasion damage was concerned). Apparently pot size is important if these assumptions apply as was shown by den Ouden (6) for tolerances of potato to *H. rostochiensis* if nematode densities are expressed per unit weight or volume of soil. The theory also explains why data of field experiments with *H. rostochiensis* on potato are in accordance with eq. (2).

The relation between nematode density and plant weight at different times after sowing or planting was also determined assuming that eq. (3) applied during the entire growing period and that the relation between plant size and time was given by the logistic curve. The latter was chosen because it is easily handled. However, the conclusions based on it would hold for any sigmoid growth curve of approximately the same shape. The relation between nematode density and plant weight then appeared to be close to those according to eq. (2) assuming that T=0. Possibly a better fit of theory to observations at low nematode densities is obtained if the size of the whole plant could be derived from that of the root system as influenced by nematode attack. For example, attack by Longidorus elongatus on Fragaria vesca and by Pratylenchus penetrans on Vicia faba increased the shoot-root ratio (9). This means that either the plant can maintain a certain rate of growth with less root than it has available without damage to the root system, or that the nematode attack apart from a damaging factor also contains a growth stimulating factor (10, 15).

Working the model out for nematode attack beginning at different times after the beginning of the development of the plant, showed that the actual tolerance of the plant is greater the later this attack begins.

A more detailed account of the results of these investigations with the help of simplified growth models will be published elsewhere.

APPLYING KNOWLEDGE OF RELATIONSHIPS BETWEEN NEMATODE DENSITY AND YIELD

Although our knowledge of the relations between nematode density and yield still is largely restricted to the recognition of some distinct patterns which seem to occur generally, important practical applications are already possible. Eq. (1) has been used extensively by Kaai (3,5) to derive density independent characterizations of the activity of systemic nematicides against stem nematode attack in onions from density dependent results of field experiments. The reasoning is as follows (3, 5, Fig. 3): If a treatment increases the proportion of healthy plants from y_u (untreated) to y_t (treated) and $y_u=z^{Pu}$ then $y_t=z^{Pt}$. Thus, $Pt/Pu = \log y_t/\log y_u$ and then Pt is the density to which Pu should have been reduced by killing nematodes in order to obtain a yield y_t. Experiments by Kaai have further shown that if a treatment is given in a field with a range of stem nematode densities the relation between initial nematode density and yield remains in accordance with eq. (1): $y_u = z_u^P$, $y_t = z_t^P$ and $\log y_t/\log y_u = \log z_t/\log z_u = Pt/Pu$. There-

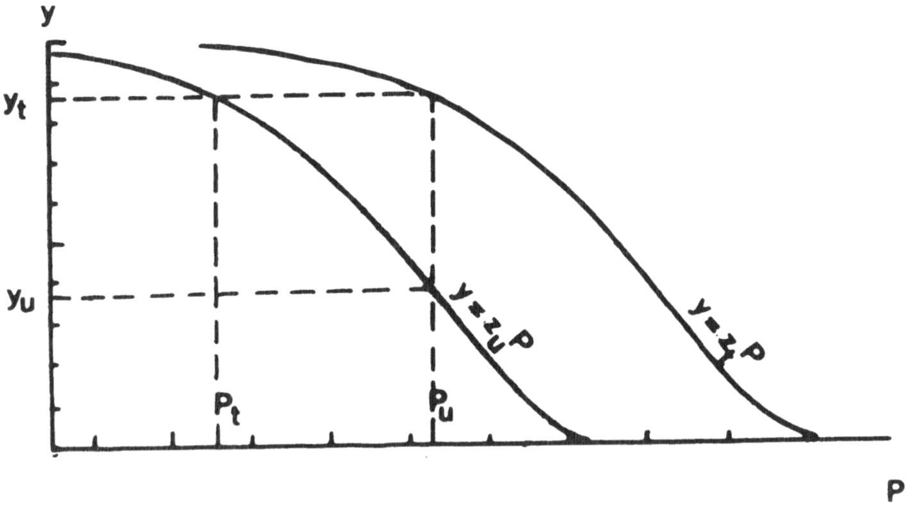

Fig. 3. Derivation of the equivalent survival, Pt/Pu, of stem nematodes after treatment of onions with a systemic nematicide, by applying eq. (1) P = nematode density, Pu = nematode density on untreated plots, Pt = equivalent density on treated plots, y = relative yield of onions. y_u, y_t = relative yield on untreated and treated plots. Further is $y = z_u^P$ the relations between nematode density and yield on untreated and $y = z_t^P$ that on treated plots with different densities.

fore, as z_u and z_t are independent of nematode density this is also the case with Pt/Pu. This ratio is called the equivalent survival, and 1-Pt/Pu the equivalent mortality. The term equivalent is used because the figures have the same meaning in relation to effectiveness of a systemic nematicide (or any treatment that protects the plants against nematode attack otherwise than by killing the nematodes before sowing or planting) as the real survival or mortality of nematodes after a nematicidal treatment. The same use could be made of eq. (1) to determine the effectiveness of treatments against virus transmission by nematodes. No example of this is available. In all cases the procedure requires an accurate estimate of the number of healthy plants in the absence of nematodes or of virus transmission. In the case of stem nematode control this standard was obtained by treating a number of plots with a high but non toxic dosage of a systemic nematicide, preferably aldicarb (5).

The effectiveness of systemic nematicides against yield losses caused by root infesting nematodes could be determined as described above using eq. (2) but much greater problems are associated with it than with determining the efficacy of these chemicals against stem nematode attack in onions. Eq (2) contains two parameters which can only be determined if a range of nematode densities is available for the experiment. Further, the yield in the absence of nematodes cannot be determined by simply using a high dosage of a good nematicide. This may influence yields in still another way than by protecting the plants against nematode attack.

DAMAGE PREDICTION FROM NEMATODE DENSITIES

Equations (1) and (2) could be used to predict the yield reduction to be expected in certain crops from nematode densities found in soil samples taken before sowing or planting. Relevant values of the parameters z (stem nematodes in onions) or T and m (crop yields to which eq. (2) applies), must then be known. Losses in onions at a given density of stem nematodes vary considerably from year to year and from place to place (2). They may already be considerable at densities of one nematode per 500 g soil (2). Such densities cannot be determined with any degree of accuracy. The same difficulties may be encountered in predicting attack by nematode transmitted viruses from densities of virus carrying nematodes in soil samples.

Densities of root nematodes causing growth reduction of plants generally are large enough to be determined with a sufficient degree of accuracy for yield predictions. Not much is known yet of the variation of minimum yields and tolerances in the field but observations on *H. rostochiensis* so far indicate that they may be limited for this nematode (7, 12).

NEMATODE MULTIPLICATION ON PLANTS

Our knowledge of population dynamics of plant parasitic nematodes has been reviewed by Seinhorst (1) who gives references to original papers. A short discussion of the main points suffices here.

Most nematode species are more or less continuously active so long as external conditions permit. In only a few species there are periods of dormancy which, to be overcome, require a period of ageing (*H. rostochiensis*) or of low temperature (*H. avenae*, *Meloidogyne naasi*). Certain *Heterodera* spp. stay inactive in the soil as second stage juveniles in eggs until they are irreversibly stimulated to activity by host roots. *Paratylenchus* needs such a stimulation to moult (1). Most other species hatch from eggs and remain active without external stimulus but growth and multiplication is possible only while food is available. Development of oocytes stops (those in an advanced stage of development even are resorbed) immediately upon withdrawal of the food source and is resumed when food is available again. As a result individual nematodes survive much longer without than with food. A common death rate of Pratylenchus in soil without plants is about 50% per year. Population changes in the absence of host plants are also slow in most other species. When a host is available multiplication depends on its qualities as a food source under the prevailing external conditions, and the activity of the nematodes under these conditions. In relation to the latter, temperature and moisture content of the soil are important but also other unknown soil characteristics which are expressed as "influence of soil type". Sometimes there is a clear correlation with soil classification as made by the soil scientists, but often there is not.

On a given host and under a given set of conditions the rate of multiplication depends on population density, and therefore on degree of intraspecific competition. The relation between rate of multiplication and population density is simple and is determined by the quantity of food that can be obtained by a single individual if there is no competition and the total amount of food available to the population. This leads to two parameters that characterize the host status of the plant: the maximum rate of multiplication (occurring at low densities) and the equilibrium density (the highest density that can be sustained by the host). The relation between nematode densities at the beginning and at the end of a certain period of multiplication on a host (Pi and Pf) generally is in accordance with the equation $Pf = aEPi/\{(a-1)\ Pi+E\}$ (eq. (4)), especially at values of Pi well below E. The relation between the densities of two subsequent genera of H. rostochiensis, Pi and Pf, is given by $Pf = a(1-q^{Pi})/-\ln q$ (eq.5). In these equations a is the maximum rate of multiplication (at low densities Pf aPi), E the equilibrium density, q a constant <1 and indicates n natural logarithm. Eq. (4) is based on the logistic curve for population increase, an arbitrary expression of the effect of intraspecific competition. Eq. (5) relates the size of the second generation to the amount of food that can be taken up by the parent generation if it is distributed randomly along the roots (according to Nicholson's competition hypothesis). Numerically both equations give very similar results (Fig. 4).

When the host is damaged by the nematode the total amount of food it produces becomes smaller the greater the nematode density. However, as this damage mostly results in a reduction of the root system a smaller proportion of the nematode population attacks it the greater its density. The nematodes that do not attack the plant decrease slowly in number. As a result, a larger proportion of the final population consists of animals of the initial population the greater the density of the latter. This keeps the relation between initial and final densities more or less in accordance with eqs. (4) and (5). Observations on population changes on a

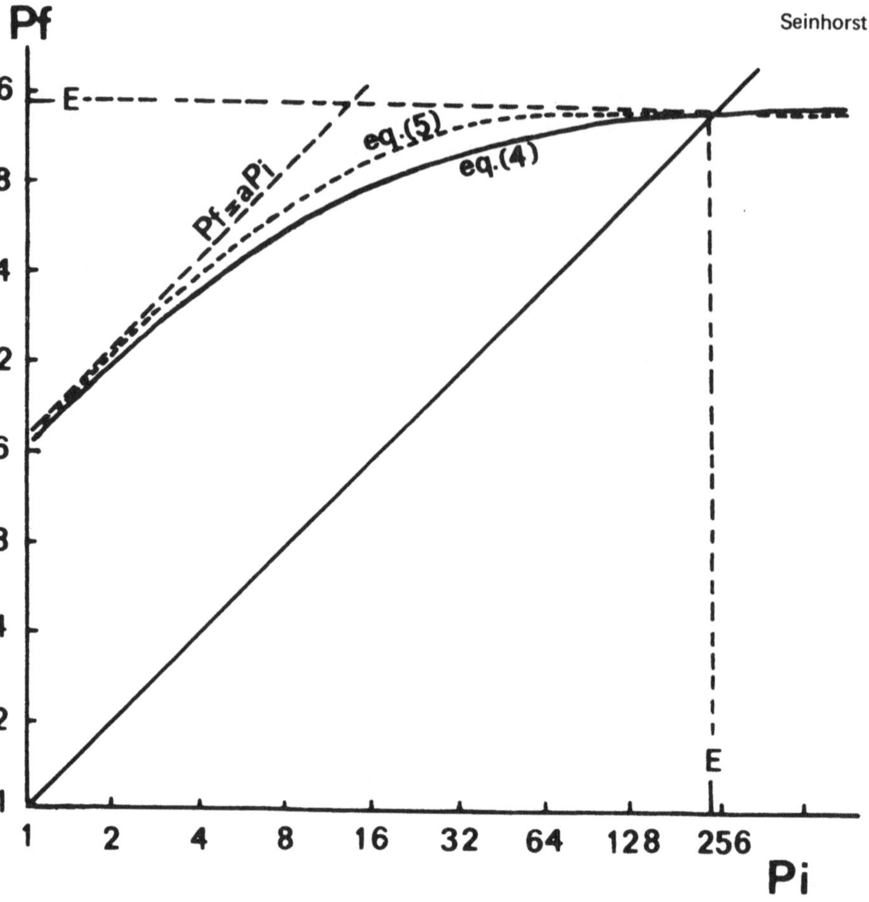

Fig. 4. Relations between initial and final densities Pi and Pf of nematode populations multiplying on a host during a certain period, according to eq. (4): P_f=aEPi (a-1)Pi+E and eq. (5): $a(1-q^{Pi})-\ln q$

host seldom reveal whether the latter was damaged by the nematode. In certain cases the tolerance limit for damage is about the same as the density at which the competition for food begins to reduce the rate of multiplication e.g. in H. rostochiensis (1-5 eggs per g soil). In other cases the former is higher than the latter and often the equilibrium density is well below the tolerance limit. The reasons for this have not yet been investigated. As an example: Trichodorus pachydermus, T. similis and T. viruliferus occur very generally on sandy soils in the Netherlands but in most instances at densities below a few hundred, if not below 50, in 500 g soil. Occasionally much higher densities are found still without conspicuous effects on the crops grown.

PRACTICAL APPLICATIONS

Rates of multiplication of nematodes on a host decrease and damage increases with increase of nematode density (Figs. 1, 2, 4). As a result there generally is a balance between cost of nematode control and economic return at a nematode density at which still some yield decrease occurs. Long term nematode control strategies must aim at striking this balance. To do this not only tolerances, but also rates of population increase at various initial densities, must be known. Eqs. (4) and (5) are of help here in devising the proper (field) experiments and the interpretation of the observations. Field observations have shown that probably the rate of multiplication of H. rostochiensis at the tolerance limit of potato is already almost at its maximum level. Therefore, in programmes of control for this nematode the maximum rate of multiplication must be counteracted especially if they aim at eradication, as in the Netherlands. However, according to observed rates at higher densities i.e. the equilibrium density of (about) 200 eggs per g soil and eq (5), direct determination of this multiplication rate can only be done at initial nematode densities below 2 eggs per g soil. This requires the investigation of large quantities of soil. However, good estimations could also be made from observations at higher densities and possibly with less effort with the help of eq. (5).

Investigations on the effects of systemic nematicides on population increase also require insight in nematode population dynamics. These effects are not only density dependent but are, most probably, in different ways for different nematode species.

REFERENCES

1. Fisher, J.M. (1966). Observations on moulting of fourth-stage larvae of Paratylenchus nanus. Austr.J.Biol.Scs.19, 1073-1079.

2. Kaai, C. (1966). Stengelaaltjes in groente-gewassen Jaarversl. Inst.Pl.ziektenk. Onderz. 1965, 99-100.

3. Kaai, C. (1967). Control of stem nematode attack in onions with O.O diethyl O.2 pyrazinyl phosphorothioate ("zinophos") and O-phenyl N,N[1] dimethyl phosphorodiamide ("Nellite"). Nematologica 13, 605-616.

4. Kaai, C. (1970). The cause of occurrence of stem nematode attack in patches. Proceedings of the IXth International Nematology Symposium. Zeszyty Problemowe Postepow Nauk Rolniczich 92, 277.

5. Kaai, C. (1972). Systemische nematiciden. Gewasbescherming 3, 33-40.

6. Ouden, H. den, (1965). Potgrootte en schade aan aardappelen door H. rostochiensis. Jaarversl. Inst. Pl. ziektenk. Onderz. 1964, 103.

7. Ouden, H. den, (1973). De tolerantie van aardappelen aangetast door H. rostochiensis. Jaarverl. Inst. Pl. ziektenk. Onderz. 1972, 98.

8. Seinhorst, J.W. (1965). The relation between nematode density and damage to plants. Nematologica 11, 137-154.

9. Seinhorst, J.W. (1966). Longidorus elongatus on Fragaria vesca. Nematologica 12, 275-279.

10. Seinhorst, J.W. (1968). A model for the relation between nematode density and yield of attacked plants including growth stimulation at low densities. Comptes Rendus du Huitieme Symposium International de Nematologie, Leiden 83

11. Seinhorst, J.W. (1970). Dynamics of populations of plant parasitic nematodes. Ann.Rev. Phytopath. 8, 131-156.

12. Seinhorst, J.W. (1974). Principles and possibilities of determining degrees of nematode control leading to maximum returns, I. Protection of one crop sown or planted soon after treatment Nematologia Mediterranea 1, 93-105.

13. Seinhorst, J.W. (1974). The relation between nematode distribution in a field and loss in yield at different average nematode densities. Nematologia 19, 421-427.

14. Seinhorst, J.W. & Ouden, H. den. (1970). The relation between density of Heterodera rostochiensis and growth and yield of two potato varieties. Nematologica 17, 347-369.

15. Wallace, H.R., (1971). The influence of the density of Nematode populations on plants. Nematologica 17, 154-166.

DISCUSSION

In replying to a question about correlation of nematode density with plant yield Seinhorst explained that there was an upper limit to the number of nematodes which would reduce the yield of a crop; once this limit was reached no further sites would be available for nematode attack. Using Heterodera rostochiensis on potato as an example it was calculated that 256 nematodes/100g soil (X axis of

Fig.1) was the upper limit and an increase above this would not cause any further loss of yield (yield curve becoming parallel with X axis). Jatala said that he had recorded much higher densities (1500 cysts/100g - 150 larvae/cyst) of H. rostochiensis affecting potato yield, but Seinhorst said he could not offer an explanation for this, pointing out that such high densities had not been recorded in Europe.

When asked about the effect of mixed populations Seinhorst said that it was difficult to analyse and that modelling studies should be confined to individual species until their biology was better understood. Mixed population effects would be complicated as some species occupied the same ecological niche and would cause accumulative damage, whilst others would perhaps compete for some factor independent of the host plant and would therefore be less damaging than if on their own.

CHEMICAL CONTROL OF NEMATODE VECTORS OF PLANT VIRUSES

I.J. THOMASON and MICHAEL MCKENRY

Department of Nematology,

University of California, Riverside, U.S.A.

INTRODUCTION

Chemical control of plant pathogenic nematodes has been practiced successfully for 30 years. Both fumigant and nonfumigant nematicides are presently used. In the production of annual crops the nematicides are applied at rates sufficient to reduce the nematode population below the economic threshold level but generally no attempt is made to eradicate the nematodes. An 80-90% reduction in the population in the upper 40-60 cm of soil is usually adequate to provide economic control.

Preplant treatments for perennial crops may require higher levels of control to greater depths for establishment of trees and vines and for their continued profitable production for a number of years. Eradication of the nematode would be desirable in the case of perennial crops attacked by nonvector nematodes but it is seldom achieved and often not economically justified.

When a crop is to be grown in soil infested with a nematode vector of nepoviruses or tobraviruses, and the nematodes are known to be viruliferous or have access to the viruses, then the problem of control becomes quite different. Some viruses will persist in nematodes for months. In addition the virus may be transmitted to a plant by a single nematode.[1] Thus it becomes necessary to achieve or approach eradication of the nematodes in the root zone of the crop to be planted.

---

[1] If one vector nematode per kg of soil occurred in the root zone of a grapevine this would be equivalent to 2250 nematodes per vine assuming a rooting zone of $2m^3$.

There are a number of interacting factors that will determine the feasibility of controlling the nematode vector with nematicidal chemicals. They are a) host plant, b) nematode, c) nematicide, d) soil, and e) weather conditions. These factors, all interacting, determine the biological and physical feasibility of controlling the nematode-vectored virus. Application techniques will also be important assuming conditions at least minimally suitable for nematode control. Additional major considerations are economic and environmental. First, we shall discuss the biological and physical factors that affect control strategy and later address the problems of costs and secondary environmental affects.

HOST PLANT

Nematode vectored viruses injure annual, biennial and perennial crops. Control of these viruses, however, is a much more serious problem on perennials, as for example trees and vines. On annual crops a lower level of nematode control may reduce the virus incidence sufficiently to escape significant yield reduction or adverse effects on quality. If effective control is not achieved the production from only one crop year is lost and a nonsusceptible crop can be grown in subsequent years.

Perennial crops represent a much greater commitment in time and money. Nematodes surviving treatment may not cause direct injury immediately or even be viruliferous. In subsequent years however they may again acquire the virus from surviving weed hosts or root fragments from the previous crop. Some strains of both nepoviruses and tobraviruses are known to be transmitted through the seed of weed hosts. Raski et al (19) reported that fanleaf virus could be detected in old grape roots 4 to 5 yr after the vines had been removed and Xiphinema index was recovered from the fallow soil of these old vineyards even after 10 yr without a host crop.

The rooting habit of the host plant also must be considered. Chemical control of vectors may be more easily achieved with shallow rooted than with deep rooted crops. There is, however, evidence that nematodes may move up toward the surface relatively rapidly when they survive at depths of 2.5 to 3m (10). Xiphinema spp. which occur to depths of 2 m or more in deep alluvial soils may be particularly difficult to kill.

NEMATODES

It is fortunate that the three genera known to transmit plant viruses, i.e. Trichodorus, Longidorus and Xiphinema, are all ectoparasitic. Thus they are exposed to the action of nematicidal chemicals in the soil. Data on the relative susceptibility of the three genera to various fumigant (e.g. alkyl halide) and non-fumigant (e.g. organophosphate and carbamate) nematicides are not available. Experience with other nematodes (3, 7, 8, 14 and 28) suggest that differences in susceptibility of various stages of a given species might exist as well as differences between species and/or genera. Van Gundy et al. (24) reported Xiphinema index adults to be more susceptible to methyl bromide in soil than X. index larvae. Trichodorus spp. have been observed to increase in numbers rapidly following soil fumigation but it is not clear whether this is due to survival of eggs and a rapid increase in the population in the absence of competitors or to other factors. The response of virus vector nematode genera, as well as their life-stages, to various nematicides is an area that needs additional research.

Species of Xiphinema and Longidorus are associated with the transmission of nepoviruses in perennial plants, including trees, vines and shrubs, whereas Trichodorus species are associated with tobraviruses of annual crops and a few perennial ornamentals. Thus chemical control of Xiphinema and Longidorus is generally more difficult and has received more attention.

NEMATICIDE

Two types of nematicides, fumigant and nonfumigant, can be distinguished based on their behaviour in soil. The fumigant nematicides such as dichloropropene, ethylene dibromide and methyl bromide have been used most widely for the control of nematode vectors of plant viruses. The important physico-chemical properties of some fumigant nematicides are given in Table 1. Methyl isothiocyanate is probably not a true fumigant nematicide in that it moves through soil with water rather than in the gaseous phase. Other nematicides which are similar in behaviour are metham-sodium (sodium-N-methyl-dithiocarbamate) and dazomet (3,5-dimethyltetrahydro-2H-1,3,5-thiadiazine-2-thione).

Nonfumigant nematicides most commonly used in attempting to control nematode vectors are either carbamates, carbamoyl oximes or organophosphates (see Table 2). The most popular formulation is a 10 to 15% granule which can be spread on the soil surface and incorporated. These nematicides are generally good soil insecticides and many have good systemic insecticidal properties. Unfortunately they may also have high mammalian toxicity. More recently, oxamyl, a systemic insecticide/nematicide, has been introduced and shows promise of being a chemical that can be applied to foliage and can control nematodes in the rhizosphere of established plants. Whether this approach would be feasible in the control of nematode vectored viruses is questionable but it might prove useful in conjunction with pre-plant treatments.

Table 1. Some important physico-chemical characteristics for the behaviour of fumigants in soil. All values for 20°C. After Goring (1962, 1967); Leistra (1972).

Compound	Vapour pressure (mm of mercury)	Solubility in water (μg/g)	Concentration ratio		Diffusion coefficient in air[3] (cm^2/s)
			water: air[1]	org. matter: water[2]	
monobromomethane	1380	16000	4.1		0.097
1,2-dibromoethane	7.69	3370	42.7	39	0.081
cis-1,3-dichloropropene	25	2700	17.7		0.074
trans-1,3-dichloropropene	18.5	2800	24.6		0.074
1,2-dibromo-3-chloropropane	0.58	1230	164	75[4]	0.070
trichloronitromethane	20	1950	10.8		0.069
methyl isothiocyanate	21	7600	92		

[1]Estimated from the solubility in water and the vapour pressure.

[2]Equilibrium ratio between the amount of fumigant adsorbed per gram of organic matter and the amount of fumigant dissolved per gram of water.

[3]Estimated values.

[4]Determination for one soil.

Table 2. Nonfumigant nematicides most frequently tested for nematode vector control[1]

Common Name	Compound	Formulation(s)
Aldicarb	2-methyl-2-(methylthio) propionaldehyde O-(methy-carbamoyl) oxime	10% Granular
Carbofuran	2,3-dihydro-2,2-dimethyl-7-benzofuranyl methyl carbamate	10 & 15% Granular, 45 kg ai/l Flowable; 75% Wettable Powder
Ethoprop	O-Ethyl S, S-dipropyl phosphorodithioate	10% Granular; 0.67 kg ai/l E.C.
Fensulfothion	O,O-Diethyl O-p-(methyl-sulfinyl)phenyl phosphoro-thioate	2 & 10% Granular 0.45 kg ai/l E.C.
Oxamyl	Methyl N'N'-dimethyl-N-[(methylcarbamoyl)oxy]-1-thiooxaminidate	Water Soluble Liquid
Phenamiphos	Ethyl 4-(methylthio)-m-tolyl isopropylphosphoramidate	5 & 10% Granular 0.4 kg ai/l E.C.[2]

[1]Does not include all possible nonfumigant nematicides or constitute a recommendation for use for those listed.

[2]E.C. = emulsifiable concentrate.

Quintozene (parachloronitrobenzene) has also been used in an attempt to control nematode vectors (15). Aqueous extracts from raspberry canes and roots were reported (23) toxic to *Longidorus elongatus*.

SOIL

In discussing chemical control of nematode vectors it is necessary to consider the soil as this is the habitat in which they are found and in which they must be killed. One must be concerned not only with the basic textural composition of the soil such as percentage clay, silt and sand but must also understand the effects of organic matter, moisture, temperature, stratification and depth. In a colder climate like that in northern Europe and Great Britain, considerable difficulty is encountered in the use of soil fumigants because the soils often are cold, wet and high in organic matter. By contrast, in the mediterranean area and in the southwestern United States the use of soil fumigants is somewhat more successful because the soils are low in organic matter and can be treated when warm and relatively dry.

There have been detailed discussion of those physical and biological factors of soil that influence the efficacy of soil fumigants (1, 7, 8, 12, 13). Leistra's (12) research in the Netherlands on the behaviour of cis- and trans-1,3-dichloropropene under cool, wet conditions and in soils high in organic matter is of special value to agriculturists in Europe who must control virus vector nematodes under such conditions. McKenry and Thomason (13) and Abdalla et al. (1) have studied critically the influence of soil moisture, temperature, and organic matter on fumigant action under the somewhat less rigorous conditions that might be encountered in the alluvial soils of California, U.S.A. They have extended the use of special techniques developed by Kolbezon and Abu-El-Haj (11) for sampling and analyzing deep field soil atmospheres. Detailed consideration of this subject is given in the literature referred to and only a brief account can be presented here to illustrate the information that has been developed.

Nematicide Movement in Soil

Nematicides in soil are in a dynamic equilibrium between the soil phase (i.e. clay, organic matter) soil solution and the soil air, The distribution of a nematicide in soil is illustrated in Fig. 1. The nonfumigant nematicides such as organophosphates, which have relatively low vapour pressures, are mostly in the soil and soil water phases. They move through the soil with soil water primarily by mass flow. Diffusion through the pore space is not of great importance.

Soil fumigants, e.g. alkyl halide nematicides, by contrast, are distributed in the soil, soil water and soil air phases. Their inherent vapour pressure and water solubility will determine to a large extent how rapidly they will move through soil. This movement will be primarily in the gas phase through the pore spaces and will be at least X1000 faster than diffusion through the soil water phase. The ratio of fumigant in the soil water phase to that in the soil air phase is a constant (referred to as Henry's constant K_h) and for a nematicide such as cis-1,3 dichloropropene the $K_h = 17.7$ at 20°C (see Table 1 for physical characteristics of some important soil fumigants).

Influence of soil moisture

If the rapid movement of soil fumigant is through the air spaces in soil it becomes apparent that soil moisture content is of primary importance. First, a significant part of the chemical is dissolved in water and as soil water content increases more chemical is in solution and not available for diffusion. Second, if a soil is saturated or near saturation, pore spaces are blocked and diffu-

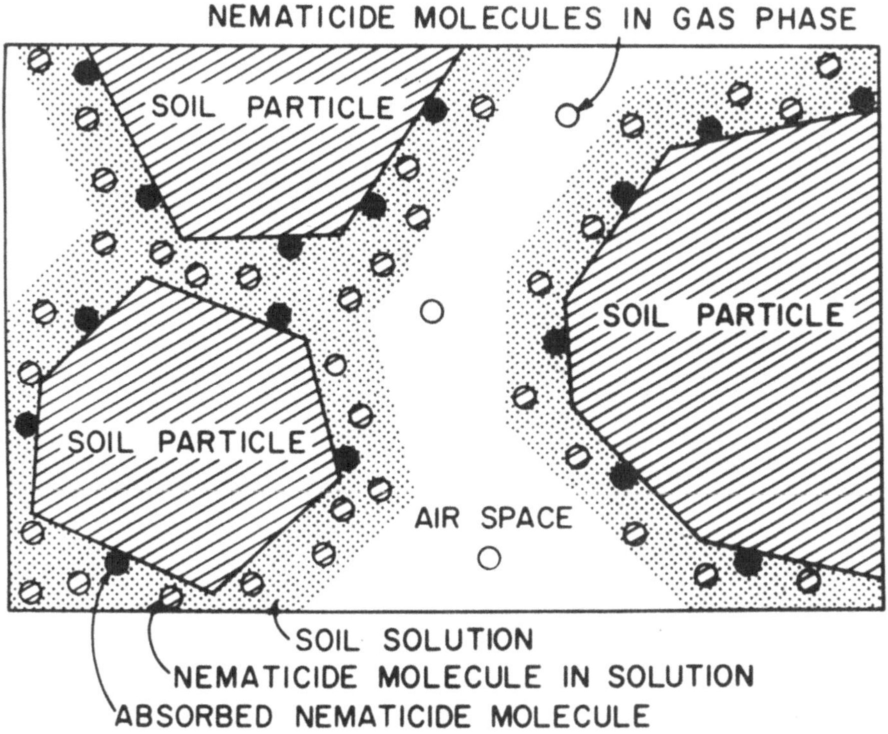

Fig.1. Diagrammatic cross section of soil particles, showing the distribution of nematicide molecules in the soil air space, soil solution and adsorbed to the soil particle.

sion is slowed down or prevented. Since clay soils have smaller soil pores it is easier for excess water to block these pores and restrict diffusion. Thus excess moisture in a clay soil, or a clay lens in a soil profile, is more likely to restrict fumigant movement than high moisture in a sandy soil.

The influence of soil moisture on the concentration of cis-1,3-dichloropropene 17.5 cm from the point of injection is illustrated in Fig. 2a. Note that the chemical reached a concentration of 3 x 10^{-5} moles/1 in the soil vapour phase within 48 h after injection in this sandy loam soil at 7.7% moisture content. At 23% moisture the highest concentration reached was less than 2 x 10^{-5} moles/1 and this was reached 96-120 h after injection.

The restrictive influence of soil moisture on movement of 1,3-dichloropropene in field soil is shown in Fig. 3. As soil moisture

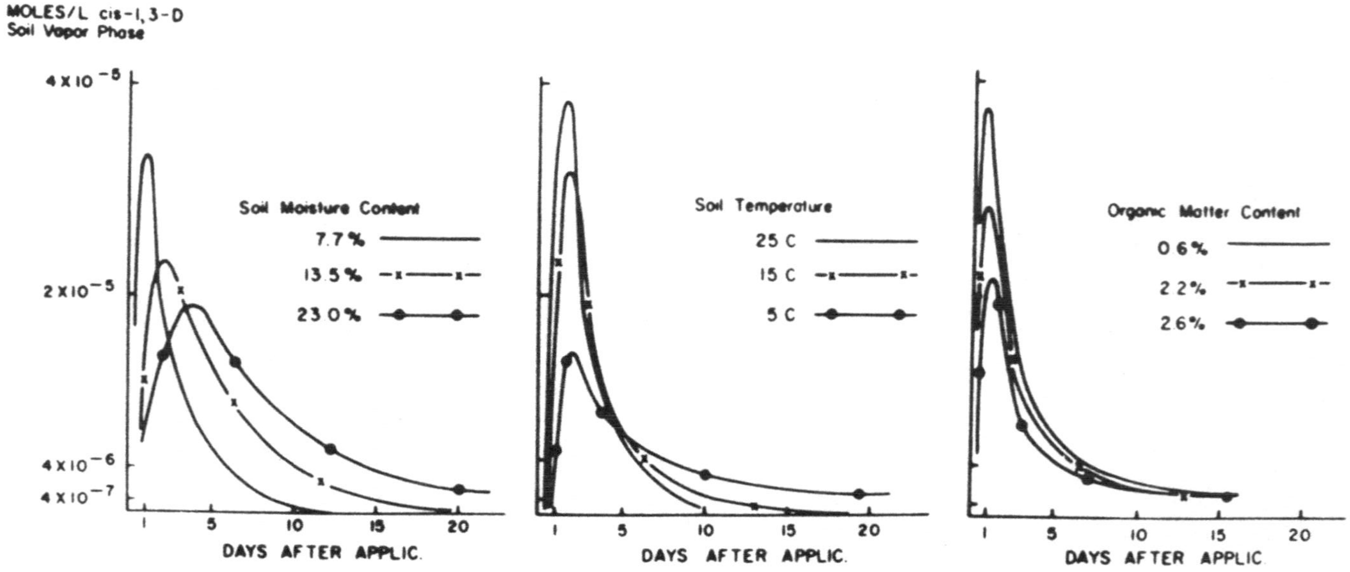

Fig. 2a, b and c. Influence of soil moisture content, soil temperature and soil organic content on the concentration of cis-1,3-dichloropropene in moles/l in the soil vapour phase 17.5 cm from the point of injection at various intervals following application.

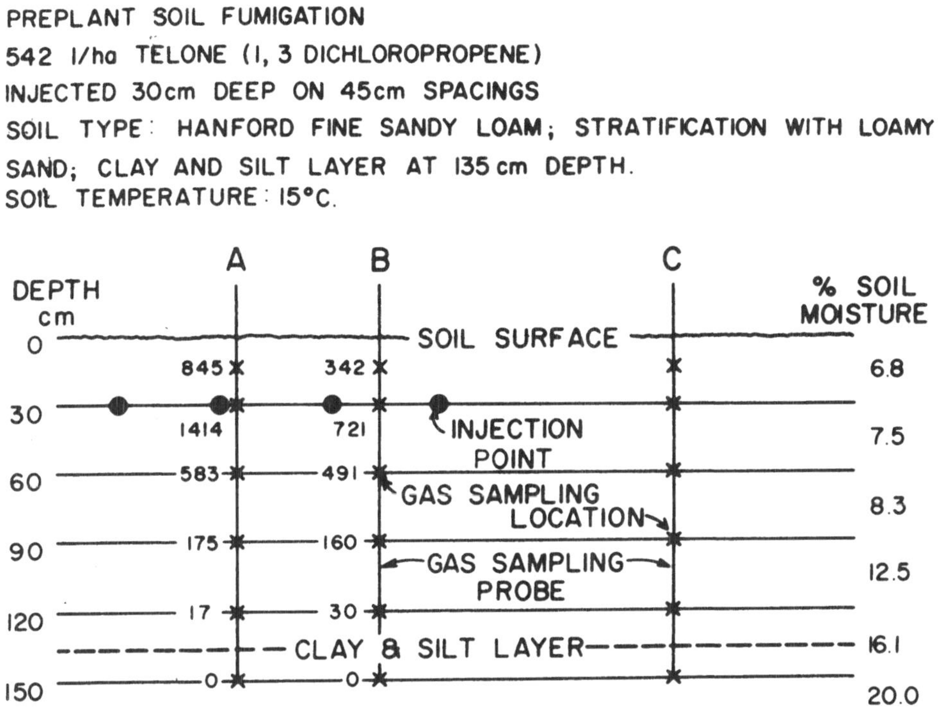

Fig. 3. Accumulated dosage of 1,3-dichloropropene at various points in the soil profile expressed as ppm.day in the soil water phase 21 days after field application of Telone.

increased with depth in this Hanford sandy loam soil the downward diffusion of nematicide was restricted. This is particularly apparent between a depth of 90–120 cm when the total accumulated dosage of 1,3-dichloropropene in the soil water phase 21 days after treatment was 160–175 ppm.day (13) at 90 cm and only 17–30 ppm.day at 120 cm.

Also apparent is the influence of soil stratification. A moist clay and silt layer between 120–150 cm depth provided an almost complete barrier to the downward diffusion of fumigant. The problems associated with soil stratification cannot be overemphasized. Stratification can result from the presence of soil particles of a different texture, organic matter or dense layers resulting from natural causes or from compaction during tillage and/or harvesting operations.

Influence of Soil Temperature

Soil temperature may function in three ways to affect the outcome of nematode control with soil fumigants. First it influences the ratio of the chemical in the water and air phases. As temperature decreases more of the chemical is in the water phase and thus not available for diffusion through pore spores. This is illustrated in Fig. 2b. At 25°C the concentration of 1,3-dichloropropene obtained at 17.5 cm from the point of injection is over twice that at 5°C and 72 hrs after treatment.

Higher soil temperatures improve the diffusion of soil fumigants but they also increase the rate at which these chemicals degrade in soil (2, 27). Finally, there is clear evidence that some nematicides e.g. ethylene dibromide, are not as effective in killing nematodes at temperatures below 10°C (3). These effects of temperature must be taken into consideration in using nematicides; experience has shown that best results are obtained at temperatures between 15 and 30°C.

Influence of Soil Organic Matter

Organic matter in soil adsorbs the nematicide and reduces the amount available for diffusion. It also appears to hasten the rate of nematicide decomposition in soil. The effect of even relatively small amounts of organic matter in soil is illustrated in Fig. 2c. The maximum concentration of cis 1,3-D reached at 17.5 cm from the point of injection was 3.5×10^{-5}, 2.75×10^{-5} and 2.0×10^{-5} moles/l after 48 h at 0.6, 2.2 and 2.6% organic matter, respectively.

It is not uncommon for soils in the Netherlands and other areas to have from 5 to 50% organic matter. In addition some soils may have stratification involving layers of organic matter. Leistra (12) gives a detailed consideration of fumigant behaviour in cool, wet soils, high in organic matter.

APPLICATION OF NEMATICIDES

Soil Fumigants

Injection with chisel applicators is essentially the standard method for application of fumigant nematicides into soil in large fields. This equipment is most often mounted on wheel or crawler tractors and the injection chisels (shanks or tines) pulled through the soil at various spacings and depths. For annual crops or shallow soil the distance between chisels is usually 30 cm and the depth of injection is 15-20 cm. With perennial crops the spacing ranges from 30-75 cm and the depth from 25-75 cm. Treatments in California vineyards which have given vector control have been made with 1,3-D (Telone or D-D) at 2.375 l/ha injected to depths of 75 cm (21).

Prior to injecting the chemical the soil may be tilled in various ways. Ploughing and/or disking is adequate for annual crops. In fields where perennial trees or vines are to be planted the soil should be deep ploughed or ripped in several directions at depths from 45 to 90 cm. Tillage is especially important in compacted soils or in soils having layers made up of clay or organic matter as indicated previously. The surface soil should be tilled to a seed bed condition. Following injection the soil surface is then smoothed and compacted. A ring roller is suitable for this operation. Control near the soil surface may not be adequate when large amounts of nematicide are injected deep into the soil in order to obtain control of nematode vectors at depths of 2 to 2.5 m. The control pattern often obtained with deep injection is shown in the middle diagram in Fig. 4. In order to overcome this problem the fumigant is injected at 2 depths and a pattern similar to that shown in the bottom diagram of Fig. 4 is obtained.

Seinhorst (22) has recommended the combined use of a fumigant injected into the soil and a nonfumigant, dazomet, applied to the soil surface. This approach might be quite effective in those areas where rainfall or irrigation water is available to redistribute nonfumigant nematicides in the surface 20-30 cm of soil.

Movement of fumigants through fine textured soils may be improved by removing excess soil moisture through cropping with a cereal or with Sudan grass in areas where rainfall is seasonal or irrigation is practiced.

Some acreage in California grape growing areas has been treated with methyl bromide. This chemical is injected into the soil and the soil surface covered with a polyethylene sheeting. Fumigant is injected and soil covered in one operation.

Non-fumigants

These nematicides are most often in granular form and can be spread on the soil surface with fertilizer spreaders or special granular applicators. They are then mixed with soil by disking or the use of a rotary cultivator. Subsequent redistribution is dependent on rainfall or irrigation water. In the production of annual crops treatment down the planting row may be adequate.

The use of nonfumigant nematicides for preplant treatment control of vector nematodes in perennial tree or vine crops has not shown great promise and may not be feasible in deep soils.

TREATMENT RATES AND ECONOMICS

Application rates for fumigant nematicides range from 190-285 l/ha for 1,3-dichloropropene nematicides on annual crops to 475-

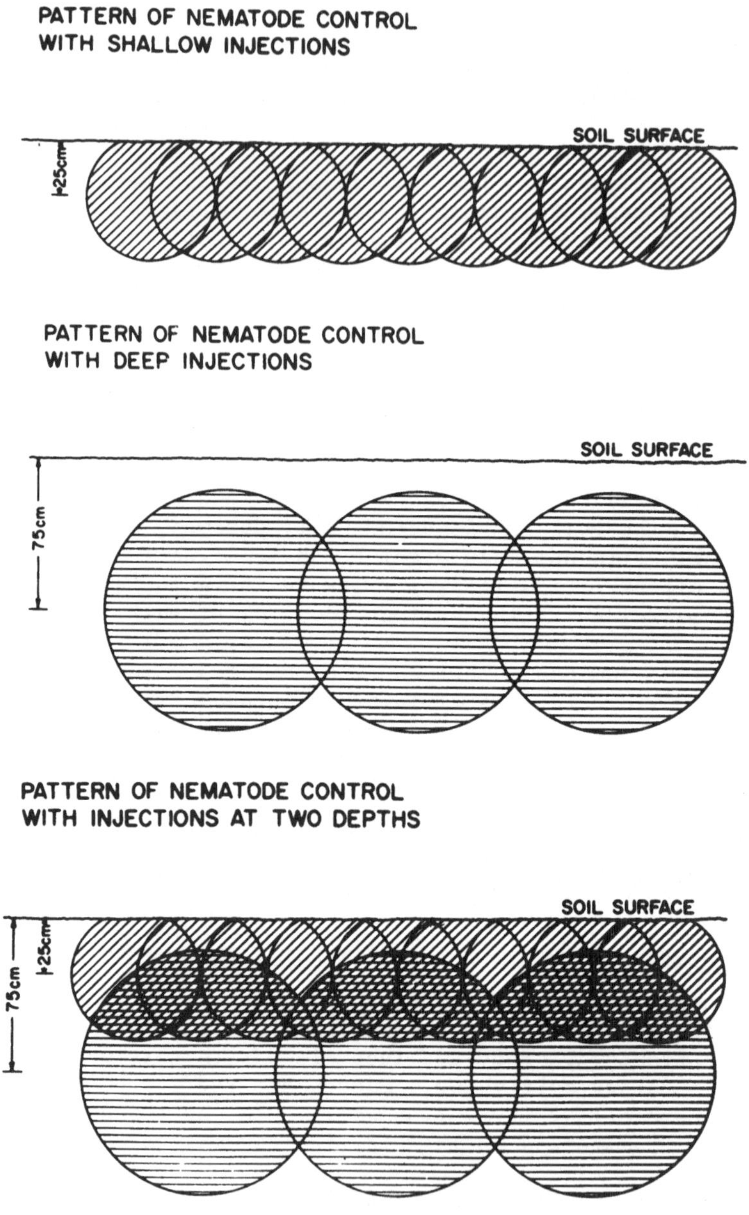

Fig. 4. Diagrammatic representation of patterns of nematode control when a fumigant nematicide is only injected at depth of 25 cm or 75 cm and where a combined application is made at 2 depths.

2,375 l/ha on perennial crops. The cost can range from approximately 100 to 1000 United States dollars per hectare. Methyl bromide at the rate of 450 kg/ha can cost 1,000 United States dollars per hectare. Thus, the attempted eradication of nematode vectors from orchard and vineyard soils by chemical means can be a very expensive operation. Great care should be taken to ensure its success.

Control of vector nematodes in annual crops by fumigants, nonfumigants or combinations can range from 50 to 100 United States dollars depending on chemical used and rate of application. Considerable literature now exists on various aspects of chemical control of virus vector nematodes. Recommendations for nematode control on small fruits and grapevines are contained in the book entitled "Virus Diseases of Small Fruits and Grapevines" (4). The published literature also provides additional information (some references are 9, 15, 17, 18, 20, 21, 25, 26).

SECONDARY EFFECTS OF NEMATICIDES

Soil fumigants applied at high dosage rates are known to interfere with nitrification (5) reduce populations of beneficial endotrophic mycorrhiza (6, 16) and persist in toxic amounts either as the parent compound or as toxic degradation products (12, 27). Effects on nitrifying bacteria have been known for some time. In those crops where ammonia nitrogen is not toxic the depressed nitrification can be ameliorated by judicious applications of nitrate forms of nitrogen.

The effects on rhizomorphic fungi, such as Endogene spp., have only recently been recognized. Elimination of these fungi is associated with reduced availability of soil phosphates to plants. In some cases minor element nutrition is also upset. Again an awareness of this problem can lead to correction through application of needed nutrients and/or the reintroduction of the rhizomorphic fungus.

Toxic amounts of fumigant nematicides usually persist in cold, wet soils. This problem can in part be solved by improving soil conditions prior to treating, reducing dosage, and treating well ahead of the time when trees or vines must be set out.

Several rather unique problems are worth mentioning. Methyl bromide applied to vineyard sites in California has resulted in toxic bromine residues in oat hay harvested from this soil. Bromine residues would be an important consideration in any soil treated with bromide containing compounds.

In the Rhine River Valley in Germany, there is a concern for the contamination of ground water under the coarse soils in the vineyards on the steep slopes in this area.

SUMMARY

Controlling nematode vectors of plant viruses presents unique and challenging problems. Much of the basic information on how and when to treat those soils harbouring nematodes has been developed in recent years. In those areas where either climate or economics or both do not limit application of this technology some success has been achieved. Much remains to be done.

REFERENCES

1. Abdalla, N., Raski, D.J., Lear, B., & Schmitt, R.V. (1974). Distribution of methyl bromide in soils treated for nematode control in replant vineyards. Pestic. Science 5, 259-269.

2. Castro, C.E. & Belser, N.O. (1966). Hydrolysis of cis- and trans-1,3-dichloropropene in wet soil. Agric. and Fd. Chemistry 14, 69-70.

3. Evans, A.A.F. & Thomason, I.J. (1971). Ethylene dibromide toxicity to adults, larvae and moulting stages of Aphelenchus avenae. Nematologica 17, 243-254.

4. Frazier, N.W., Fulton, J.P., Thresh, J.M., Converse, R.H., Varney, E.H. & Hewitt, W.B. (eds.) (1970). Virus diseases of small fruits and grapevines. Univ. Calif. Div. Agric. Sci. Berkeley, U.S.A. 290 pp.

5. Gasser, J.K.R. & Peachey, J.E. (1964). A note on the effects of some soil sterilants on the mineralization and nitrification of soil-nitrogen. J.Sci. Fd. Agric. 15, 142-146.

6. Gerdemann, J.W. (1970). The significance of vesicular-arbuscular mycorrhiza in plant nutrition. In Root Diseases and Soilborne Pathogens. T.A. Toussoun, R.V. Bega and P.E. Nelson, eds. U. of Calif. Press, Berkeley, 125-129.

7. Goring, C.A.I. (1957). Factors influencing diffusion and nematode control by soil fumigants. ACD Inform. Bull. 110 (Dow Chemical Co., Midland, Mich., Nov. 1957).

8. Gording, C.A.I. & Youngson, C.R. (1957). Factors influencing nematode control by ethylene dibromide in soil. Soil Science 83, 377-389.

9. Harrison, B.D., Peachey, J.E. & Winslow, R.D. (1963). The use of nematicides to control the spread of arabis mosaic virus by Xiphinema diversicaudatum (Micoletsky). Ann. Appl. Biol. 52, 243-255.

10. Johnson, P.W. & McKeen, C.D. (1973). Vertical movement and distribution of Meloidogyne incognita (Nematodea) under tomato in a sandy loam greenhouse soil. Can. J. Plant Sci. 53, 837-841.

11. Kolbezon, M.J. & Abu-El-Haj, F.J. (1972). Fumigation with methyl bromide. II. Equipment and methods for sampling and analyzing deep field soil atmospheres. Pestic. Sci. 2, 73-80.

12. Leistra, M. (1972). Diffusion and adsorption of the nematicide 1,3-dichloropropene in soil. Agric. Res. Reports No. 769 Laboratory for Research on Insecticides, Wageningen, The Netherlands, 105 pp.

13. McKenry, M.V. & Thomason, I.J. (1974). 1,3-dichloropropene and 1,2-dibromoethane compounds: I. Movement and fate as affected by various conditions in several soils. II. Organisms-dosage response studies in the lab with several nematode species. Hilgardia 42, 393-438.

14. Moje, W. & Thomason, I.J. (1963). Toxicity of mixtures of 1,2-dibromo-3-chloropropane and 1,2-dibromoethane to the root-knot nematode, Meloidogyne javanica. Phytopathology 53, 428-431.

15. Murant, A.F. & Taylor, C.E. (1965). Treatment of soils with chemicals to prevent transmission of tomato black-ring and raspberry ringspot viruses. Ann. Appl. Biol. 55, 227-237.

16. Possingham, J.V. & Groot Obbink, J. (1971). Endotrophic mycorrhiza and the nutrition of grapevines. Sonderdruck aus der Zeitschrift "Vitis". 10, 120-130.

17. Raski, D.J. (1972). Nematode control: some things will work, such as deep fumigation. Wines and Vines 53, 30-31.

18. Raski, D.J., Hart, W.H. & Kasimatis, A.N. (1973). Nematodes and their control in vineyards. Calif. Agric. Expt. Station Circular 533, 20 pp.

19. Raski, D.J., Hewitt, W.B., Goheen, A.C., Taylor, C.E. & Taylor, Taylor, R.H. (1965). Survival of Xiphinema index and reservoirs of fanleaf virus in fallow vineyard soil. Nematology 11, 349-352.

20. Raski, D.J., Hewitt, W.B. & Schmitt, R.V. (1971). Controlling fanleaf virus-dagger nematode disease complex in vineyards by soil fumigation. California Agriculture 25, 11-14.

21. Raski, D.J., Schmitt, R.V., Luvisi, D.A. & Kissler, J.J. (1973). 1,3-D and methyl bromide for control of root-knot and other nematodes in vineyard replants. Plant Disease Rptr. 57, 619-623.

22. Seinhorst, J.W. (1973). The combined effects of dichloropropane-dichloropropene mixture injected into the soil and of dazomet applied to the soil surface. Neth. J. Pl. Path. 79, 194-206.

23. Taylor, C.E. & Murant, A.F. (1966). Nematicidal activity of aqueous extracts from raspberry canes and roots. Nematologica 12, 488-494.

24. Van Gundy, S.D., Munnecke, D., Bricker, J. & Minteer, R. (1972). Response of Meloidogyne incognita, Xiphinema index and Dorylaimus sp. to methyl bromide fumigation. Phytopathology 62, 191-192.

25. Vuittenez, A. (1958). Actvité compareé des fumigants des insecticides, et divers products appliqués en traitment du sol sur les contaminations par la dégénérescence infectieuse de la vigne. C.R. Acad. France 44, 901-907.

26. Vuittenez, A. (1960). Nouvelles observations sur l'activité des traitements chimiques du sol pour l'eradication des virus de la dégénérescence infectieuse de la vigne. C.R. Acad. Agr. France 46, 89-96.

27. Williams, I.H. (1968). Recovery of cis- and trans-1,3-dichloropropene residues from two types of soil and their detection and determination by electron capture gas chromatography. J. Econ. Ent. 61, 1432-1435.

28. Youngson, C.R. & Goring, C.A.I. (1963). Nellite nematicide-laboratory and greenhouse studies on a new phosphorus material for root-knot nematode control. Down to Earth 18,

DISCUSSION

A query was raised about the loss of fumigant through the chimney formed by the injector chisel and whether this was critical. Thomason replied that there were no data on how much fumigant escapes from the soil by this route but pointed out that at low dosages 5 h is required for the nematicide to completely volatilize from the liquid to the gaseous phase. Thomason went on to explain that the gas in the soil was sampled by pushing into the soil to a depth of 2-5 cm a 2 mm stainless steel tube, with a bronze screen at the lower end to prevent the ingress of soil; gas samples were then drawn off through the tube by means of a 50 ml syringe. On the subject of cost effectiveness of chemical control for nematodes, he said that in California the rate of application of D-D would be 5 gal/acre at a cost of 2 USA dollars per gallon. Such treatment before replanting vines could result in obtaining a crop one year earlier than from an untreated plantation, and taking raisins selling at 750 dollars/ton this would more than pay for the cost of treatment.

Thomason explained that there was no State control over the contractors who applied the nematicides, but advice was given to farmers by the State Advisers about the necessity or otherwise for chemical treatment and also Thomason and his colleagues could undertake investigations with their equipment after treatments to ascertain whether they had been undertaken correctly and effectively.

Asked about the time required for a nematicide to break down in the sub-soil, Thomason commented that this should be ascertained by bio-assays; redworms were used for bioassay if 1-3-D, to which they were very sensitive. In reply to another question Thomason said that it did not matter whether a nematode was exposed to 10 ppm for 100 days or 100 ppm for 10 days, the effect was the same. The accumulated dose (or CT product) was the important quantity to express in terms of nematode control. Seinhorst referred to his experiments which showed that a fumigant nematicide like D-D tended to build up its concentration in the lower layers of the soil whereas dazomet applied to the surface as granules was most concentrated in the surface layer. Hence he had found that 97-99% kill was obtained by a combination of D-D and dazomet. The action of dazomet was, however, dependent on rainfall.

Short Report

CHEMICAL CONTROL OF TRICHODORIDS AND OF TRANSMISSION OF TOBACCO RATTLE VIRUS

T.J.W. Alphey[1], J.I. Cooper[2] and B.D. Harrison[1] ([1]Scottish Horticultural Research Institute, Invergowrie, Dundee, Scotland; [2]Commonwealth Forestry Institute, Oxford, England.)

Field trials were undertaken in two consecutive years with some systemic nematicides for the control of trichodorid nematodes and the transmission of tobacco rattle virus (TRV) to potatoes. In replicated plot experiments oxamyl (Vydate) at 5.6 or 7.8 kg a.i./ha, aldicarb (Temik) at 3.4 kg a.i./ha, phenamiphos (Nemacur) at 15.0 or 30.0 kg a.i./ha and dichloropropane-dichloropropene (D-D) at 224 kg/ha were applied to soil containing trichodorids carrying TRV. The plots were planted with a spraing-susceptible potato cultivar, Pentland Dell. Numbers of trichodorids were estimated, at about 6-weekly intervals throughout the growing season, by wet-sieving soil samples (200 g) taken from each treatment.

In the 1972 trial all nematicides tested significantly ($P = 0.05$) decreased trichodorid numbers in comparison to the control treatment, by mid summer. Nematode control was maintained until the autumn harvest. In 1973, oxamyl (5.6 kg a.i./ha) caused a significant decrease in nematode numbers, for a limited time, but aldicarb and phenamiphos did not. In both years trichodorid numbers at harvest in all treatments, including the controls, were smaller than at planting time.

The incidence of spraing disease was assessed by recording the percentage of tubers showing internal symptoms at harvest. In 1972 the incidence of spraing was decreased from 14% (in the untreated control plots) to 0.4% by phenamiphos (15.0 kg a.i./ha), 1.4% by oxamyl (7.8 kg a.i./ha), 1.8% by phenamiphos (30.0 kg a.i./ha) and 3% by dichloropropane-dichloropropene (224 kg/ha). In 1973 oxamyl (5.6 kg a.i./ha) and aldicarb (3.4 kg a.i./ha) significantly decreased the incidence of spraing but phenamiphos (15.0 kg a.i./ha) did not.

None of the nematicides was phytotoxic or diminished tuber yield at the rates used.

Short Report

CHEMICAL CONTROL OF CORKY RINGSPOT DISEASE ON POTATOES IN FLORIDA, U.S.A.

D.P. Weingartner[1], J.R. Shumaker[1] and G.C. Smart, Jr.[2] ([1]Agricultural Research Center, Hastings, Florida 32045, U.S.A.;[2] University of Florida, Gainsville, Florida 32611, U.S.A.)

Corky ringspot (CRS) believed to be caused by tobacco rattle virus (TRV) is an important disease of potatoes in northeastern Florida, U.S.A. Although potato cultivars such as Pungo possessing resistance to CRS are available, other varietal characteristics such as susceptibility to bacterial wilt (caused by Pseudomonas solanacerarum E.F.Sm.), tendency to produce dark potato chips, and low yield potential in Florida's sub-tropical climate, have limited their use. Since 1970 most potato fields in northeast Florida have been fumigated with chemicals such as DD for control of Belonolaimus longicaudatus Rau, Meloidogyne incognita (Kofoid & White) Chitwood, and Tylenchorhynchus claytoni Steiner, as well as nematodes in five other plant parasitic genera. There were reports from other sections of Florida that following soil fumigation, numbers of Trichodorus christiei Allen (the vector of TRV) on other crops sometimes increased to levels exceeding those in non-treated controls. Experiments were therefore initiated during 1970 to determine whether numbers of T. christiei would increase on potatoes following soil fumigation and whether incidence and severity of CRS would be affected by any observed increases; and, whether non-volatile carbamate and organophosphate nematicides would reduce incidence and severity of CRS.

All chemicals, except foliar applications of oxamyl 2L, were applied in-row because overall treatments, particularly with soil fumigants, are impracticable on the highly ridged rows used in the area. Most experiments were performed on the cultivar Sebago which is highly susceptible to CRS. The cultivar Pungo was included in most, but not all experiments, as a resistant control. In addition to T. christiei, most soils were also inhabited by B. longicaudatus, M. incognita, T. claytoni, Criconemoides sp., and Helicotylenchus spp. Except in rate studies, 1,3D was applied at 75 l/ha in-the-row. Non-volatile chemicals applied to soil were used at 3.4 kg a.i./ha in-the-row. When oxamyl 2L was applied to foliage, three separate applications, each at 1.1 kg a.i./ha overall, were made. Control data were based on percent tubers showing typical concentric surface lesions and arcs of internal necrosis associated with CRS in potato tubers. The disease was severe in all experiments with the mean percentage of tubers showing internal CRS in non-treated controls ranging from 17.6 - 40.9% in the several experiments.

No significant increases in T. christiei populations were observed following in-row applications of five different rates of ethylene di-bromide (7.5-18.7 l/ha), DD (56-132 l/ha), or 1,3D (56-132 l/ha). None of these fumigants effectively controlled tuber symptoms of CRS at the rates tested. The fumigant DD-MENCS (28 l/ha) was also ineffective. When compared to non-treated controls, carbofuran 4F and 10G, aldicarb 10G, and ethoprop 10G applied to soil at planting significantly reduced incidence and severity of CRS symptoms on Sebago tubers. In one experiment incidence and severity of CRS were significantly reduced by carbofuran 4F when used alone or in various combinations with 1,3D and foliar applications of oxamyl 2L, and by at plant applications of oxamyl 2L followed by foliar applications of oxamyl 2L; but not by oxamyl 2L applied at planting, 1,3D applied 21 days before planting, or foliar applications of oxamyl 2L used in combination with a preplant treatment with 1,3D. Under Florida conditions soil fumigants applied in-row do not control CRS. Application of carbofuran, aldicarb or ethoprop to soil at time of planting or use of those chemicals in combination with 1,3D or DD can provide economical control of CRS. For example, using 1974 prices of chemicals and the 1973 average market price of potatoes, the increases in saleable crop resulting from nematode control and reductions in CRS in the above experiment would have produced the following cost benefit ratios (i.e., cost of treatment/net return): carbofuran 1/18, 1,3D 1/10, and 1,3D + carbofuran 1/8.

Short Report

NEMATODE CONTROL IN ESTABLISHED RED RASPBERRY PLANTINGS

F.D. McElroy (Canada Agriculture, 6660 NW Marine Drive, Vancouver, B.C., Canada.)

Red raspberry (*Rubus idaeus*) is an important crop in southwestern British Columbia and ranks second in the world in terms of yield per acre. Plantings free of disease remain productive for 10-15 yr and yield 4-5 ton/acre. However, in recent years plantings declined considerably, with yields in many fields down by 40-50% and plantings becoming unproductive after only 7-8 yr. No single factor appears responsible for this decline and in fact it may be a complex of biotic and abiotic factors. It is estimated that in 1973 there was a 31% loss to the potential 9 million dollar industry due to this slow decline problem.

In a recent survey of raspberry plantings in British Columbia to determine the involvement of nematodes in this decline, 16 stylet-bearing nematode genera were recovered. *Pratylenchus penetrans* was the most prevalent species, found in 90% of 350 samples, followed by *Xiphinema bakeri* and *X. americanum*. Highest populations of the first two species occurred under raspberries growing in loam soils and were consistently associated with declining plants. *X. americanum* occurred only in clay loam soils and was associated with tomato ringspot virus on raspberry.

Several tests were conducted to determine possible methods of control. In greenhouse tests oxamyl was applied as a foliar spray to raspberries growing in nematode infested soil in pots. Rates of 0.5, 1.0 and 2.0 lb/100 gal decreased *P. penetrans* soil populations by 86%, 97% and 99% and root populations by 96%, 99% and 99% respectively; *X. bakeri* numbers were decreased by 85%, 97% and 99% respectively. Plant growth was also increased by 25%, 21% and 37% for the above rates, respectively.

In another test four sprays were applied at 2-wk intervals and nematode counts were made one week after the second and fourth sprays. After two sprays, *P. penetrans* soil and root populations were decreased by 68% and 87% respectively, and by 96% and 92% respectively after four sprays. Similarly *X. bakeri* numbers were decreased by 67% and 96% after two and four sprays respectively.

In the field, oxamyl at 0.5, 1.0 and 2.0 lb/100 gal was applied to a 4 yr old planting, four times at 2-wk intervals as a foliar spray beginning at full leaf in May. Numbers of both nematode species under raspberries treated with all rates of oxamyl declined the year of treatment. The year following, however, only populations under the two highest rates remained significantly lower than the checks. All treatments significantly increased yields in the treatment year and the year following. The highest rate of treatment resulted in an increase of 4246 lb/acre over non-treated plants, or a gross return of $ 1132/acre over a two year period.

These treatments gave good control of nematode species causing direct damage and resulted in increased growth and yield. However, such control measures would undoubtedly be inadequate to prevent vector species from transmitting virus. Control never reached 100% in my tests even with the highest rate and therefore a small percentage would remain to act as vectors.

Short Report

CHEMICAL AND CULTURAL CONTROL OF NETTLEHEAD AND RELATED VIRUS DISEASES OF HOP

R.S. Pitcher (East Malling Research Station, Maidstone, England)

Special features of hop-growing practice and the epidemiology of the diseases associated with the hop strain of arabis mosaic virus (AMV-H) transmitted by Xiphinema diversicaudatum complicate the problems of their control (see previous report by Pitcher in these Proceedings). Attempts to eliminate the vector by soil treatment with fumigants or other nematicides have little chance of success, as even a 99% control of the minimal population readily detectable by routine soil sampling (i.e. 1/200 ml ≡ 10,000/plant) will have 100 survivors, many of which may be viruliferous and thus capable of infecting a healthy hop. This task is made more difficult, in fact virtually impossible, by the heavy-textured soil in which hops are usually grown and by the deep and extensive root system and the long-term perennial nature of hops, which allow small foci of infection to expand year by year.

Better and more permanent success has been achieved by cultural means involving the provision of hop-free intervals of 2 yr(1). This can be done either by bare-fallowing or by planting virus-immune hosts ('break crops'). During a hop-free interval the vectors gradually lose their infection, either by moulting or, presumably, by erosion or inactivation of the virus particles adhering to the cuticular oesophageal lining (see Taylor and Robertson in these Proceedings). If conditions allow the grower to apply hop-free break periods for 2 yr a permanent control will result, provided that care is taken to ensure that AMV-H is not re-introduced to the field, either in planting material or by the growth of infected hop seedlings from the same or an adjoining field. Further research may allow the hop-free interval to be reduced to little over a year, virtually halving the loss of crop resulting from a 2 yr break. Fallowing techniques are less likely to succeed with other strains of AMV, which have a wider weed host range and would therefore demand a much more rigorous fallow regime.

If a 2 yr hop-free interval is not practicable, thorough cultivation and as long a hop-free interval as possible, followed by soil fumigation with 1-3 dichloropropene (at approx. 500 l/ha) is the most efficient treatment yet devised and may delay serious reinfection with AMV-H for periods up to 5 yr. Dazomet is moderately effective and the oximecarbamates less so.

REFERENCE

1. McNamara, D.G., Ormerod, P.J., Pitcher, R.S. & Thresh, J.M. (1973). Fallowing and fumigation experiments on the control of nettlehead and related virus diseases of hop. Proc. 7th Br. Insectic. Conf. (1971). 314-318.

Short Report

CONTROL OF *XIPHINEMA* IN SOUTH AFRICA

P.C. Smith (Plant Protection Research Institute, Stellenbosch 7600, South Africa)

The data presented are taken from normal nematode control experiments in established vineyards infested with various plant parasitic nematodes. The experiments were not specifically aimed at virus or virus vector control. The soils were sands or sandy loams where the highest nematode populations seem to occur.

Applications of liquid nematicides were by hand injection or flood irrigation. In the latter case the nematicide was applied during the full irrigation period. Granular nematicides were broadcast on the soil surface and then incorporated into the top 10 cm of soil. Treatments were followed immediately by an irrigation of about 10 cm water to seal the soil surface. At irregular intervals nematodes were extracted from soil samples, using the Oostenbrink elutriator/cotton wool filter technique and counted. These data were analysed statistically, using a log n + 1 transformation. Coefficients of variation were generally considered acceptable - not more than 30% and usually much lower.

Results so far have been variable. In one experiment excellent, statistically highly significant (1% level) control was achieved and maintained for almost 2 yr (96 wk) of several phytoparasitic nematode species including *Meloidogyne incognita*, *Tylenchulus semipenetrans* and *Macroposthonia xenoplax*. However, the treatments - two E.C. formulations of DBCP at 58 kg a.i./ha - did not significantly decrease the numbers of two *Xiphinema* species.

In another experiment a similar treatment significantly (5% level) depressed the numbers of *X. elongatum* after 20 weeks. Treatments with granular formulations of DBCP (58 kg a.i./ha), 'Vydate' (oxamyl) (9 kg a.i./ha) and 'Nemacur' (phenamiphos) (20 kg a.i./ha) had no significant effect on the *Xiphinema* population.

Treatments with an E.C. formulation of DBCP at dosage rates between 29 and 116 kg a.i./ha all gave highly significant (1% level) control of a *Xiphinema* species after 23 weeks in a third experiment. The degree of control obtained with 58 and 116 kg was also significantly (5% level) greater than with 29 kg.

All observations are being continued and will be reported on in more detail at a later date.

LIST OF LECTURERS

E. Cohn,
Division of Nematology, ARO, The Volcani Center, Bet-Dagan, P.O. Box 6, Israel.

A. Coomans,
Instituut voor Dierkunde, Laboratorio voor Morfologie en Systematick, Ledeganckstraat 35, B9000 Gent, Belgium.

A. Dalmasso,
Station de Recherches sur les Nématodes, 123 Boulevard du Cap, Antibes (06), France.

D.J. Hooper,
Nematology Department, Rothamsted Experimental Station, Harpenden, Herts. AL5 2JQ., England.

F. Lamberti,
Laboratorio di Nematologia Agraria applicata ai Vegetali, Via G. Amendola 165/A, 70126 Bari, Italy.

P. Lippens,
Instituut voor Dierkunde, Laboratorio voor Morfologie en Systematick, Ledeganckstraat 35, B9000 Gent, Belgium.

P.A.A. Loof,
Laboratorium voor Nématologie, Landbouwhogeschool, Wageningen, The Netherlands.

M. Luc,
Laboratoire de Nematologie, ORSTOM, Boîte Postale 1386, Dakar, Senegal.

G.P. Martelli,
Instituto di Patologia Vegetale, Universita di Bari, Via G. Amendola 165/A, 70126 Bari, Italy.

R.S. Pitcher,
East Malling Research Station, East Malling, Maidstone, Kent, ME19 6BJ, England.

W.M. Robertson,
Scottish Horticultural Research Institute, Invergowrie, Dundee. DD2 5DA., Scotland.

D. Roggen,
Eenheid voor Dierkunde Fakulteit Wetenschappen, Urÿe Universiteit Brussel, Triomflaan, Brussels, Belgium.

J.W. Seinhorst,
Institut voor Plantenziektenkundig Onderzoek, Binnenhaven 12, Wageningen, The Netherlands.

C.E. Taylor,
Scottish Horticultural Research Institute, Invergowrie, Dundee. DD2 5DA., Scotland.

I.J. Thomason,
Department of Nematology, University of California, Riverside, California 92502, USA.

B. Weischer,
Biologische Bundesanstalt, Institut für Hackfruchtkrankheiten und Nematodenforschung, 44 Münster (Westf.), Toppheideweg 88, Germany.

A.L. Winfield,
Agricultural Development and Advisory Service, Block C, Government Buildings, Brooklands Avenue, Cambridge. CB2 2DR., England.

U.R. Wyss,
Institut für Pflanzenkrankheiten und Pflanzenschutz der Technischen Universität Hannover, 3 Hannover-Herrenhausen, Herrenhäuser Strasse 2, Germany.

LIST OF PARTICIPANTS

T.J.W. Alphey,
Scottish Horticultural Research Institute, Invergowrie, Dundee. DD2 5DA., Scotland.

Miss L. Ambrogioni,
Istituto Sperimentale per la Zoologica Agraria, Via Romana 15-17, 50125 Firenze, Italy.

Miss M. Arias,
Instituto Español de Entomologia, J. Gutierrez Abascal 2, Madrid - 6, Spain.

Miss K. M. Atkinson,
ADAS, Lawnswood, Leeds 16, England.

Mrs T. Bleve-Zacheo,
Laboratorio di Nematologia Agraria applicata ai Vegetali, Via G. Amendola 165/A, 70126 Bari, Italy.

B. Boag,
Scottish Horticultural Research Institute, Invergowrie, Dundee. DD2 5DA., Scotland.

D.J.F. Brown,
Scottish Horticultural Research Institute, Invergowrie, Dundee. DD2 5DA., Scotland.

F. Callieris,
Laboratorio di Nematologia Agraria applicata ai Vegetali, Via G. Amendola 165/A, 70126 Bari, Italy.

Mrs A. Capusso-Tosi,
Laboratorio di Nematologia Agraria applicata ai Vegetali, Via G. Amendola 165/A, 70126 Bari, Italy.

F.E. Caveness,
International Institute of Tropical Agriculture, PMB.5320, Ibadan, Nigeria.

Mrs B. Choleva,
Institute of Plant Protection, Kostinbrod, Sofia-district, Bulgaria.

Miss M. Coiro,
Laboratorio de Nematologia Agraria applicata ai Vegetali,
Via G. Amendola 165/A, 70126 Bari, Italy.

J.I. Cooper,
Unit of Invertebrate Virology, Commonwealth Forestry Institute,
South Parks Road, Oxford, England.

J. Cotten,
Plant Pathology Laboratory, Hatching Green, Harpenden. AL5 2BD,
England.

F. D'Errico,
Istituto di Entomologia Agraria, Universita degli Studi,
Faculty Agraria, 80055 Portici (Napoli), Italy.

F. Elia,
Laboratorio di Nematologia Agraria applicata ai Vegetali,
Via G. Amendola 165/A, 70126 Bari, Italy.

A.M. Emechebe,
Department of Crop Science, Makerere University, P.O. Box 7062,
Kampala, Uganda.

K.B. Erikkson,
Department of Plant Pathology and Entomology, Agricultural
College of Sweden, S-750 07 Uppsala 7, Sweden.

R.H. Estey,
Department of Plant Pathology, Macdonald College,
McGill University, P.O. Box 219, Quebec, Canada.

N. Greco,
Laboratorio di Nematologia Agraria applicata ai Vegetali,
Via G. Amendola 165/A, 70126 Bari, Italy.

Mrs S. Grimaldi di Zio,
Istituto di Zoologica e Anatomia Comparata,
Via G. Amendola 165/A, 70126 Bari, Italy.

V. Lo Giudice,
Istituto Sperimentale per l'Agricultura, Corso Savoia 165,
95024 Acireale (Catania) Sicily.

R.W. Hackney, University of California, College of Biological
Agricultural Sciences, Department of Nematology, Riverside,
California 92502, USA.

J. Jakobsen,
Statens Plantepatologiske Forsøg, Lottenborgvej 2,
2800 Lyngby, Denmark.

P. Jatala,
International Potato Centre, Apdo Postal 5969, Lima, Peru.

Miss M.E.John,
ADAS, Shardlow, Derby. DE7 2GN., England.

D. Kaplan,
Department of Plant Pathology, University of Massachusetts, Amherst, Mass. 01002, USA.

Z.A.S. Katcho,
Plant Pathology Division, Abu-Ghraib, Iraq.

Mrs J. Kozlowska,
Zaklad Ekologie, Polska Akademia Nauk, Rembertow A1, Sztandarow 11, Warszowa, Poland.

F. McElroy,
CDA Research Branch, 6660 North West Marine Drive, Vancouver, 8, B.C., Canada.

L.I. Miller,
Department of Plant Pathology and Physiology, Virginia Polytechnic Institute, Blacksburg, Virginia 24061, USA.

Mrs R. Morone di Lucia,
Istituto di Zoologia e Anatomia Comparata, via G. Amendola 165/A, 70126 Bari, Italy.

D. Nevo,
Plant Protection Department, P.O. Box 15030, Jaffa, Israel.

A.J. Ordosgoitti,
c/o Laboratorio di Nematologia Agraria applicata ai Vegetali, Via G. Amendola 165/A, 70126 Bari, Italy.

J. Rau,
Institut für Pflanzenkrankheiten und Pflanzenschutz, Herrenhäuser Strasse 2, Hannover, Germany.

R. Roca,
Laboratorio di Nematologia Agraria applicata ai Vegetali, Via G. Amendola 165/A, 70126 Bari, Italy.

Mrs M. Sabova,
Helmintologicky ustav SAV, Duklianskych hrdinov 11, 04000 Kosice, Czechoslovakia.

S. Sharafat Ali,
Laboratorio voor Morfologie en Systematick Rijksuniversiteit, K.L. Ledeganckstraat 35, 9000 Gent, Belgium.

H. Shoeffling,
Zentralstelle für Klonenselektion der Landes, Lehr versuchsanstalt für Weinbau, Gartenbau und Landwirtschaft, Egbertstrasse 18-19, 55 Trier, Germany.

A. Siniscalco,
Laboratorio di Nematologia Agraria applicata ai Vegetali, Via G. Amendola 165/A, 70126 Bari, Italy.

P.C. Smith,
Plant Protection Research Institute, Stellenbosch 7600, S.Africa.

W.C. Snyder,
Department of Plant Pathology, University of California, Berkeley, California, USA.

M. Støen,
The Norwegian Plant Protection Institute, Vollebekk, Norway.

A. Szczygiel,
Sadowniczy Zaktad Doswiadczalny, Brzezna, pow. Nowy Sącz, Poland.

D.P. Weingartner,
Agricultural Research Center, P.O. Box 728, Hastings, Florida 32045, USA.

K.J.O. Williams,
Commonwealth Institute of Helminthology, The White House, 103 St. Peter's Street, St. Albans, Herts., England.

A. Vuittenez,
Centre de Recherches de Colmar INRA, Station de Pathologie Vegetale, 28 Rue de Herrlisheim, B.P.384, 68021 Colmar, France.

A.M. Yassin,
Gezira Agricultural Research Station, Wad-Medani, Sudan.

SUBJECT INDEX